T0185165

NATIONAL MILITARY ESTABLISHMENTS AND THE ADVANCEMENT OF SCIENCE AND TECHNOLOGY

BOSTON STUDIES IN THE PHILOSOPHY OF SCIENCE

VOLUME 180

NATIONAL MILITARY ESTABLISHMENTS AND THE ADVANCEMENT OF SCIENCE AND TECHNOLOGY

Studies in 20th Century History

Edited by

PAUL FORMAN

Smithsonian Institution, Washington, D.C.

and

JOSÉ M. SÁNCHEZ-RON

Universidad Autónoma de Madrid

KLUWER ACADEMIC PUBLISHERS

DORDRECHT / BOSTON / LONDON

Library of Congress Cataloging in Publication Data

National military establishments and the advancement of science and
technology : studies in 20th century history / edited by Paul Forman
and José M. Sánchez-Ron.
     p.   cm. -- (Boston studies in the philosophy of science ;   v.
180)
   Includes index.
   ISBN 0-7923-3541-4 (hb : alk. paper)
   1. Military research--History--20th century. 2. Discoveries in
science--History--20th century. 3. Technological innovations-
-History--20th century.   I. Forman, Paul.   II. Sánchez Ron, José M.
(José Manuel)   III. Series.
U390.N19 1996
355'.07'0904--dc20   95-17492

ISBN 0-7923-3541-4

Published by Kluwer Academic Publishers,
P.O. Box 17, 3300 AA Dordrecht, The Netherlands

Kluwer Academic Publishers incorporates the publishing
programmes of D. Reidel, Martinus Nijhoff, Dr. W. Junk and MTP Press.

Sold and distributed in the U.S.A. and Canada
by Kluwer Academic Publishers,
101 Philip Drive, Norwell, MA 02061, U.S.A.

In all other countries, sold and distributed
by Kluwer Academic Publishers Group,
P.O. Box 322, 3300 AH Dordrecht, The Netherlands.

Printed on acid-free paper

Printed in the Netherlands

TABLE OF CONTENTS

R. S. COHEN

FOREWORD

To some philosophers, seeking to understand the human condition, technology is a necessary guide. But to think through the complex human phenomenon of technology we must tackle philosophy of science, philosophy of culture, moral issues, comparative civilizational studies, and the economics of specific industrial and military technologies in their historical contexts. The philosopher wants to grasp the technological factor in this troubled world, even as we see it is only one factor, and that it does not speak openly for itself. Put directly, our human troubles to a considerable extent have been transformed, exaggerated, distorted, even degraded, perhaps transcended, by what engineers and scientists, entrepreneurs and politicians, have wrought. But our problems are ancient, problems of dominations, struggles, survival, values in conflict, greed and insane sadisms. To get some conceptual light on the social reality which seems immediately to be so complicated, a philosopher will need to learn from the historians of technology.

A few years ago, the philosopher Elisabeth Ströker concluded that "a historical philosophy of technology [is required] since history – and history alone – provides all those concepts that form part of the repertoire of the philosophical analysis of technology". And she added that this goes far beyond the triviality that like other cultural achievements technology has its historical development. Now historical comprehension is no substitute for a logical methodology in the analysis of technological problems. What the engineers do, and why and how they succeed or fail, seem to be questions that parallel those of the philosophers of science: the still illuminating distinction drawn by Reichenbach between contexts of discovery and contexts of justification; by Carnap among the syntactic, semantic, and pragmatic components required for adequate clarity about the language of science; and by the conceptual historians (Canguilhem, Fleck, Kuhn) between basically differing frameworks of meaning, institutionally established. Historical case studies have typified these works, and so it is with technology too as historians investigate its influence and massive development in the modern world.

The philosophical problem is not directly a matter of epistemology, of some distinction to be found between allegedly 'pure' science and applied knowledge, between science and engineering know-how, a distinction which in any case is deeply muddy. 'Man the maker' has always been active, rarely a passive observer who then acts. But in recent centuries there has been an historical explosion of industrial technology within our industrial society.

vii

P. Forman and J. M. Sánchez-Ron (eds.), National Military Establishments and the Advancement of Science and Technology. Studies in the 20th Century History, vii–viii.
© 1996 Kluwer Academic Publishers. Printed in the Netherlands.

David Landes rightly calls his study *The Unbound Prometheus* (1969), and he points to the European stress, after medieval times, upon rational adaptation of means to ends. Technology is central to understanding means.

What then do we know about the means? Among the several factors of production and transformation of human lives, among the great changes in agriculture and the growing dependence upon machines and other technologies of power, we find always present a military factor. The history of technology tells of sources of new powers, and of competitions among social forces, a continuing development. Is technology determining, or not? (We always recall medieval military power after the introduction of the stirrup.) Surely it will be seen as underdetermining. But in the social context of warfare and its technologies, a philosophy of science and technology must confront the grave issues of the philosophy of war and peace, and of technology at war.

The massive phenomena of the past 150 years have smashed the previous mere continuity of growing technology and its increasing influence upon social and political history by what may usefully be seen as plain discontinuities: monstrous human destruction in the American Civil War, followed by the unmeasurable disasters of wars ever since; the new giant metropolitan cities; repeated unemployment of skilled as well as increasingly deskilled workers; major TV visual literacy replacing literacy of the written word, which in turn becomes popular illiteracy; the cybernetic hardware revolution; the growth of AI and the information technology software revolution; birth control nearly everywhere, and perhaps genuine population management; severe and ominous planetary limitations upon resources; and the vast apparatus of nuclear warfare. Among these, to what extent have military sciences, military technologies, military options dominated the science and technology policies of governments, economic establishments, and research enterprises? The questions cannot be avoided by historians, economists, or philosophers of modern times. Might there yet be alternatives among technologies? How has this science-technology-military-industrial-political complex developed? What indeed is the political economy of war, and what is the political economy of technology?

Paul Forman and José Sánchez-Ron present a collection of ten detailed historical studies, a fascinating beginning which may lead the reader to the question they raise so carefully: do you like what you see? Forman and Sánchez-Ron speak as philosophers who feel the complexity at the century's end, our time "deprived of a simple structure of geo-political polarization and deprived of transcendent values..."

Boston University   Robert S. Cohen
March 1996

PAUL FORMAN AND JOSÉ M. SÁNCHEZ-RON

INTRODUCTION

We, the editors, admit to being ourselves not avid readers of the obligatory 'introductions' to volumes of collected essays. Taking such a volume in hand, we neither find ourselves inclined to linger over lengthy precis when the turning of some pages opens to us the several works themselves, nor have we, as critical readers, any expectation that editors of such collections would be such fools or knaves as now, at this stage, to appear as critics of the papers they have so laboriously brought together. We editors, having rendered our criticisms in confidence – in contrast, say, with the commentators at a scholarly meeting – necessarily come forward now as our authors' 'advance men,' smoothing their way into the reader's favor. How can we, then, in good conscience, claim the attention of those other critical readers who may take this volume in hand? This moral dilemma being once admitted, the implication is clear: spare the reader (and the publisher, and the forests) those inevitably self-congratulatory summaries, or, at least, spare them all lengthy precis of the several papers' arguments and principal contributions to knowledge. What, however, we owe our readers, and do here attempt to provide, is a very brief indication of ways in which each of the papers included in this volume bears upon its theme, viz., that national military establishments – and not merely that of the United States, but those of many countries – have been a major factor in the advancement of science and technology in the twentieth century.

It is with a conscience all the clearer that we restrict ourselves to noticing so very briefly the thematic unity of our volume inasmuch as our own "Introduction" is, on the one side, preceded by Robert Cohen's welcome preface making evident the pertinence of this collection of historical essays to the philosophy of science, and, on the other side, is followed immediately by David Edgerton's forceful essay demonstrating historically – to be sure, with respect to Britain only – the urgent necessity of just such studies as the other contributors to this volume provide.

Throughout this century, Edgerton argues, "British scientific intellectuals," – i.e., British scientists speaking about science in its wider social relations, speaking not as mere laymen but with claims to authority – have generally displayed very little of that independence of judgement and respect for fact upon which, as scientists, they prided themselves. Their dicta regarding the relations of science, technology, and war have either been expressions of the social-political attitudes typical of the British intellectual class at that historic moment, or else they have been endeavors, specific to the scientists,

ix

P. Forman and J. M. Sánchez-Ron (eds.), National Military Establishments and the Advancement of Science and Technology. Studies in the 20th Century History, ix–xiv.
© 1996 Kluwer Academic Publishers. Printed in the Netherlands.

to evade any conscious awareness of the extensive connections between their
national scientific establishment and their national military establishment.
The British scientific intellectuals thus succeeded in misleading the British
public – including, Edgerton emphasizes, British historians of science and
technology – even as they misled themselves. They wished to believe, and led
others to believe, that, at any rate in Britain, these two enterprises, science and
war, have, historically, had only an intermittent and no intimate relationship,
and that, more particularly, the rate and direction of scientific and technologic
progress may be discussed without close attention to society's investment in
its war-making capabilities.

Certainly ours is not the first volume to be devoted to the correction of
this misapprehension.[1] However, we believe ours to be distinctive in that
it is not overwhelmed, as previous collections have generally been, by the
conspicuous military 'forcing' of the physical sciences in the United States
of America in and since World War II. Rather, this collection of essays
makes it possible to see that which arose in the United States in the middle
decades of this century in a longer and broader historical perspective, to see
it, namely, as but one of many variants of a fundamental symbiosis between
national scientific and military establishments characteristic of all advanced
and advancing countries in the twentieth century.

Turning, then, to the first of our 'national' groupings, "Mainly in Germany,"
and to the first of the papers within it, that by Helge Kragh on "Telephone
technology and its interactions with science and the military, ca. 1900–1930,"
we must immediately acknowledge that this paper, as also the paper by
Michael Eckert which follows it, declines to limit itself to a single nation.
Indeed, Kragh's paper constitutes, along with Edgerton's, a second 'overture'
in that the period considered is the opening decades of this century, and,
while focusing chiefly upon Germany, Kragh also draws our attention to
the military and political significance accorded to long-distance telephony in
France, Britain, and the U.S.A. He shows us a nominally civilian technology
– one whose peace-time employment was indeed overwhelmingly civilian –
whose shape and extension was largely determined, in Germany especially,
by consideration of its employment for military purposes in time of war (and,
after the National Socialists seized power, for control of the civilian populace
at all times).

Eckert, following the paths taken into war-related research in the early
1940s by two students of the German theoretical physicist Arnold Sommerfeld – one, Hans Bethe, who emigrated to the United States in the 1930s,
and one, Heinrich Welker, who remained in Germany – shows how the more
complete integration of American physicists as a research community, and
of the American physicists within American society, brought about a greater
reorientation of Bethe's research direction (i.e., to accord more nearly with
immediate military needs). For Welker, making his way in a national sci-

entific establishment characterized by specialized, autonomous, monocratic research institutes, the military establishment was primarily a resource, the indispensable protector and supporter for his continued efforts along established lines of research. Eckert points out that nearly all comparisons of the contributions of German physicists and American physicists to their respective war efforts fail to take into account this difference in social integration and research orientation, i.e., fail to see that the same measuring rod cannot be applied to scientific work and workers in both countries.

With Herbert Mehrtens' important paper, "Mathematics and war: Germany 1900–1945," we come to a study of these relationships in just one country – as indeed are all following papers in this volume as well. Mehrtens is endeavoring to grasp and to interconnect the social integration of mathematicians, on the one hand, and the nature of mathematics and of its hold on reality, on the other hand. What, Mehrtens asks, is the relation between the mathematician as operative of mathematics – of mathematics as system of 'arbitrary' signs and rules – and the practical operativity of that system of signs and rules as seen, say, in the shape of an armor-piercing projectile? How does mathematics so defined come to be, get to be, "the" means of control over the physical and the social environment, and, more particularly, to be such for Germany's military-technical enterprises? His essay thus addresses the theme of this volume at a more fundamental level than does any other paper here included.

Mehrtens' analysis of the carriage of German mathematicians during the Second World War differs somewhat from Eckert's in his greater stress on the mathematicians' 'self-mobilization.' Yet at bottom both authors are in agreement that this 'self- mobilization' was indicative of the distress that scientists felt at the absence, under National Socialism, of an effective symbiosis between the military and scientific establishments (with the technologists it was quite a different matter), and that, moreover, this 'self-mobilization' did indeed advance the interests of the scientific establishment and make resources available for significant scientific accomplishment even as the Third Reich was collapsing.

"Three Latin Countries," our second main grouping of the essays here collected, contains studies of France, by Lewis Pyenson, of Argentina, by Eduardo Ortiz, and of Spain, by Javier Ordoñez and José M. Sánchez-Ron. The first two of these begin their examinations of the relation between science and the respective national military establishment well before the twentieth century – in Pyenson's essay on France, even before the nineteenth century. Pyenson, highlighting a few French military men who occupied key positions, and played key roles as administrators of scientific institutions and enterprises through two centuries, draws our attention to circumstances to some extent characteristic of our two other 'Latin' countries as well: the French military establishment was of first importance for the advancement of science and technology in that country because (a) military institutions had so large a role in the educational formation and professional careers of the nation's scientists and technologists, and (b) the 'civilizing mission' in the context of colonialism gave heightened importance to scientific expeditions and outposts, and placed a premium upon the skills and resources of military men. As a consequence of the French scientific establishment being based so largely on the military establishment, the French military contained an unusually high proportion of men with active scientific interests, and such men came especially readily into influential positions in the French scientific establishment.

In "Army and science in Argentina: 1850–1950," Ortiz gives us a sweeping, but detailed and connected, history of the exploitation, and the promotion, of science and technology by the Argentine army. Because the military was long the dominant institution in Argentine society, it was inevitable that through fostering of science and technology to advance its own interests, it would come also to play a leading role in creating – and eventually crippling – that nation's scientific establishment.

The third essay in this group, "Nuclear energy in Spain," covers that period in Spain's history when the structure of its society bore strong similarities to that of Argentina, viz., during the regime of Generalissimo Franco. In 1945, when the atomic nucleus burst upon the world, Spain too was still an underdeveloped country, scientifically speaking – notwithstanding significant, but isolated, researches such as those of Miguel Catalán in atomic spectroscopy. In Spain too the military was the dominant social institution. This circumstance gave the military an even larger role in nuclear energy than the nature of that subject alone could lead us to expect. Nonetheless, the situation that they describe is a complex one, in which Spain's need for acceptance by an Atlantic alliance of democratic states moderated the secretiveness of a military establishment preoccupied with retaining its hold upon state power.

The last group of papers in this volume approaches our theme over the now relatively familiar ground of military-scientific relationships in mid-century America. But though that which Hevly, DeVorkin, and Forman present will not be altogether new and surprising to those acquainted with the extensive literature on The Cold War and American Science – to borrow the title of S.W. Leslie's recent book[2] – even the cognoscenti will have encountered few studies that attempt quite what these essays do. Whereas most studies of the relationship between military establishments and scientific practice have problematized and elucidated the institutional connections, few have asked after the specific character of the science done, of the knowledge produced (and its relation to technology), at research institutions within the military or under military sponsorship. These last three papers show how science was being redefined even as it was being advanced.

Bruce Hevly, in his essay, examines a research laboratory in and of the U.S. military, the Naval Research Laboratory, before and after the Second World

War, seeking in its research program organized around "tools of science" that which is specific to such an institution. David DeVorkin describes the actual creation of a field of scientific research by the military, that is by its provision of the V–2 rocket as vehicle, and its provision of support to those scientists – not, originally, specialists in 'space science,' but merely the possessors of appropriate techniques – who would give themselves the very considerable trouble of preparing experiments for, and risking them upon, rocket launches. And, finally, Paul Forman examines a military-funded research laboratory at a university, describing how, out of the difficult reconciliation of academic postures and military objectives, there arose a fundamentally new sort of device and a new set of scientific concepts. In concluding, Forman returns us to Edgerton's 'overture,' drawing attention to a compartmentalization of knowledges in the scientists' minds that helps account for the disparity Edgerton highlights between the pronouncements of scientists qua intellectuals and the real terms on which they have lived their scientific lives in the twentieth century. As Gunnar Myrdal observed towards the end of his life:

When people try to deny, to the outside world and to themselves, that they live in moral compromise and that they ceaselessly and habitually violate their own ideals, they are customarily brought to falsify their perception of reality in order to conceal this from themselves and others.[3]

The value of any product of research and scholarship is to be measured not only by the information it contains and the questions it answers, but also by the questions it raises. Every essay in this volume will be found to meet that test. Yet so little as we presumed to summarize each of our contributors' evidence and argument, even less do we presume to suggest what questions they should raise in our readers' minds. One question, however implicit, obtrudes in every one of these essays, and we will not forbear to raise it here: 'do you like what you see?'

A decade ago, when most of us were still unaware that the advent of postmodernity was radically transforming our evaluative standards – indeed, the very possibilities of valuation – we would have answered with a firm and unequivocal 'no.' We stood then in the midst of the Reaganite revivification of the Cold War, in which advanced military technology, based upon scientific research correspondingly oriented, was given a leading role. This 'Star Wars' turn of research priorities – and the general conformance of the U.S. scientific establishment to them – was itself largely responsible for the critical attention that, beginning in the early 1980s, has been given to the role of 'the military' in shaping those modes of cultural production which, as historians of science and technology, we had theretofore been inclined to celebrate. As we can now more clearly see from the perspective of postmodernity, our indignation at that reckless intensification of the Cold War in a scientific-technical direction was all the greater as it was so striking a demonstration of the demotion of the cultural elite, with which we associated ourselves, from the role of 'legislators.'[4] But even as postmodernity has marginalized intellectuals to the role of mere 'interpreters,' it has liberated them to make their own moral judgements – and, curiously, elevated their importance (though not the esteem in which they are held) by putting culture, in place of the Cold War, at the center of the political arena.

Decades in the making, postmodernity was concealed from us in its incipient stages by the 'old world order' that the Cold War kept frozen in place. The sudden, unexpected liquidation of that 'old world order' at the very end of the eighties has allowed a general widening and lengthening of historical perspectives, including that achieved in this volume on the relations between science, technology, and the military. Our answers to the question 'do you like what you see?' are no longer so categoric as a decade ago – as our world itself, our 'new world *dis*order', is no longer so categoric. The debate over the 'correct' science policy, over the 'good' direction for research and technology to take, has become much more complex in a world deprived of a monolithic structure of geo-political polarization and deprived of transcendent values upon which valuations of historical circumstances can be surely based. Nonetheless, no reader will fail to sense that every author contributing to this volume is struggling to find – to create – grounds for making just such judgements. We hope that these essays will help our readers to shape their own.

Smithsonian Institution's National Museum of American History, Washington
Departamento de Fisica Teorica, Universidad Autónoma de Madrid

NOTES

[1] The principal collections of papers on this subject are: Merritt Roe Smith, ed., *Military enterprise and technological change* (MIT Press: Cambridge, Mass., 1985); Robert W. Seidel, ed., *Historical Studies in the Physical Sciences 18*, part 1 (University of California Press: Berkeley, 1987); Everett Mendelsohn, Merritt Roe Smith, and Peter Weingart, eds., *Science, technology, and the military*, 2 vols. (Kluwer: Dordrecht, 1988); Michelangelo De Maria, Mario Grilli, and Fabio Sebastiani, eds. *The restructuring of physical sciences in Europe and the United States, 1945–1960* (World Scientific: Singapore, etc., 1989); Peter Galison and Bruce Hevly, eds., *Big science: the growth of large-scale research* (Stanford University Press: Stanford, Calif., 1992). See, also, the introductory discussion in Lewis Pyenson's paper in this volume, and especially his note 6.

[2] Stuart W. Leslie, *The Cold War and American science: the military-industrial-academic complex at MIT and Stanford* (Columbia University Press: New York, 1993).

[3] Gunnar Myrdal, *Hur styrs landet?* (Stockholm, 1982), 263, as quoted and translated in Sissela Bok, "Introduction," *Daedalus* (Winter 1995), pp. 1–15 (=*Proc. American Acad. of Arts and Sciences 124*, nr 1), on p. 6.

[4] Zygmunt Bauman, *Legislators and interpreters* (Blackwell: Oxford, 1987); Zygmunt Bauman, *Intimations of postmodernity* (Routledge: London and New York, 1992).

D.E.H. EDGERTON

BRITISH SCIENTIFIC INTELLECTUALS AND THE RELATIONS OF SCIENCE, TECHNOLOGY AND WAR[1]

Scientists, humanists, historians and sociologists of science, even proponents of a 'science of science', have long held that scientists' expertise did not extend to systematic reflection on the place of science in the world. Scientists speaking of anything but their own work have been seen as everything from mere laymen to self-serving ideologues. However, scientists ignored these injunctions to silence and spoke and wrote, as scientific intellectuals, on the social relations of science. Historical studies of what they have said are rare; much such work is, in any case, sympathetic to the arguments of the particular scientists studied, if not to scientific intellectuals in general.[2] Studies of futuristic literature include discussion of the work of scientific intellectuals, but these studies often place a premium on the supposed prescience of their subjects.[3] Despite the lack of systematic critical attention, the influence of scientific intellectuals has been, and probably still is, greater than that of qualified analysts of the social relations of science. It is worth studying for that reason. Furthermore, as our independent historical knowledge of the micro- and macro-social, economic, and political relations of science has grown, we are in a position to compare scientitific intellectuals' commentary with our own.

In this paper I will examine an important subset of British scientific intellectuals' writings: their reflections on the relations of science, technology and war in twentieth century Britain. I will show that there is remarkably little correspondence between our historical picture of what they said and our historical picture of the actual relations between science, technology and war. This applies to the funding of British science and technology, to the identification of key scientific and technological institutions, and to the roles of individual scientists and technologists in war. In discussing the relations of science, technology and war in twentieth century Britain, scientific intellectuals typically invoked images and arguments which distanced science from war. Science, and the scientist, were seen as internationalist, universal, moral, civil, peaceful and peace-creating; nationalism, immorality, capitalism, militarism and war would wither away with the advance of science. Scientific advance would make war impossible, either because it would create a 'global village' or because war would become so destructive that societies would abolish it. Scientists were more intelligent and forward looking than serving officers, and science would be corrupted if it came under military control. Sci-

1

P. Forman and J. M. Sánchez-Ron (eds.), National Military Establishments and the Advancement of Science and Technology. Studies in the 20th Century History, 1–35.
© 1996 Kluwer Academic Publishers. Printed in the Netherlands.

entists were not only intellectually superior to others, they were also morally superior; they tried and usually failed to make war more humane and efficient. Had their methods been applied war could have been avoided altogether. The overall picture is that science and war were antithetical to each other or, at least, radically different enterprises.

These stories, which were intended to be taken literally, were fairy stories, but ones that have bewitched many students of the relations of science, technology and war. The divergence between this picture and the most straightforward empirical analysis of the relations of science and war is depressing yet sobering; sometimes it will astonish even the most hard-bitten cynic. Although many details of the relations of science, technology and war were not known by contemporaries, and much remains to be discovered today, the divergence cannot be explained by secrecy. There was always sufficient information in the public domain to yield a very different picture, and some scientific intellectuals did indeed produce a different picture. Furthermore, lack of information did not prevent large amounts of literature being produced, full of empirical detail. In any case, most of the scientific intellectuals who discussed the relations of science and war had direct personal involvement in scientific work in wartime, and often in peacetime too. Examining the divergence over time shows it to be systematic and longstanding. More interestingly, there was an unacknowledged repetitiveness in the literature of staggering scale. There were also, however, important differences across historical periods, again unacknowledged by scientific intellectuals themselves. Both change and continuity need to be explained.

A historical analysis reveals a broad trend: in the twentieth century scientific intellectuals' picture of the relations of science, technology and war became increasingly mythological. In order to defend science, scientific intellectuals created a paradigmatic if ironic example of almost pure apriorism, lack of empirical justification, and indeed lack of cumulation of knowledge. The charge that scientitific intellectuals often do not apply the canons of scientific method to the discussion of the place of science in society – for example, that scientific intellectuals are peculiarly prone to teleological thinking about science and about society – is a familiar one, but it has rarely been made empirically, much less historically. Doubtless this is because academic analysts are not keen on the systematic study of the banalities of the powerful.

And yet, the case is worth making empirically and historically. As indicated above, only such a study will bring out certain key features of scientific intellectuals' discourse. This paper seeks to establish in particular the sheer repetitiveness and lack of originality of the scientific intellectuals' discourse on science and war. It also seeks to show that it is almost entirely derivative from established political views. But this paper also claims that historians who have used these discourses as a source for histories of science and war, and state-science relations, have often been seriously misled. Scientific intellectuals' views of the relations of science and war live on in the historical literature.[4] Uncritical readings, I suggest, have led to a historiography of science-state-war relations written almost exclusively in terms of state-funded *civil* science and technology.[5] In contrast to the US case, the relations of science, technology and the military have been systematically neglected.

That British scientific intellectuals should not have produced an empirically sound account of the relations of science and war would not come as any surprise to students of the post-Second World War military-scientific complex of the United States.[6] Forman writes of American scientists' "false consciousness, which succeeded so well in what it was intended to do, to mislead others even as it blinded themselves", about the scale of the links between postwar physics and the military.[7] In explaining this "false consciousness" Forman has argued that the scientific community has fostered a structural amorality associated with the commitment to a trancendental view of science.[8] I would point more than Forman does to the importance of political thought, and the political climate, national and international, in shaping what scientific intellectuals wrote about the relations of science, technology and war. The arguments scientific intellectuals made on this topic were analogous to general arguments about society, indeed fairly commonplace ones. There was no ideologically distinctive scientific voice. Scientific intellectuals were very much of their time and place. I suggest that, before the First World War, British scientific intellectuals shared in the then prevalent nationalist militarism; in the 1930s the dominant voice was in the liberal to marxist range; while in the 1950s and 1960s, a desiccated, and profoundly unhistorical, liberal humanism took over. But in some ways there was something distinctive about the case of scientific intellectuals; there was an increasing decadence in their reflections on the world, at least into the 1960s. There is hardly a better case to be found to refute the belief that knowledge necessarily increases its scale and scope over time.

I

Before moving on to look at what scientific intellectuals make of the relations of science, technology and war in twentieth century Britain, it is worth giving a brief outline of those relations, derived largely from my own work.[9] Perhaps the most important single point is that throughout the twentieth century, in peace and in war, warlike R&D has been significantly higher than state-funded civil R&D. Indeed, in the interwar years the Air Ministry was easily the largest R&D spending institution in Britain. In the mid–1920s it was spending £1.34 million on R&D, compared with £0.98 million by the Admiralty, and £0.49 million by the War Office. The largely civil Department of Scientific and Industrial Research (DSIR) was spending a mere £0.38 million.[10] For the mid–1930s we can make a comparison with industrial R&D spending. In

1935/36 the Air Ministry spent £1.25 million on R&D, while British industry spent, though the figure is an underestimate, £2.7 million. ICI, the largest industrial R&D performer, spent £0.59 million.[11] In the 1940s and 1950s defence R&D was much larger than state-funded civil R&D and larger also than industry-funded civil R&D. Something like half of total national R&D spending was devoted to warlike purposes, a proportion which dropped to about 30% by the 1960s. A very crude estimate for the interwar years yields a comparable figure. Thus, for the twentieth century as a whole defence has accounted for over 30% of total national R&D expenditures.

The key agencies for funding warlike R&D were the service departments and special supply ministries. The service ministries before the Great War were the Admiralty (the navy ministry) and the War Office (the army ministry); they were joined by the Air Ministry in 1918. It was not until 1964 that these ministries were merged to form the modern Ministry of Defence. Before and after the Great War these ministries directly funded warlike R&D in their own laboratories and in private firms. However, in the Great War the R&D and procurement functions for the army and the air services were taken over by a supply ministry, the new Ministry of Munitions. In the Second World War the R&D functions of the Army and Air Force were taken over by the Ministry of Supply and the Ministry of Aircraft Production respectively, and these were merged after the war into a Ministry of Supply. This became the Ministry of Aviation in 1959, which itself became the largest part of the Ministry of Technology in 1967. Only in 1973 did responsibility for all warlike R&D go to the Ministry of Defence, where it remains today. The largest government spenders at any one time on R&D were successively: the Admiralty, the Ministry of Munitions, the Air Ministry, the Ministry of Aircraft Production, the Ministry of Supply, the Ministry of Aviation, the Ministry of Technology, and the Ministry of Defence.

These service and supply ministries funded R&D within their own laboratories as well as in industry. Warlike R&D establishments were many and various, but many kept their identity and location for long periods; some of the largest were created before the Great War. By the interwar years the principal research (rather than development and design) establishments were the Royal Aircraft Establishment, the Admiralty Research Laboratory and the War Office's Research Department at Woolwich. The Second World War saw further expansion and the creation of new laboratories, notably the Telecommunications Research Establishment (TRE) for radar research. After the Second World War more new centres were created, especially but not only for the development of nuclear weapons. The scientists and technologists in these establishments were typically civil servants; they were members of what became known after the Second World War as the Scientific Civil Service.

During the war, many scientists and technologists, including many holders of university chairs, were recruited into the civil service on a temporary basis. In contrast to the situation in the United States, universities did not acquire large military-funded research laboratories, and this remained the case in Britain after the war. Another important way in which academic scientists entered the warlike R&D system was as seconded advisers, either to advisory positions or to advisory committees. One of the most significant of these committees was founded as the Advisory Committee on Aeronautics in 1909, but was known for most of its life as the Aeronautical Research Committee or Council. Some scientists became advisors not on weapons development but on operational matters, hence 'operational research'; these advisors, and associated operational research sections, formed part not of the supply organisations, but of individual commands of the armed services themselves, for example, the RAF's Bomber Command.

The private manufacturers of arms themselves employed a very substantial proportion of all the scientists and technologists engaged in warlike R&D. Like the structure of the government machine, that of the arms industry was very complex. In the Edwardian years the great armourers like Vickers, Armstrong and Whitworth were rightly recognised as particularly science-oriented firms. Their work on armour plate, on the design of heavy guns and mountings, and on warships themselves, required huge inputs of scientific and technological work. So did the development of explosives by such makers as Nobel. In the interwar years the airframe, and above all aero-engine firms (notably Rolls Royce and Bristol), came to the fore as major industrial performers of R&D. During and after the war, industrial R&D for warlike purposes came to be dominated by the now very large airframe, aero-engine and electronics firms. In the 1950s such firms as Hawker-Siddeley, Rolls-Royce, Bristol, English Electric, and Associated Electrical Industries dominated British industrial research.

Finally, some speculative comparative remarks may be helpful. Before the Great War Britain almost certainly devoted no less resources to warlike research, development and invention than did Germany or France, and almost certainly more than the other large powers; the USA, Russia and Austria-Hungary. In the 1920s and early 1930s Britain almost certainly held the lead. After the Second World War Britain had easily the largest military R&D complex outside the USA and the USSR. Especially before 1941, British scientists and engineers made perhaps the single largest contribution to the development of weapons of war; even after that date Britain's contribution was one to be reckoned with.

II

In the nineteenth century scientific intellectuals tended to believe that 'militant' societies were being replaced by peaceful 'industrial' societies.[12] This general argument, rightly associated with liberalism, insisted on an antithesis between scientific, technological, industrial and commercial progress on the one hand, and war on the other. It was to prove longlasting, as many examples to be given below will testify. However, by the beginning of this century a new nationalistic and militaristic spirit was in the air. Scientists were by no means immune. In 1902 the astronomer and former War Office civil servant Sir Norman Lockyer made a famous speech, "The Influence of Brain Power in History" – the title was an allusion to the famous American naval theorist Admiral Mahan's *The Influence of Seapower on History*. Lockyer argued that:

Every scientific advance is now, and will be in the future more and more, applied to war. It is no longer a question of an armed force with scientific corps; it is a question of an armed force scientific from top to bottom. Thank God the navy has already found this out. Science will ultimately rule all the operations both of peace and war, and therefore the industrial and the fighting population must both have a common ground of education. Already it is not looking too far ahead to see that in a perfect State there will be a double use of each citizen – a peace use and a war use; and the more science advances the more the old differences between the peaceful citizen and the man at arms will disappear. The barrack, if it still exists, and the workshop will be assimilated; the land unit, like the battleship, will become a school of applied science, self-contained, in which the officers will be the efficient teachers.[13]

The speech argues, quite contrary to the nineteenth century liberal tradition, that the scientisation and militarisation of society are parallel and desirable processes. Lockyer's speech was an important and influential one: it was intended to help create a new scientific lobby and in this it succeeded. The British Science Guild was, in Frank Turner's words, "clearly a conservative, social imperialist pressure group seeking to combine the intellectual prestige of science with the political attraction of efficiency and empire". As Turner argues, these scientists, before 1914, "were defining the character of modern warfare and the manner of its preparation so as to enhance their own possible contribution to it and subsequent recognition through it".[14]

The Great War did indeed enhance the status of the scientist, and secured permanent government support for scientific research on a scale unknown before the war. While this increased state support was welcomed by scientists there was also more than a twinge of regret at the devastation caused by the war, and by the new weapons that came out of belligerent's laboratories. However, the general view was that the corruption of science was something the Germans, not the British, had brought about. Lord Justice Moulton, a scientifically-trained judge who was responsible under the Ministry of Munitions for the supply of explosives, said in his 1919 Rede Lectures that while Germany had been the great scientific nation, "Yet we find Germany during a period of at least twenty years consciously and deliberately making preparations for a war to be waged upon its neighbours solely for the purpose of self-aggrandizement". The Germans attacked civilians brutally and "they introduced the use of asphyxiating gas and all the tortures of so-called chemical warfare... It is the monster which the Frankenstein of Mary Wollstonecraft created – a human being with his powers magnified to those of a giant but destitute of moral sense".[15] This sentiment was by no means unusual. As citizens of the neutral United States were told in 1916 by Sir Edward Grey, the British Foreign Secretary: "All their scientific genius has been dedicated to wiping out human life. They have forced these things into general use in war... Will the outstanding contribution of Kultur disclosed in this War be such efficiency in slaughter as to lead to wholesale extermination?"[16] The image of Germans as scientific giants but moral dwarfs corrupted by militaristic doctrines had been a common one from the beginning of the war among British intellectuals.[17] During the war, however, a number of British scientists, notably Sir William Ramsay, went further by claiming that German science had not been that good after all.[18] Furthermore, scientists bearing German names faced calls, even from prominent scientists, for their removal from official posts.[19] After the war, German, Austrian and Bulgarian scientists were kept out of international scientific organisations until 1926; neutrals faced special restrictions. Even as a scientific ideology, scientific internationalism ceased to be universalistic.

During the war the journal *Nature* gave little attention to the war as it was waged, but a great deal to the scientific war on the home front. Even so, the emphasis was not on what scientists were actually doing for the war effort; rather it was on the view that scientists knew better than the politicians and the military leaders how the war should be run, that scientific education was held back by the grip of the classics, that there were too few scientists in Parliament, and that partisan politics was not in the national interest. As Turner has argued, this was merely a continuation into wartime of pre-war themes.[20]

During the 1920s, too, the relations of science and war appear to have been barely touched on by *Nature*. But when they were touched on they seemed to cause little worry. An editorial commented in 1926:

The intelligent man in the street must see that to abandon investigations on lethal weapons while other nations pursue them is to court disaster, if not extinction. However much we may detest these weapons, the calls of home, of country, and of Empire must come first. In these circumstances the responsibility of the scientific workers concerned appears to be confined to the observance of strict secrecy in their work.[21]

Another indication of lack of moral concern in the 1920s is that J.B.S. Haldane, the nephew of the great pre-war army minister R.B. Haldane, published in a popular series of books a defence of chemical warfare. The argument was that being killed by gas was no worse than being killed by other means, and that gas warfare would lead to more decisive, and therefore more humane, wars.[22] Although scientific intellectuals' attitudes in the 1920s (as opposed to the 1930s), require much more research, it seems that there was little to trouble the conscience.

III

The 1930s saw a radicalisation of a minority of British academics and students in elite institutions, in the face of mass unemployment and the threat of war in the early 1930s, and above all with the increasing power and aggression of fascism in the late 1930s. The outbreak of new wars, especially in the Far East and South America, and the Geneva disarmament conference, gave the issue of war a new salience. This concern preceded the rise of Hitler, and any serious danger of war in Europe. Anti-war groups sprang up, including the Cambridge Scientists' Anti-War Group, founded in 1932, which was the training ground for many in the social relations of science movement.[23] As J.D. Bernal put it in 1939, by which time the world situation had changed dramatically: "More than anything else the question of science and war has made scientists look beyond the field of their own enquiries and discoveries to the social uses to which these discoveries are put".[24] The claim is an interesting one which should be taken seriously, and the more so because war was not the only candidate for making scientists reflect on the social relations of science: unemployment and the waste of the Great Depression was another, and one closer to home for British scientists. Perhaps the reason for the salience of war was that the early 1930s did indeed see a general revulsion against war, and a worry that public concern with the use of new science-based weapons had the potential to discredit science. As *Nature* put it in 1931:

There is a widespread tendency to hold science, and possibly chemistry in particular, responsible for many of the worst evils of modern warfare, which is perhaps the more dangerous to society because it is apt to discredit the voice of science.[25]

Nature's defence of science against this charge would echo down the years: science had so transformed war that any distinction between combatant and non-combatant had all but disappeared; this challenged societies to abolish war. It was on the basis of science, of the scientific study of the causes of war, that society would abolish war.[26]

The 1930s did indeed see a great deal of reflection by leading scientists on the relations of science and warfare. It took many forms. None, however, celebrated the linkages of science, the state and war, in the way Lockyer had done. Some were quite relaxed about the association of science with war, but even they echoed the themes that science and war, and the military, were in some sense antithetical. For Professor Low, a former teacher of Royal Artillery officers, and a noted populariser of science, the involvement of science with war was inevitable given that human societies tended to go to war, even though "To the scientist, war should have absolutely nothing to recommend it, for the purpose of science is to create not to destroy".[27] The biologist Julian Huxley, writing in 1934, recognised the extent to which the state supported warlike research, but did not disapprove because he felt defence should be scientific. He suggested the establishment of a War Services Research Council along the lines of the civil research councils to give scientists a greater say in warlike research policy.[28] Huxley was prepared to be quite frank. Science could be directed:

If you are willing to pay for more men and more facilities in war research than in, say, medical research, you will get more results adapted to killing people and less adapted to keeping them alive. And it is when we look at the amounts of expenditure in different fields that we begin to realise what a large share of the nation's scientific brains is occupied with war.

It is very difficult to obtain exact figures; but I have attempted to reach rough estimates which I think are not too far from actuality to be of service. I submit them with reserve, and as subject to an error of at least 15 to 20 per cent. For research in industry, and in the sciences mainly basic to industry, like physics and chemistry, the country spends perhaps 2 1/4 or 2 1/2 millions a year. War research comes next to this, with certainly over a million pounds, perhaps a million and a half. Research in agriculture and the agricultural side of biology take somewhere around three-quarters of a million; research in health and the physiological side of biology about half a million or probably less. And research in the specifically human sciences like psychology and sociology probably accounts for well under a hundred thousand. Money talks; and these figures tell a tale. Science is being applied on a large scale to the ends of destruction, not because science is essentially destructive or scientists particularly militarist, but because the nation, through its appointed government, is paying handsomely to secure that it shall be so applied.[29]

For Huxley, there was a case for applying science to disarmament, and especially through the study of psychology, to study the causes of war. Appropriately funded, and directed, science could serve peace, as in the early 1930s, it served war. If Huxley proposed studies of the causes of war; J.B.S Haldane studied some aspects of war. In a book on air raid protection, published in 1938, Haldane wrote quite matter-of-factly about air raids, arguing against the like of H.G. Wells, who held up a vision of aerial gas war bringing a country to its knees in days. Gas would not be used for military and technical reasons, he argued, and good protection against high explosives and incendiaries could be given.[30] Haldane's patrician marxism, and his familiarity with war and defence policy, gave him the confidence to argue that:

I am so dastardly a Red that I had rather a quarter of a million miners won the next war, or even better averted it, by digging holes [air raid shelters], than that the young gentlemen of the Royal Air Force lost it by bombing the cities of Europe.[31]

Haldane was criticising Britain's commitment to a bomber fleet as both morally and technically wrong. Haldane explained this commitment in terms of old-fashioned militarism and the lack of scientific culture among Britain's ministers and civil servants: science suggested a very different defence policy.[32] Haldane was extremely unusual in discussing the relation of science to strategy, even if he did so very briefly.

Huxley and Haldane argued for a new peace policy, and a new defence policy, based on science. More common was the equation of science with pacifism, to stress the pacific uses of science, and to see scientists as pacifists. An article in *Nature* in 1937 argued that:

Already in the time of the Great War the prostitution of science to warfare was deeply felt by many British men of science, some of whom refused altogether to participate either as soldiers or scientific workers. The same attitude would again be taken by many, should we be cursed by another war; even in the present time of uneasy peace, the activity in rearmament is renewing in the minds of many men of science the same tension of feeling, the same problems of conscience.[33]

No evidence was provided of any British man of science who was a conscientious objector or refused to put his scientific expertise at the service of the state (though there were some), or that many "deeply felt" the prostitution of science. Certainly the pages of the wartime *Nature* (which was edited by Norman Lockyer) reveal no such scruples. The identification of science with moral concern about the application of science to war was a product of the 1930s.

Scientific intellectuals acknowledged the links of science, technology and war, but these were often felt to require a contingent explanation. A common way, perhaps the dominant one, of explaining, indeed explaining away, the linkages, was the invocation of a 'cultural lag', though it was not called this. As Sir Alfred Ewing, a former Director of Naval Education and codebreaker from the Great War, put it in his 1932 Presidential Address to the British Association for the Advancement of Science "... the engineer's gifts have been and may be grieviously abused... Man was ethically unprepared for so great a bounty. In the slow evolution of morals he is still unfit for the tremendous responsibilities it entails. The command of Nature has been put into his hands before he knows how to command himself".[34] However, for Ewing the radio was an unmitigated blessing "which even the folly of nations cannot pervert... Can you imagine any practical gift of science more indispensable as a step towards establishing the sense of international brotherhood which we now consciously lack and wistfully desire?"[35] The faith in science and technology as essentially civil and liberating phenomena remained undaunted, even if faith in humanity proved premature.

The idea of a general 'cultural lag' was generally used to shift blame onto something so large and abstract that little could be done about it. However, more refined versions of the thesis pinned the blame on something more specific, and proposed or implied particular courses of action. The philosopher R.G. Collingwood wrote in the late 1930s that:

I knew that for sheer ineptitude the Versailles treaty surpassed previous treaties as much as for sheer technical excellence the equipment of twentieth-century armies surpassed those of previous armies. It seemed almost as if man's power to control 'Nature' had been increasing *pari passu* with a decrease in his power to control human affairs. That, I dare say, was an exaggeration... I seemed to see the reign of natural science, within no very long time, converting Europe into a wilderness of Yahoos.[36]

The answer was neither more good will and human affection (as most exponents of the cultural lag argued) nor was it to be found in the extension of the methods of natural science to the social and the human (as most scientific intellectuals argued, notably Huxley). A new kind of history, informed by a thorough philosophy of history, could teach "man to control human situations as natural science had taught him to control the forces of Nature".[37] J.G. Crowther, the science correspondent of the *Manchester Guardian*, complained that "aviation has been frustrated by its submission to military needs, by the possessors of real estate, by nationalistic governments; and its influence on human culture has been frustrated because it has been developed for the use of soldiers and sportsmen, and not for the service of the working civilian, whose intellectual and physical labour creates the most durable human values".[38] The remedy proposed is obvious, but the diagnosis is not: it was precisely nation states, the military, and the rich that had created and sustained aviation; a purely civil and civilian aviation would have been a puny thing. What is interesting is that Crowther cannot reject the aeroplane, or indeed any technology; they all have potential for good in the right hands. Such arguments could easily slide into the belief that modern technologies themselves could be actively used to destroy nationalism and militarism. The image of a new world state which would enforce world discipline through technology, in particular through an 'international air police', was a 1930s commonplace. A *Nature* reviewer quoted the following with approval: "Hand over all those recent and future applications of science to warfare to your new international police force, and let domestic armies be content with pre-War armaments".[39] The 1930s saw a whole series of novels about an international force based on high technology making war to end war, the most important of which was that by H.G. Wells, *The Shape of Things to Come: the Ultimate Revolution* (1933).[40]

The marxism of the Second and Third International was itself in many ways a cultural lag theory: capitalist society was acting as a fetter, not only on social development, but on the productive forces themselves. For marxists, nationalism, militarism, and the rich were not themselves the problem, but were seen as the inevitable product of a capitalist economic structure. Thus, as a thoughtful marxist, the physicist J.D. Bernal saw the causes of war, and thus the frustration of science, in capitalism, and not in human nature or in nationalism as such. War, he wrote, "cannot effectively be combatted unless its social and economic nature is fully understood, and scientists are a long way yet from this understanding".[41] His argument about the relations between science and war is fundamentally an attack on capitalism.[42] He wanted to create a social system in which science could be free to flourish, and that for him was socialism. In his famous *Social Function of Science* he wanted to show the extent to which science was systematically corrupted under capitalism. A key example was the extent to which British science was funded by the military. Bernal highlighted this expenditure, getting the picture broadly right: "It would not be unfair to say that something between one-third and one-half of the money spent on scientific research in Britain is spent directly or indirectly on war research... And this in peace time".[43] At the same time, however, he was keen to highlight that the civilian scientist was a superior being to the soldier, even when it came to winning wars. Writing in 1935, he argued that even from a military point of view warlike R&D was inefficiently performed: in the Great War, civilian scientists drafted from the outside made the major contributions; the "military mind is naturally averse to innovations which make war so much nastier for all concerned", he wrote.[44] In 1939 he highlighted the vile new weapons of the Great War, and condemned the chemist Sir William Ramsay for his aggresive nationalism. But he also condemned the military for not giving H.G.J. Moseley a scientific job, and again insisted that warlike research had been done wastefully and inefficiently.[45]

The idea that civilian science was superior to military science, even in winning wars, was a common one; we have already seen the point made by Huxley and Haldane. A basic assumption was that modern wars were industrial and technological wars, rather than wars of men imbued with the martial spirit. Bernal and others implicitly went further: instead of arguing that military technology – the aeroplane and so on – had made the decisive contribution to the Great War, they insisted that industrial technology was critical: Germany had been successful at the beginning of the Great War because its *industry* had made greater use of science than British *industry*. Bernal argued that "the War, and only the War, could bring home to Governments the critical importance of scientific research in the modern economy. This was recognised in Britain by the formation of the Department of Scientific and Industrial Research".[46] Bernal, who recognised the important role of the Air Ministry, Admiralty and War Office in British science in the 1930s, did not mention their wartime predecessors, the Ministry of Munitions or the service departments. During the war, these ministries, and especially, the Ministry of Munitions, had a scientific and technical effort which dwarfed that of the DSIR, just as in peacetime. The identification of government-funded science, even in the Great War, with the DSIR, has proved particularly enduring, as we shall see.

IV

The argument that civilian and civil science was critical in modern warfare proved important to the anti-fascist scientific left. Two weeks after the Munich agreement *Nature* published an editorial entitled "Science and National Service". The anonymous author – in fact Bernal – argued that the mobilisation of science for war would not be successful "unless its ultimate aim – the utilisation of science for human welfare in times of peace – is kept steadily in view"; indeed, "Preparation... for war requires not only a much more thorough organisation of science, but also a much closer integration between scientific research and other activities of the community, particularly those of industrial production, agriculture and health".[47] These themes were fully endorsed in an influential, anonymously produced, Penguin Special called *Science in War*, which was written and published in a few weeks in 1940 by a group of progressive scientists, including Bernal. Its central argument was for a greatly expanded use of science by the British state and industry. It barely discussed the most obvious use of scientists in war, the design and development of new offensive weapons. Attention was firmly fixed on protection of the civil population, care of the wounded, and the rational organisation of the armed forces, economy and society. In a book of 144 pages only twelve were devoted to warlike technologies (aeroplanes, tanks and magnetic mines), compared with 38 pages dealing just with food. Furthermore, argued *Science in War*:

the world of scientists until now, and still, is a world which has little or no say about the uses to which scientific advances are put. Had it been otherwise, had scientific method played its part in the conduct of social and international affairs, wars and other forms of aggression might have been wiped from the face of the earth.[48]

If science could no longer be associated with pacifism, it need not be associated with militarism or with the military. Even in war science should be a fundamentally civil and progressive activity. But the reality was, needless to say, rather different. Scientists were recruited into government laboratories, industry, and the forces themselves, primarily to design and develop weapons, and to advise on their use.

In the Second World War British scientific intellectuals saw themselves as defending both science and democracy. Whereas Wilhelmine Germany could not easily be portrayed as an enemy of science, Nazi Germany could: it had dismissed many scientists from their positions, and endorsed a specifically Aryan science. In the Great War, British scientific intellectuals were not notable for their defence of democracy, but in the Second World War their position was rather different. *Science in War* argued that "Science and democracy are no longer merely desirable goals: they are conditions of survival".[49] Furthermore, scientists presented themselves as being especially prescient about the dangers of the Nazis and the nature of modern war: in September 1941 the British Association for the Advancement of Science held a conference on the subject of "Science and the World Order". A report of the conference, another Penguin, argued that "The men of science, more plainly than anyone, had foreseen what war, made terrible by the abuse of science, would mean... They had seen science, which must be international... fettered as 'Nazi science' or 'Fascist science'".[50] Just at the moment when science was fully mobilised to national war purposes, the internationalism of science was invoked; just as scientists turned their full attention to the making of weapons, they retroactively became prophets of peace.

The question arises: did scientists play any role during the war in resisting the encroachment of the state on science, or the lives of citizens; did they

speak out against any abuses of science in the service of war? There is little
or no evidence that they did so. Many years after the war it was claimed
that scientists stood out against 'strategic' bombing. But during the war there
were hardly any public stands against such bombing, even though it was in a
sense the perfect test case for supporting internationalism (defending German
civilians) and objecting to the corruption of science. An exception was A.V.
Hill, a veteran of Great War naval research and a leading figure in the Royal
Society, who was scathing about it in Parliament, but not on the grounds
that it was immoral.[51] It might be objected that no opposition was possible,
but in fact it did exist.[52] In view of their claims to a particularly powerful
internationalism, and a higher morality, scientists were noticeable by their
absence from the ranks of the dissenters.

V

Once the war was over, the scientific intellectuals argued that science had been
very successfully applied to winning the war. In contrast to what had been
said in the 1930s there was no hint of inefficiency or wastefullness of warlike
research, nor indeed was there any particular objection to the way science had
been used in the war. There was, nevertheless, some sensitivity to the issue.
The socialist trade union, the Association of Scientific Workers, which had
boomed in membership during the war, noted in its 1947 manifesto, *Science
and the Nation*, that

it is easy to curse science and scientists. But the real lesson which the development of "scientific
weapons" should make as clear as daylight is: when sufficiently large resources of finance,
organisation and scientists are used there are few problems of the control and exploitation of
Nature that cannot be solved, and often with unexpected speed.[53]

Some scientists complained directly about the uses to which science had been
put in war. For example, Sir Henry Dale, President of the Royal Society,
wrote in 1946 that: "The use of science as an aid to war is a perversion from
its proper purposes, and its rapidly extending misuse in the recent war, as a
direct agent of violence and destruction on a stupendous scale, creates a threat
to the survival of civilization".[54] But that statement was made in a foreword
to a celebratory book called *Science at War* commissioned by the Scientific
Advisory Council to the British Cabinet, of which Dale had been a member.
Another blistering attack came from a former Director of Medical Research
at the War Office, F.A.E. Crew. He wrote in 1952:

The scientist, aware of the expanding power of the weapons of war that are being fashioned out
of scientific knowledge in application, is required by his sense of social responsibility to protest
against such misuse of scientific knowledge. To the biologist, the pattern that war is now taking
is completely horrible. He must in his own life work actively to oppose it. He must make his
contributions to a deeper understanding of the actual causes and consequences of modern war,
and he must play his part in the creations of a world in which war will be unknown. Peace
will not be achieved... as long as there are among us so many who are warped, embittered,

and frustrated by the conditions and circumstances of the social environment in which they were raised. [It was necessary to harness the rebellion of the young] against things as they are to the tasks of overcoming ignorance, superstitition, disease, of preventing the exploitation of human beings and the fettering of science and technology for national aggrandizement and the expansion of the dominion of the unproductive.[55]

The tone and style are reminiscent of the 1930s, but Crew was writing in the early 1950s, in the context of massive rearmament, and the Korean War. He noted the aim of both was officially stated to be to demonstrate that aggressive action did not pay. But he said: "I do not know, and I have no means of knowing" whether this argument was valid. "Possibly war has come to have a really new purpose; maybe war of this kind is something that can be justified". "I must", he concluded, "as an ordinary citizen, uneasily place my trust in those whose responsibility it is to lead".[56] Scientists, he seems to suggest, should, as scientists, oppose war, but as citizens, trust political leaders. The genuine doubts of the citizen have no place in the bombastic rhetoric of the scientist.

Thus scientists did not engage with the issues raised by actual wars; neither did they engage with the question of what war did to science. Indeed, I know of only one argument that the mobilisation of science and technology in the Second World War was not a good thing. It dates from 1956 and comes from the hand of the engineer Sir William Stanier, a member of the wartime Engineering Advisory Council. After giving many examples of the way in which the two world wars advanced British engineering, he wrote:

The foregoing may suggest that the influence of war on the advancement of engineering is wholly beneficial. But it should be noted that though war stimulates advances it does so only in restricted fields. In other fields advance is brought almost to a halt not merely 'for the duration' but for long afterwards... during the war, the thoughts of many brilliant men had to be turned away from the creation of things beneficial to the human race and concentrated upon devising new means of destruction or new means of averting an enemy's destructive intentions... But little was learnt from much of that activity which can be applied in peace, certainly not nearly as much as could have been learnt had all the brains and insight been applied to the advancement and construction of peaceful appliances. In short, the influence of war upon engineering advancement is to distort rather than to further it. The benefit, so very apparent in certain fields, is, in this writer's opinion, more than over-balanced by the setbacks suffered in other fields and the wastage of talent inherent in the design of destructive instead of constructive things.[57]

Stanier's moderate and plausible analysis that the net effect of the war was to retard civil technical development is remarkable for its rarity.

War research continued in Britain after the war, indeed it hardly fell from its wartime peak.[58] Scientific intellectuals, however, while celebrating the successes of wartime warlike R&D, ignored its continuance in peacetime. *Science and the Nation* (1947), for example, made only a passing reference to services research.[59] The 1956 Reith Lecturer, Sir Edward Appleton, former Secretary of the DSIR, felt that he had to explain his title, "Science and the Nation" because science was obviously international. His chapter on science and war, implies that before the war scientists were all civil; his chapter on government and science is mostly concerned with civil science.[60] This postwar neglect of warlike R&D must be seen in the context of the very significant deradicalisation of intellectual politics in the face of 'Natopolitan culture'.[61] The militarisation of research was an embarrassment to the scientists, but to be a pacifist was to be seen as a communist; consequently, to draw critical attention to the extent to which R&D was funded for military purposes was to brand oneself a subversive. This would prevent one working in the higher civil service, and obtaining an American visa.[62]

Those who did speak out on the extent of warlike R&D were indeed typically communists. The AScW, which came under increasing Communist Party influence, constantly pointed to the high level of warlike R&D in Britain.[63] J.D. Bernal, who was AScW President between 1947 and 1949, was the key spokesman. In a 1949 pamphlet written with Maurice Cornforth he showed clearly the importance of warlike research. They argued that war increasingly had the object of "blind slaughter without even military excuse"; this was "foreign to the whole tradition of science" but if accepted would lead to scientists losing "all sense of social responsibility and moral value in science". The development by scientists of weapons of mass destruction gave "apparent justification to the non-scientific public to associate science with war in its most horrible aspects". The imperatives of military secrecy would destroy free scientific communication.[64] They did not reflect that exactly the same objections could be made against war research during the Second World War. Moreover as a supporter of the Soviet Union, Bernal rejected calls for international control of atomic weapons.[65]

In the 1950s Bernal developed the traditional liberal argument: war had become irrational because science produced greater wealth than conquest:

The wealth that could be available to us now, through the application of the amount of science we know already, is far greater than anything that could be obtained from the conquest of the most fruitful territories or by winning the most exclusive controls of raw materials, oil or coal.[66]

The advance of science had rendered the old marxist category of imperialism obsolete. Nuclear energy would help usher in a world of plenty; Bernal regretted that military nuclear development was delaying civil use.[67] Science for Peace, established in 1951, the principal platform for the much diminished scientific left, also adopted a very liberal view of science and its social relations.[68]

VI

One of the many ironies of the early postwar era was that the atomic bomb featured in many arguments calling for a radically new engagement of scientists in politics, and a new scientific ethics. *Science and the Nation* argued that

"Particularly among scientists... a new and more vivid awareness has arisen of the need for the social control and social use of science and the dangers which threaten if it is not so controlled and used", as a result of the atom bomb.[69] A.V. Hill argued that

The only hope... of averting the disaster which science, misapplied, could inflict on humanity is an international brotherhood of scientific men, with a common ethical standard by which potential crimes of this character would be exposed and prevented. For, if political isolationism and aggresive nationalism are to exploit science and its applications, not for the benefit of mankind but in order to prepare in secret for mutual destruction, they are very likely to succeed.

Scientists, he went on, "must be allowed to work together in mutual confidence and sincerity. Ethical standards... must be restored... so that misuse of scientific knowledge... will be regarded – like cowardice in a soldier or dishonesty in a banker – as the unforgivable sin".[70] Despite this British scientists seemed less concerned about the relations of science and war than in the 1930s. The prospect of gas warfare, or aerial bombing, had had more effect than the reality of strategic bombing or the atomic bomb. Indeed, the bomb caused more concern than strategic bombing, even though more civilians were killed by conventional bombing raids on Japan than by the atomic bomb: a single large conventional raid did as much immediate damage as the first atomic bombs.

Even so, rhetorical heartsearching about the bomb did not translate into action against the bomb. In 1946 an Atomic Scientists's Association was formed, with Neville Mott as President and Patrick Blackett, John Cockcroft and Lord Cherwell as Vice Presidents. It was made up largely of scientists who had worked on atomic weapons during the war, and acted as an information service and a forum for the discussion of atomic issues. In its early years, the main topic was international control of atomic energy. But even Mott, who was hostile to a British atomic bomb, was not in principle against atomic weapons. He suggested that peace might come from the bomb: "we can only hope that, with this sombre bomb in the background, war between nations will come to be regarded like war between Canada and the United States, as something unheard of, against which no preparation need be made. This, or a world state, seem the only alternatives before civilisation". Mott (unusually) made connections between fear of the bomb and fear of air power before the war.[71] He was, with Kathleen Lonsdale, the only 'vociferous' opponent of the British bomb to be found among the contributors to *Atomic Scientists' News*.[72] There was, however, little if any recognition that talk of world states, and international control of weapons, had been a staple of 1930s, and indeed wartime, arguments about aviation. When transferred to nuclear weapons they appeared original.

The emphasis of scientific intellectuals on the bomb as a moral question, and as a world transforming technology, obscured what it most obviously was: a weapon of war. Patrick Blackett, who wrote a very remarkable book on atomic weapons published in 1948, argued that atomic weapons had changed war, but had not utterly transformed it. He put atomic bombing in the context of wartime strategic bombing, and argued still more unusually that the dropping of the bomb on Japan was intimately connected with the entry of the Russians into the war.[73] For Blackett this political and military contextualising was essential; without it scientists' idealistic calls for international control were nothing but a cover for power politics.[74] Blackett was to continue to challenge official policy, and the arguments of scientists. In his Lees-Knowles Lectures of 1956 he argued that the 'First and Second World Wars were won by the Allies primarily by their superiority in numbers over Germany, certainly not by any overall technological superiority'.[75] The Baruch Plan for the international control of atomic energy

was the illegitimate offspring of the idealism of conscience-stricken scientists and the conservative realism of hard-bitten statesmen. It was based on the absurdity that a few dozen atomic bombs could defeat a continental power cheaply and quickly, and so nourished – for many the fear and for a few the hope – of a preventive war.[76]

Blackett neatly turned a favourite argument about the backwardness of soldiers on its head: "It is an old crack, however unjust, that the natural conservatism of soldiers makes them plan the next war with the concepts of the last war but one. It seems to me that, historically speaking, the extreme advocates of air power have often planned on the basis of the next war but one". In an age when war would once more become limited "perhaps now the soldiers should really plan for the last war but one".[77]

VII

Most of the scientific intellectuals discussed above believed, strongly, that scientists did have something to say about how science should be used. They stressed the utility of science, but downplayed what was in fact one of its key uses – its contribution to the preparation for and waging of war. Nevertheless, they argued that civil science served the nation in war. Many argued for the systematic planning of the entire scientific and technological resources of the nation. Planning meant the direction of these resources towards *civil* objectives. The high point for this kind of leftist scientific intellectualism was the late 1930s and early 1940s. But from the 1940s there was an organised opposition to the scientific left's equation of science with socialism and planning, and the failure to distinguish science from technology. The Society for Freedom in Science, believed 'Bernalism' to be a fundamental challenge to pure university science. In defending it some were led to a critique of scientism. While they stressed the social benefits of pure research, they did not concern themselves with the research that went on in government and industrial laboratories, which they acknowledged was and should be directed.

In effect they consigned the vast bulk of scientific activity to the domain of technology, where there was an obvious connection to society, politics, economics and warfare. To that extent they did not oppose the planning of most of science.[78] Thus Michael Polanyi could write in 1962: "the movements for guiding science towards a more direct service of the public interest, as well as for coordinating the pursuit of science more effectively from a centre, have all petered out".[79]

Polanyi's implicit definition of 'science' as pure academic science lay behind much writing in the 1950s and 1960s. Thus John Ziman, in a book on 'the social dimension of science': "... as I read popular journals about 'Science', I cannot help being disturbed by the use of the word to cover almost any conscious rational technique". For Ziman, science was not technology, although to "the vulgar eye" the distinction "must seem perverse".[80] The "vulgar conception" of Science as the mastery of Man's environment, "confounds Science with Technology" and "confuses ideas with things".[81] He finished his book with: "The subject of "rational technique" is even vaster and more diffuse than Science as I have delimited it, and calls into question the whole of our way of life, our ethical and moral standards, and the poetical quality of our earthly existence";[82] "Science", by implication, did not. But the vulgar conception that Ziman complained of was as much the product of scientific intellectuals efforts as of popular misunderstanding. As Ravetz was later to comment, "... for the preservation of the innocence of science, the term is understood to include only that research where both the research-worker and the supporting agency have the purest of internal criteria of significance".[83] The upshot of this was clear; on this definition 'science' could have nothing to do with war.

The equally legitimate conclusion that if a scientific activity was related to war in any way it could not be called 'science' was not stressed, except in an inverted form. Military irrelevance was used as a criterion of 'purity', and strong efforts were made to show that the history of science and the history of war barely intersected, and that the external links of science were with philosophy. A few examples will suffice. Robert Oppenheimer was invited to give the 1953 Reith Lectures. The public and press naturally expected them to be about science, politics and war; instead they got history and philosophy of science.[84] The historian of science Rupert Hall claimed that in the scientific revolution ballistics and mechanics had developed entirely separately.[85] Sir Gavin de Beer wrote a book largely concerned with Humphrey Davy's visit to France in the Napoleonic Wars to receive a prize, in order to show that science was deeply internationalist: it was called *The Sciences were never at War*.[86] More remarkable still was the fact that some outspoken critics of the interelationship between science and war should argue that the relationship was recent. Robert Jungk's *Brighter than a Thousand Suns* presented a picture of interwar physics as pure: it asked why certain scientists had fallen from grace, and argued that, paradoxically, the German physicists were the ones that fell least.[87] In 1960 C.P. Snow gave a lecture entitled 'The Moral Un-neutrality of Science', in which he claimed that:

The scientific world of the Twenties was as near a full fledged international community as we're likely to get... the atmosphere of the Twenties in science was filled with an air of benevolence and magnanimity which transcended the people who lived in it... The discovery of atomic fission broke up the world of international physics.[88]

At least the history of science was safe from the taint of war.

The implied breakpoint in the history of science was 1939/40, perhaps Hiroshima itself. After it, science, politics and war were inextricably linked; innocence had been lost. The focus on the bomb shifted attention away from the whole military-scientific complex of the second world war. Furthermore the bomb-centered story obscured the fact that the military-scientific complexes of the war did not appear out of thin air: all the Great Powers had strong military-scientific complexes going back decades. The really significant break in destructive capacity came with the development of the H-bomb, rather than the A-bomb. But seeing a decisive change in the 1950s would have pointed to the need for action, and to the fact that this escalation was carried out in time of peace. Better in a sense to have already known sin, than to fall into it.

If science had known sin, that did not mean that science was to blame, or that science did not provide guidance for the human spirit in the new and dangerous atomic age. Many scientific intellectuals insisted, repeatedly, that science itself provided the only way out. Jacob Bronowski, for example, blamed the relationship between science and war on divisions within society: there were no panaceas, 'there is no cure in high moral precepts'. Only science could cure that division in society which caused war.

This is a remarkable age of science. It is for us to use it to broaden and liberate our culture. These are the marks of science: that it is open for all to hear, and all are free to speak their minds in it. They are the marks of the world at its best, and the human spirit at its most challenging.[89]

This sort of scientific humanism needed heroes. Oppenheimer was not the only key member of the military-industrial-scientific complex remodelled as a moral hero. C.P. Snow tried to do the same for Sir Henry Tizard. One of Snow's main purposes in his *Science and Government* lectures was to show that science had been on the side of good in the Second World War. A great part of these lectures were devoted to an account of the so-called Tizard/Lindemann dispute over strategic bombing in 1942. In Snow's account Tizard was a goody who opposed strategic bombing and Lindemann a baddy who supported it; while Tizard was the "English of the English", Lindemann was "quite un-English"; Tizard was poor, Lindemann rich; Lindemann was strange, Tizard normal; and, to cap it all Tizard was the better scientist.[90] Strategic bombing was "an unrationalised article of faith"; "Lindemann had always believed in this faith with characteristic intensity".[91] The burden of

Snow's case was that Tizard and Blackett had demonstrated that Lindemann's calculations, made in early 1942, were wrong; but that neither Lindemann nor the Government accepted this and therefore a disastrous strategic bombing policy was stepped up. Blackett was to vouch for the accuracy of the Snow story, and wrote:

I confess to a haunting sense of personal failure, and I am sure Tizard felt the same way. If we had only been more persuasive and had forced people to believe our simple arithmetic, if we had fought officialdom more cleverly and lobbied ministers more vigourously, might we not have changed this decision'.[92]

But, in fact, the argument was over a simple arithmetical error Lindemann had made, and accepted that he had made; neither Blackett nor Tizard noticed another simple error Lindemann had made.[93] The decision had not depended only on estimates of effects alone, nor did Blackett's own figures prove accurate. Furthermore, Lindemann's notorious paper was based, at least in part, on a scientific analysis of the effects of bombing on Hull and Birmingham, carried out by Solly Zuckerman and J.D. Bernal. Indeed, Zuckerman had suggested that such an analysis be done to Lindemann himself![94]

Blackett and Snow came to believe that there was a moment of decision when reason was overlooked. Michael Sherry has noted the importance of this kind of moral thinking about air warfare and nuclear weapons, but comments that decisions "certainly resulted from choices but not from a moment of choice".[95] Snow, especially, did not produce any historical context for the "dispute"; he did not make clear that Tizard had been the senior scientific adviser to the Air Ministry since 1933, or that he had been a member of the ruling Council of the Ministry during the early part of the war. Snow later admitted that he was not really sure whether Tizard had any fundamental objections to strategic bombing, without realising the damage that this did to his whole case.[96] Blackett, a former naval officer and wartime scientific adviser to the Admiralty, was an opponent of strategic bombing, but he to came to forget his history: "So far as I know, it was the first time that a modern nation had deliberately planned a major military campaign against the enemy's civil population rather than against his armed forces. During my youth in the Navy in World War I such an operation would have been unthinkable".[97] Blackett forgot that the civilians were bombed in the Great War, and that the major task of his Royal Navy was to impose a blockade against Germany which aimed to cause hardship to the civilian population.

The Tizard/Lindemann dispute was in effect a dispute involving 'operational research', which after the war would become a much celebrated example of the use of scientists in war.[98] This was not surprising for OR was 'clean'; it did not involve the making of deadly new weapons; rather it was concerned with the analysis of operational data and recommending changes in operations on the basis of the results. It was in a sense a pure 'scientific method'. The story of operational research was often told in a way which suggested that soldiers, sailors and airmen did not know how to conduct operations rationally – the scientist, on the basis of some figures, could. What was not stressed was that the US equivalent was carried out by lawyers, economists and businessmen, or that in Britain the techniques were pioneered in business, and in the pre-war Air Ministry. Furthermore, operational research was not unambigously on the side of saving life and humanising war. Most attention was given to the operational research sections of the RAF's Coastal Command, which was largely concerned with anti-submarine warfare, and with that of the Admiralty. Hardly any attention was given to one of the very largest operational research sections; that attached to RAF Bomber Command. But we are fortunate to have a recent memoir by Freeman Dyson, published in 1979, of his time as a junior member of the section. He recalled how his advice on reducing losses of aircrew, by taking out guns, and making escape hatches better, was ignored. In a scathing attack on Bomber Command he said "it might have been invented by some mad sociologist as an example to exhibit as clearly as possible the evil aspects of science and technology". The root was the doctrine of strategic bombing which had been "attractive to political and military leaders", though not, it is implied, to scientific leaders. "Bomber Command was an early example of the new evil that science and technology have added to the old evils of soldiering. Technology has made evil anonymous. Through science and technology, evil is organised bureaucratically so that no individual is responsible for what happens". Even he himself had no feeling of personal responsibility.[99]

Dyson is unsparing in criticism of himself, but nevertheless he exonorates himself to some extent by his insistence on the lack of individual responsibility in a bureaucracy, and above all because it is the scientist that is trying to save life, while the soldier is profligate with it. The wartime commander of Bomber Command, much criticised by Dyson as an antediluvian militarist, was by any standard a great admirer of science, and the head of one of the most scientific branches of the armed forces. As he wrote in 1947: a new Defence Force

must obviously become more and more dependent on science, and must make it its main business, as the only condition of winning the next war, to exploit the best contemporary weapons that science has to offer, with no more regret when it relinquishes an older weapon than a scientist shows when a hypothesis is exploded, or when he finds a quick and easy method to replace one that was inefficient and laborious.[100]

It is clear that the image scientific intellectuals wanted to portray of commanding officers as stupid, bloodthirsty, and wasteful, in contrast to the humane but powerless scientist simply will not wash.

Just how powerful, and sickeningly absurd, mythmaking about scientists and war could become is well illustrated by the BBC's famous thirteen part television series, *The Ascent of Man*, broadcast in 1973. This history of science was written and presented by Jacob Bronowski. Bronowski discussed the atomic bomb not through the life of say, Oppenheimer, or Teller, or General

Groves, but through that of Leo Szilard. Commenting on Szilard's 1934 chain reaction patent he says that, characteristically of scientists at that time, Szilard "wanted to keep the patent secret. He wanted to prevent science from being misused. And, in fact, he assigned the patent to the British Admiralty, so that it was not published until after the war".[101] In the 1930s, Bronowski went on, Szilard tried and failed to get others not to publish; later he failed to get the U.S. government to merely demonstrate the power of the atom bomb: "Szilard failed, and with him the community of scientists failed," but Szilard redeemed himself by giving up physics for biology. Again we have a story of a scientist trying and failing to save humanity at the centre of a story more obviously told in other ways. But even the story of Szilard could have been told very differently: Szilard was an active promoter of the idea of atomic weapons, he was fascinated with the idea of world government and with weapons of mass destruction.[102] Szilard was among the very first to put forward the idea of 'preventive' atomic war against the Soviet Union.[103]

VII

It is time to leave the never-never land of the postwar scientific intellectuals to ask how historians have dealt with the relations of science, technology and war in twentieth century Britain. We will find many echoes of the arguments of scientific intellectuals. Clark's *The Rise of the Boffins* gives the appearance of covering the whole field of the relations of science and the Second World War, but in fact concentrated on radar, radio, and operational research, with emphasis given to the work of academic scientists.[104]

Rose and Rose's widely read *Science and Society*, published in 1969, was a good summary of the arguments and evidence presented up till then. They covered the period 1914 to 1945 in two chapters. "The Chemists' War" argued, in summary, that in 1914, due to the neglect of science by government and industry, Britain was short of key materials. Government belatedly responded, especially through the creation of the Department of Scientific and Industrial Research. The rest of the chapter is devoted to the industrial research associations, the DSIR laboratories, the Medical Research Council and the scientists' trade union. There is but a fleeting mention of defence research!

"The Physicists' War" centred on the Second World War, has some discussion of pre–1939 defence research, which, before 1914, it is claimed: was "not of a level which could really be dignified by so grandiose a title".[105] They mention the development of poison gas and the scientific work of the Admiralty in 1914–1918, but they say that by the early 1920s, "the innate conservatism of the military and the genuine hopes of disarmament ... certainly helped inhibit active war research in Britain".[106] In the 1930s however, war research picked up, notably in radar. Research in wartime is discussed in terms of operational research, atomic power, the mobilisation of civil scientists, and the higher organisation of the scientific effort. Only for the postwar years do they clearly recognise the importance of defence R&D. Similarly, Gummett's standard text on science and British government, was heavily biased towards civil science.[107]

This kind of picture of relations of science, technology and war in Britain was filled out by detailed studies by professional historians of science. Cardwell, in a paper on science in the First World War discussed only the essentially civil DSIR.[108] MacLeod and Andrews, as well as Pattison, who considered warlike departments of the First World War, saw the application of science to war as essentially new and brought about by civil scientists entering the war machine in war. Furthermore they did not adequately contextualise their studies of particular institutions within wartime R&D as a whole.[109] The major recent general work on science-state relations in the period 1850–1920, Alter's *The Reluctant Patron*, reproduces the neglect of the military to an extraordinary degree. Alter's study ignores the warlike departments of the state, in peace and war, without specifying that his treatment covers only civil R&D. He ignored therefore the most important element of the patronage of science and technology by the British state.[110] In none of these works would one get any sense of what the most significant ministries for the support of science and technology were, either in peace or war. One way this could be summed up is by noting that aviation barely figures.

The historiography is thus faithful to the public pronouncements of scientists (and to a much lesser extent engineers), except in two telling cases. On the one side, the treatment of Norman Lockyer, who, given his importance, appears in many of these works. In his biographical sketch of Norman Lockyer, Alter does not mention that most of Lockyer's working life was spent at the War Office. The militaristic content of "The Influence of Brain Power on History" is noted but the issue is seen as one of scientific strength being another element in defence, to be added to naval power.[111] But at least Alter went part of the way. North, in a biographical sketch of Lockyer, published in 1969, inferred from that 1903 speech: "He deplored the miserably small sums of money which the state advanced for the universities just as he deplored the substitution of battleships for intellectual values".[112] On the other side, the failure to reproduce Bernal (and Huxley's) analysis of spending on science in the interwar period which revealed the importance of warlike R&D.

This sort of misreading of the history of science-state relations, and of the politics of science is all the more surprising because the late 1960s and early 1970s saw the emergence of critical commentary on science, and a new historiography of science, which rejected the scientific humanism of the Cold War as an analysis of science. Critical to the development of this new conception of science was the United States's technological destruction of Vietnam. Technocracy, warfare, social control were now seen as inextricably and sytematically linked. However, critical work was concentrated on the biological sciences, and such issues as IQ, race, class and gender.[113] It is notable, for example, that Albury and Schwartz's critical textbook on the politics of science and technology barely mentioned the military.[114] It is significant too that British authors looked to the United States for questionable cases of relations of science and war, as in Brian Easlea's stimulating book on [American] nuclear weapons as expressions of masculine values.[115] Just as noteworthy is the fact that the key, early work on the international relations of British science around the Great War, which demolished the myth of scientific internationalism, was done by non-British scholars.[116] And here again, the British contributor was a biologist.[117]

That a focus on science, and traditions of writing about science, was a factor explaining the shaping of the historiography of the relations of science and war is suggested by the fact that historians without such an orientation have produced a different picture. Margaret Gowing's official history of the wartime British bomb project noted clearly the lack of British scientific opposition to the atomic bomb.[118] Her two volumes on the postwar British bomb project also noted the lack of opposition by scientists to the bomb, with the exception of Blackett and the left-wing Association of Scientific Workers.[119] Particularly striking was Guy Hartcup's study of invention in the Great War. Hartcup looked at the most important inventions and technical developments of the war: the DSIR barely figures in the story.[120] Mary Kaldor hardly touched on the institutions and arguments of the scientific intellectuals: battleships, aeroplanes, and tanks were her concern.[121] Maurice Pearton has explored the consequences of the 'industrialisaton' of war, again focussing on military technology.[122] But even some of these studies reproduce the images scientists themselves presented. Thus Hartcup stresses that the military in the Great War were resistant to new technology. Mary Kaldor explains what she sees as the baroque character of military technology as resulting from the technical dynamism of civil capitalist corporations interacting with the conservative military; in brief the military corrupt technology. Pearton sees the origins of the industrialisation of war as purely civil. This comment applies, though to a lesser extent, to two superb recent studies of British military and naval technology by Sumida and by Travers.[123]

VII

In a famous letter to the Grand Duchess Christina, Galileo set out a straightforward argument for the independence of scientific inquiry from literal readings of the Bible. In certain passages, the Bible spoke of the physical world in specific ways. But God had not intended these statements to be taken literally: they were designed for the common people. Intellectuals, who knew better, were free to enquire into the natural world.[124] But science, too, has its stories for the masses; its propaganda. Clearly the student of science cannot rely on the literal truth of the statements of scientific intellectuals, even when they deal with mundane aspects of science and of society.

British scientific intellectuals generally reworked standard liberal progressive beliefs in their accounts of the relations of science, technology and war. They were seemingly untouched by the history of a profoundly militant century. But some scientific intellectuals, in particular periods, with particular arguments to make, and influenced by particular political ideologies, subverted at least parts of this picture. A nationalist militarist like Lockyer, arguing for strong armed services before 1914, saw science and militarism marching together. A marxist like Bernal, arguing against the scientific and socio-political status quo, showed just how dependent British science and technology were on the military, as part of an argument for the abolition of the capitalism which, in his view, visited these horrors on science. A former career naval officer like Blackett, a man of the left, deeply hostile to Britain developing nuclear weapons, contextualised nuclear weapons in the history of strategy, and had no respect for conscience-striken scientists.

It needs stressing that these arguments and evidence were lost from the discourse of scientific intellectuals. This was due above all to the political climate, and changes in political ideology, but it was perhaps in part because there was no tradition of debate among scientific intellectuals on these issues, only of ex-cathedra pontification – it is striking how little reference there is to others' arguments in all the scientific and technological literature on science and war. There was no cumulation of knowledge. One result was that the age of the hydrogen bomb and the permanent mobilisation of R&D for warlike purposes produced the vacuous, and profoundly ahistorical, moralising of C.P. Snow and Jacob Bronowski. They influenced, unfortunately, a whole generation of "irritating know-alls".[125] Gil Eliot captured something very important about the Snow-Bronowski kind of rhetoric, which he described as the "Scientific Discovery of Original Sin": he asked "Why is it that the structure of this argument is so unoriginal? If science is going to offer us values, are there no original scientific values which can be of benefit to us?"[126] Scientists, so original in scientific and technological work, could only maintain a pretense of originality for their views on the social relations of science; worse, scientific intellectuals' accounts of the relations of science and war became ever more disengaged from the world they purported to describe. Taken in total, the pronouncements of scientific intellectuals on the relations of science and war display a combination of arrogance and ignorance which is as difficult to credit as it is to understand.

Sociologists of knowledge of the 1960s and 1970s influentially pointed to what they called a lack of symmetry in older sociology of knowledge: sociological explanation was only required to explain error; truth required no explanation other than the removal of barriers to its revelation. It would have been tempting, not least for polemical reasons, to apply this older 'sociology of error', which was indeed common to many scientific intellectuals, to the scientific intellectuals' reflections discussed here. Indeed, highlighting and explaining error and omission in past work is a standard device in historical and sociological studies which seek to remedy an error, or make good an omission. Thus Perry Anderson used an informal "sociology of ignorance", to explain the lack of marxism in Britain, as a prelude to the creation of a revitalised British Marxism.[127] But knowledge, like ignorance or error, is a social phenomenon, indeed knowledge and ignorance are symbiotically linked.[128] The explanation of knowledge and of ignorance should therefore have the same general form; in other words explanation should be symmetrical. This does not mean, however, that knowledge and ignorance should be equally valued. In this paper I have argued that both highlighting and ignoring the actual relations of science technology and war, are explained by political factors. But I do not suggest that some political ideologies are inherently more conducive to knowing, or less conducive to ignoring.[129] Historical context was clearly important.

This paper has looked at scientific intellectuals' reflections on aspects of science external to the immediate practice of science, and indeed how some came to define science so narrowly that it had no external aspects beyond being a beacon of goodness, an argument reflected in 'internalist' historiography of the 1950s and 1960s. However, the 1960s saw echoes of the Bernalist programme in histories of science policy, sociologies of scientists and so on. Such studies rarely engaged with theoretical issues concerning the nature of scientific practice, and became increasingly distant from the social history of scientific knowledge and practice, which developed a local and microscopic tendency.[130] We have seen a shift from an identification of social history of science with the macro-history of science to an identification of the social with the micro.[131] Especially for the study of twentieth century science, that is a perverse outcome. Particularly unfortunate is the recent tendency to create macro-stories out of micro-social studies; to move from a new view of the hard case of physics, to a new view of the world, ignoring and or misrepresenting the vast literature on the world.[132] This is indeed an ironic reflection of scientific intellectuals' arguments for the centrality of science and the scientific method in the world and their lack of engagement with history and the social sciences.

Both the macro- and the micro-worlds of science have been rendered invisible by scientific intellectuals and by historians and sociologists of science. Indeed making the macro-world of science invisible is in many ways a greater achievement: it is of necessity in some way public already. Bauman has recently analysed a compelling and powerful example of how the obvious can be rendered harmless. In his study of the place of the Holocaust in the social sciences, he shows that this event so well-known, and so universally regarded as a watershed in human history, barely figures in sociological accounts of modernity. The strategies used to keep it out were: seeing it as the work of individual criminals; seeing it as a recrudescence of barbarism; or, simply ignoring it. There is, he argues, a real incompatibility between the realities of the Holocaust, especially its modernity, and the central assumptions of social theory.[133] Scientific intellectuals's reflections on the relationship between science and war, as also many historical accounts of science and technology, show many similarities to such blinkered reflections on the relationship between modernity and the Holocaust.

Imperial College of Science, Technology and Medicine, London

NOTES

[1]   Earlier versions of this paper were presented at a seminar organised by the Peace Studies Program and the Department of Science and Technology Studies, Cornell University, at the conference in Madrid on which this volume is based, and at the EASST/4S meeting at Gothenburg in August 1992. I am grateful to the participants for their comments, especially Judith Reppy, José Manuel Sánchez-Ron, Paul Forman and Herbert Mehrtens. I am grateful to John Pickstone, Jonathan Harwood and Andrew Warwick for their comments on a later draft.
[2]   Werskey, P.G., 1978. *The Visible College: A Collective Biography of British Scientists and Socialists of the 1930s* (London: Allen Lane); McGucken, W., 1984. *Scientists, Society and the State* (Columbus: Ohio State University Press); Horner, D.S., 1986. "Scientists, trade unions and the labour movement policies for science and technology, 1947–1964", PhD thesis, University of Aston. This paper, and the bibliography, is confined almost exclusively to British scientific intellectuals. Since this paper was written, the following relevant works have appeared: Pick, D., 1993. *War Machine: The Rationalisation of Slaughter in the Machine Age.* (New Haven: Yale UP); MacLeod, R.M., 1993. "The Chemists go to War: The Mobilisation of Civilian Chemists and the British War Effort, 1914–1918," *Annals of Science* Vol. 50; Crook, D.P., 1994. *Darwinism, History and War: The Debate over the Biology of War from the 'Origin of Species' to the First World War*, (Cambridge: CUP); Garrett, S.A., 1993. *Ethics and Airpower in World War II: The British Bombing of German Cities*, (New York: St. Martin's Press).
[3]   See Armytage, W.H.G., 1968. *Yesterday's Tomorrows: A Historical Survey of Future Societies* (London: Routledge and Kegan Paul); Clarke, I.F., 1992/1979. *Voices Prophesying War: Future Wars, 1763–3749*, 2nd edn (Oxford: OUP; 1st edn 1966); *The Pattern of Expectation: 1763–2001* (London: Cape).
[4]   Turner, F.M., 1980. "Public Science in Britain, 1880–1919", *Isis, 71*: 589–608. Turner correctly noted that the "uncritical reading of the documents of public science has lead historians to take at face value speeches, reports, and essays by scientists about the condition of science when in fact those writings simply repeated arguments and polemics that had been long in the public domain" (p. 608). As I will show this is no less true of the period after 1918.
[5]   I have given a very different account of science-state relations in my "Liberal militarism and the British State", *New Left Review*, No 185 (1991), pp. 139–169; *England and the Aeroplane: An Essay on a Militant and Technological Nation* (London: Macmillan, 1991); "Whatever Happened to the British Warfare State? The Ministry of Supply, 1945–1951," in: Mercer, H. et al. (eds), 1992. *Labour Governments and Private Industry: The Experience of 1945–1951* (Edinburgh Univ. Pr.), pp. 91–116; "British R&D after 1945: a reinterpretation", *Science and Technology Policy* (April, 1993); "The 'White Heat' Revisited: the British government and Technology in the 1960's," *Twentieth Century British History* (forthcoming 1996). Military

support of R&D is not the only major source of funds neglected by scientific intellectuals and historians: business funding for R&D has suffered the same fate. See my "Science and Technology in British Business History", *Business History*, 1987, *29*: 84–103; and Edgerton, D.E.H. and Horrocks, S.M., 1994. "British Industrial Research and Development before 1945", *Economic History Review 44*: 213–238). In brief, attention has been focussed on the least significant element of R&D, state-funded civil R&D. In the above works I also criticise the 'declinism' which characterises the literature on twentieth century British science and technology.

[6] That there is much more literature on the American case than on the British one is probably related to the fact that 1) the military-scientific complex was very large and visible within the American university, while in Britain the complex was organised largely out of sight in government and corporate laboratories, 2) that Britain had no campus unheaval comparable to that which accompanied the latter stages of the Vietnam war, in which protests were often directed precisely at the presence on campus of the military-scientific complex.

[7] Forman, P., 1987. "Behind Quantum Electronics: National Security as Basis for Physical Research in the United States, 1940–1960", *Historical Studies in the Physical and Biological Sciences, 18*: 149–229, on p. 228.

[8] Forman, P., 1991. "Independence, Not Transcendence, for the Historian of Science", *Isis, 82*: 71–86.

[9] See[5] . For a comprehensive bibliography covering works published between 1968 and 1986 see Edgerton, D.E.H. and Gummett, P.J. "Science, Technology, Economics and War in the Twentieth Century" in: Jordan, G. (ed), 1988. *A Guide to the Sources of British Military History* (New York: Garland), pp. 477–500. Particularly important books are Postan, M.M. et al., 1964. *Design and Development of Weapons* (London: HMSO); Gowing, M.M., 1974. *Independence and Deterrence: Britain and Atomic Energy, 1945–1952*, 2 Vols. (London: Macmillan); Hackmann, W.D., 1984. *Seek and Strike: Sonar Anti-Submarine Warfare and Royal Navy, 1914–1954* (London: HMSO); Haber, L.F., 1985. *The Poisonous Cloud: Chemical Warfare in the First World War* (Oxford: Clarendon Press).

[10] MacLeod, R. and Andrews, A., 1969. 'The Committee of Civil Research: Scientific Advice for Economic Development, 1925–1930', *Minerva, 7*: on p. 699.

[11] Edgerton, D.E.H., 1987. "Science and Technology in British Business History", *Business History, 29*: 100–102; McKinnon Wood, R., 1935. *Aircraft Manufacture*, (London: New Fabian Research Bureau).

[12] See Edgerton, D.E.H., 1990. "Science and War" in R.C. Olby et al (eds),*The Companion to the History of Modern Science* (London: Routledge); Hinton, J., 1988. *Protests and Visions* (London: Radius); Howard, M., 1978. *War and the Liberal Conscience* (London: Temple Smith); Semmel, B., 1986. *Liberalism and Naval Strategy* (London: Allen & Unwin).

[13] Lockyer, N. "The Influence of Brain Power on History", extracted from Poole, J.B. and Andrews, K. (eds), 1972. *The Government of Science in Britain* (London: Weidenfeld and Nicolson), p. 39.

[14] Turner, F.M., 1980. "Public Science in Britain, 1880–1919", *Isis, 71*: 602, 603.

[15] Quoted in Moulton, H.F., 1922. *The Life of Lord Moulton* (London: Nisbet), pp. 177–8.

[16] Grey, Sir E., 1916. Statement to *Chicago Daily News*, in: *Speeches*, pp. 197–8.

[17] See for example, Murray, G., 1914. *Thoughts on the War* (Oxford Pamphlets; Oxford Univ. Pr.); Sadler, M.E., 1914. *Modern Germany and the Modern World* (London: Macmillan).

[18] Badash, L., 1979. "British and American Views of the German Menace in World War I", *Notes and Records of the Royal Society of London, 34*: 91–122, pp. 98–102.

[19] Badash, L. "British and American views", pp. 93–8.

[20] Turner, "Public Science", pp. 603–7.

[21] *Nature*, Vol. 117, 23 January 1926, p. 110.

[22] Haldane, J.B.S., 1925. *Callinicus, or A Defence of Chemical Warfare* (London: Kegan, Paul,

Trench, Trubner); see Clark, R., 1968. *J.B.S.: The Life and Work of J.B.S. Haldane* (London: Hodder & Stoughton). The humanisation of war through science was also a theme developed in the United States as was ably shown in: Slotten, H., 1990. "Humane Chemistry or Scientific Barbarism? American Responses to World War I Poison Gas, 1915–1930", *The Journal of American History*, 77: 476–498.

[23] Wersky, P.G., 1988. *The Visible College* (London: Free Association), pp. 217 ff; Burhop, E.H.S. "Scientists and Public Affairs", in Goldsmith, M. and Mackay, A. (eds), 1966. *The Science of Science* (Harmondsworth: Penguin), pp. 32–46.

[24] Bernal, J.D., 1939. *The Social Function of Science* (London: Routledge and Kegan Paul), p. 186.

[25] *Nature* Vol. 127, 7 March 1931, p. 333.

[26] *Nature* Vol. 127, 7 March 1931, pp. 333–4; see also review of Lefebure, V. 1931. *Scientific Disarmament* in: *Nature*, Vol. 127, pp. 513–5.

[27] Low, A.M., 1939/1943. *Modern Armaments* (London: Hutchinson), p. 256; *Musket to Machine Gun* (London: Hutchinson).

[28] Huxley, J.H., 1934. *Scientific Research and Human Needs* (London: Watts), pp. 152–3, 167.

[29] Huxley, J.H. "Peace through Science", in: Noel-Baker, P., et al., 1934. *Challenge to Death* (London: Constable), pp. 292–3.

[30] In contrast to Wells, and much other commentary, he had faith in people: "Any brave man can stand one raid. To stand a number requires military discipline, mass heroism as is found in Madrid, or a philosophy which makes it clear why these things happen, and why the final victory of fascism is impossible. A dying Marxist can say in perfect confidence: "Exoriare aliquis nostris ex ossibus ultor" [Out of our bones let some avenger arise. Virgil]." Haldane, J.B.S., 1938. *ARP* [Cambridge Scientists Anti-war Group, *Air Raid Protection*] (London: Gollancz), p. 57.

[31] Haldane, *ARP*, p. 242.

[32] Haldane, *ARP*, p. 247.

[33] *Nature*, Vol. 139, 1937, p. 980, quoted in Wersky, P.G., 1971. "The Perennial Dilemma of Science Policy," *Nature*, vol. 233, pp. 529–532, on p. 531.

[34] Ewing, Sir A., 1932. "An Engineer's Outlook", Supplement to *Nature*, p. 350

[35] Ewing, Sir A. "Engineer's Outlook", p. 349.

[36] Collingwood, R.G., 1939. *An Autobiography* (Oxford: Oxford University Press), p. 91.

[37] Collingwood, *Autobiography*, p. 106.

[38] Crowther, J.G. "Aviation" in: Hall, Sir D., et al., 1935. *The Frustration of Science* (London: Allen and Unwin), pp. 40–1.

[39] Marvin, F.S., 1931. Review of Davies, Major D., 1930. *The Problem of the Twentieth Century: A Study in International Relationships* (London: Benn), in: *Nature*, Vol. 127, p. 85.

[40] Clarke, I.F., 1961. *The Tale of the Future from the Beginning to the Present Day* (London: Library Association), is an invaluable list of works published in Britain from which the following titles and comments are taken. Newman, B., 1931. *Armoured Doves*, "An international league of scientists uses secret weapons to put an end to war"; Arlen, M., 1933. *Man's Mortality*, "Revolt against the world authority in the days when International Aircraft and Airways is the only government"; Vierck, G.S. and Eldridge, P., 1933. *Prince Pax*, "Capital cities bombed encourage peace"; Stuart, F., 1933. *Glory*, "Love and adventure in the days when TransContinental Aero-Routes control the skies of the world"; Gloag, J., 1934. *Winter's Youth*, "By 1960 the power of modern weapons has forced peace on the world"; Edmonds, H., 1935. *The Professor's Last Experiment*, "An inventor is able to stop a war"; Tunstall, B., 1936. *Eagles Restrained*, "How the International Air Police put an end to a war between Germany and Poland".

[41] Bernal, J.D., 1939. *The Social Function of Science* (London: Routledge and Kegan Paul), p. 187.

[42] See Teich, M., 1990. "Reflecting on the Golden Jubilee of Bernal's *The Social Function of Science*", *History of Science, 28*: 411–8.

[43] Bernal, *The Social Function of Science*, p. 173.

[44] Bernal, J.D. "Science and Industry", in: Hall, Sir D. et al., 1935. *The Frustration of Science* (London: Allen and Unwin), pp. 45–8, 48.

[45] Bernal, *The Social Function of Science*, pp. 171–2, 182. The Moseley myth is refuted in Heilbron, J.H. 1974. *H.G.J. Moseley* (Berkeley: Univ. of California Pr.).

[46] Bernal, *The Social Function of Science*, p. 172.

[47] Quoted in Werskey, "Science Policy", p. 532.

[48] Anon, 1940. *Science in War* (Harmondsworth: Penguin), p. 14.

[49] Anon, *Science in War*, p. 144.

[50] Crowther, J.G., Howarth, O.J.R. and Riley, D.P., 1942. *Science and World Order* (Harmondsworth: Penguin), p. 10.

[51] House of Commons, 24 February, 1942, reproduced in *The Ethical Dilemma of Science* (London: Scientific Book Club, 1962), pp. 288–295. Hill was the member of Parliament for Cambridge University. The anachronism of university seats was abolished in the late 1940s.

[52] Irving, D., 1985. *The Destruction of Dresden* (London: Papermac; first published 1963), pp. 50–53.

[53] Ibid, p. 243.

[54] Dale, Sir H., 1947. Foreword to J.G. Crowther and R. Whiddington, *Science at War* (London: HMSO).

[55] Crew, F.A.E., 1952. *Must Man Wage War? The Biological Aspect* (London: Thrift Books), pp. 92–3.

[56] Crew, *Must Man Wage War? The Biological Aspect*, pp. viii–ix.

[57] Stanier, Sir W., 1956. "The Influence of War", *The Engineer*, Centenary Number, p. 172.

[58] See Edgerton, "Whatever happened to the Warfare State?".

[59] Association of Scientific Workers, *Science and the Nation* (Harmondsworth: Penguin, 1947), p. 166.

[60] Appleton, Sir E., 1957. *Science and the Nation: the BBC Reith Lectures for 1956* (Edinburgh: Edinburgh Univ. Pr.). Sir John Reith, the first Director-General of the BBC, had been an engineer with the armament firm Beardmore! The annual Reith lectures on BBC radio were a very prestigeous platform for intellectuals.

[61] Thompson, E.P., 1960. "Outside the Whale", reprinted in: *The Poverty of Theory and Other Essays* (London: Merlin, 1978), pp. 134. There is a striking contrast between the flourishing marxist historiography of the 1950s, in which E.P. Thompson played a leading part, and the weakness of Marxist studies of science. It is also interesting to note that the marxist historians did not engage with the history of science and technology, despite the pioneering work of J.D. Bernal. See Hobsbawm, E. "The Historians' Group of the Communist Party" in: Cornforth, M. (ed), 1978. *Rebels and their Causes: Essays in Honour of A.L. Morton* (London: Lawrence and Wishart), pp. 21–48.

[62] In 1948 seventeen scientists were purged from the Civil Service, largely from warlike R&D laboratories: Jones, G., 1986. "The Mushroom-Shaped Cloud: British Scientists' Opposition to Nuclear Weapons Policy, 1945–57," *Annals of Science, 43*: 1–26.

[63] Horner, thesis[2], chapter one.

[64] Bernal, J.D. and Cornforth, M., 1949. *Science for Peace and Socialism* (London: Birch).

[65] Ibid. For the official Soviet position, see "Dr Einstein's Mistaken Notions: An Open Letter from Sergei Vavilov, A.N. Frumkin, A.F. Joffe and N.N. Semyonov" (1947) reprinted in: Einstein, A., 1956. *Out of My Later Years* (New Jersey: Citadel Press), pp. 161–8.

[66] Bernal, J.D., 1958. *World Without War* (London: Routledge & Kegan Paul), p. 2.

[67] Ibid., p. 47.

[68] Werskey, *Visible College*, p. 307.

[69] *Science and the Nation*, pp. 243–4.

[70] "Science and Secrecy", *The Spectator*, 17 August 1945; reprinted in: *The Ethical Dilemma of Science*, pp. 303–5.

[71] Mott, Sir N., 1986. *A Life in Science* (London: Taylor & Francis), pp. 80–94, p. 92. See also Peierls, R., 1985. *Bird of Passage: Recollections of a Physicist* (Princeton: Princeton University Press), ch. 12. Jones, G., 1988. *Science, Politics and the Cold War* (London: Routledge), ch. 3, 5.

[72] Gowing, M., 1974. *Independence and Deterrence: Britain and Atomic Energy, 1945–1952*, Vol. 1 (London: Macmillan), p. 184.

[73] Blackett's lone opposition to a British bomb among government advisors is clearly brought out in Gowing, *Independence and Deterrence*, Vol. 1, passim. Blackett's argument that the first atom bombs were dropped to warn the Soviet Union, rather than because they were necessary to defeat Japan was taken up in the 1960s by Gar Alperovitz.

[74] Blackett, P.M.S., 1948. *The Military and Political Consequences of Atomic Energy* (London: Turnstile).

[75] Blackett, P.M.S., 1956. *Atomic Weapons and East-West Relations* (Cambridge: Cambridge University Press), p. 4.

[76] Ibid., p. 92.

[77] Ibid., pp. 99, 100. Cf. Lord Zuckerman, 1984. "Scientists, Bureaucrats and Ministers", *Proceedings of the Royal Institution of Great Britain*, Vol. 56, pp. 205–229. "Some scientists are more ignorant for the work of others than would be a well-read layman"; "a scientific education, and still less a technological obsession, is not a sufficient condition for wise judgement"; "Ministers and civil servants not only need dispassionate scientific advice, but maps of the hidden rocks of vested and prejudicial scientific opinion" (p. 221); "Except in the short term, so-called 'operational requirements' are always formulated around technological promises which, as often as not, turn out to have been overambitious" (p. 223).

[78] See McGucken, W., 1978. "On Freedom and Planning in Science: The Society for Freedom in Science, 1940–1946", *Minerva, 16*: 42–72; Baker, J.R., 1942. *The Scientific Life* (London: Allen & Unwin) and Polanyi, M., 1962. "The Republic of Science: Its Political and Economic Theory", *Minerva, 1*: 54–73, both of which ignore, not always explicitly, the vast bulk of scientific activity.

[79] Polanyi, "Republic of Science", p. 66.

[80] Ziman, J., 1968. *Public Knowledge: An Essay Concerning the Social Dimension of Science* (Cambridge: Cambridge University Press), p. 147.

[81] Ibid., pp. 2–3.

[82] Ibid., p. 147.

[83] Ravetz, J.R. "'...et augetibur scientia'", in: Harre, R. (ed), 1973. *Problems of Scientific Revolution* (The Herbert Spencer Lectures, Oxford: Clarendon Press), p. 46.

[84] See Oppenheimer, J.R., 1989. *Atom and Void: Essays on Science and Community* (Princeton: Princeton Univ. Pr.), see also the Preface by Freeman Dyson.

[85] Hall, R., 1952. *Ballistics in the Seventeenth Century: A Study in the Relations of Science and War with Reference Principally to England* (Cambridge: Cambridge Univ. Pr.).

[86] Beer, Sir G. de, 1961. *The Sciences Were Never at War* (Cambridge: Cambridge Univ. Pr.).

[87] Jungk, R., 1958. *Brighter than a Thousand Suns: The Moral and Political History of the Atomic Scientists* (London: Gollancz).

[88] "The Moral Un-Neutrality of Science", speech, 1960, to the American Association for the Advancement of Science, in: Snow, C.P., 1981. *The Physicists: A Generation that Changed the World* (London: Macmillan), Appendix 3.

[89] Bronowski, J., 1961. *The Common Sense of Science* (1951, Harmondsworth: Penguin), pp. 153–4. Ravetz pointed to the hollowness of scientific ethics in the post war period, and to Bronowski as a particular example (Ravetz, J.R., 1973. *Scientific Knowledge and its Social*

Problems (1971, Harmondsworth: Penguin), p. 65. Bronowski was a wartime member of the branch of the Ministry of Home Security which would work for the Air Ministry on strategic bombing and was also a member of the British Scientific Mission send to Japan to study the effects of the atomic bombs.

[90] Snow, C.P., 1961. *Science and Government* (London: Oxford Univ. Pr.), pp. 8–12.

[91] Snow, *Science and Government*, p. 47.

[92] Blackett, P.M.S., 1962. "Science in Government", review of Snow, reprinted in: Blackett, P.M.S., 1962. *Studies of War* (Edinburgh: Oliver & Boyd), p. 126.

[93] Webster, C. and Frankland, N., 1961. *The Strategic Air Offensive against Germany* (London: HMSO), Vol. 1, p. 234.

[94] Zuckerman, S., 1978. *From Apes to Warlords* (London: Hamish Hamilton), pp. 139–48.

[95] Sherry, M., 1987. *The Rise of American Air Power: The Creation of Armageddon* (New Haven: Yale University Press), p. 363.

[96] Snow, C.P., 1962. *Postscript* (London: Oxford Univ. Pr.), p. 27.

[97] Blackett, *Studies*, p. 123.

[98] See Air Ministry, 1963. *Origins and Development of Operational Research in the RAF* (London: HMSO); Waddington, C.H., 1973. *OR in World War 2: Operational Research against the U-Boat* (London: Elek); Rosenhead, J., 1989. "Operational Research at the Cross-roads: Cecil Gordon and the Development of Post-war OR", *Journal of the Operational Research Society*, 40: 3–28.

[99] Dyson, F., 1979. "The Children's Crusade" in *Disturbing the Universe* (New York: Harper & Row), pp. 29–30.

[100] Harris, Marshall of the RAF Sir A., 1947. *Bomber Offensive* (London: Collins), p. 278.

[101] Bronowski, J., 1973. *The Ascent of Man* (London: BBC), p. 369.

[102] Rhodes, R., 1988. *The Making of the Atom Bomb* (1986, Harmondsworth: Penguin), pp. 21–8, 753–4.

[103] Trachtenberg, M., 1991. *History and Strategy* (Princeton: Princeton Univ. Pr.), pp. 103–4.

[104] Clark, R.W., 1962. *The Rise of the Boffins* (London: Phoenix House).

[105] Rose, H., and Rose, S., 1969. *Science and Society* (Harmondsworth: Penguin), p. 59.

[106] Ibid., p. 60.

[107] Gummett, P.G., 1980. *Scientists in Whitehall* (Manchester: Manchester Univ. Pr.). Gummett has since become one of the leading British students of warlike R&D.

[108] Cardwell, D.S.L., 1975. "Science and World War I", *Proceedings of the Royal Society A*, 342: 447–56.

[109] MacLeod, R.M., and Andrews, E.K., 1971. "Scientific Advice in the War at Sea, 1915–1917: the Board of Invention and Research", *Journal of Contemporary History*, 6: 3–40. Pattison, M., 1983. "Scientists, Inventors and the Military in Britain, 1915–19: The Munitions Invention Department", *Social Studies of Science*, 13: 521–68.

[110] Alter, P., *The Reluctant Patron: Science and the State in Britain, 1850–1920* (Oxford: Berg).

[111] Alter, *Reluctant Patron*, pp. 129–130, 136.

[112] North, J. "Sir Norman Lockyer" in: Harre, R. (ed), 1969. *Some Nineteenth Century British Scientists* (Oxford: Pergamon), p. 199.

[113] See, for example, Rose, H., and Rose, S. (eds), 1976. *The Radicalisation of Science* (London: Macmillan).

[114] Albury, D., and Schwartz, J., 1982. *Partial Progress: The Politics of Science and Technology* (London: Pluto).

[115] Easlea, B., 1983. *Fathering the Unthinkable: Masculinity, Scientists and the Arms Race* (London: Pluto). See also Easlea, B., 1980. *Liberation and the Aims of Science* (Edinburgh: Scottish Acad. Pr.), ch. 7, 10, 11, which deal with the relations of science and the military in the United States.

[116] Schroeder-Gudehus, B., 1978. *Les scientifiques et la paix: La communauté scientifique internationale au cours des années 20* (Montreal: Les Presses de l'Université de Montreal); Alter, P., 1980. "The Royal Society and the International Association of Academies 1897–1919", *Notes and Records of the Royal Society of London, 34*: 241–264; Kevles, D., 1971. "'Into Hostile Political Camps': The Reorganisation of International Science in World War I", *Isis, 62*: 47–60; Badash, L., 1979. "British and American views of the German Menace in World War I", *Notes and Records of the Royal Society of London, 34*: 91–122.

[117] Cock, A.G., 1983. "Chauvinism and Internationalism in Science: The International Research Council, 1919–1926", *Notes and Records of the Royal Society of London, 37*: 249–288.

[118] Gowing, M., 1964. *Britain and Atomic Energy, 1939–1945* (London: Macmillan).

[119] Gowing, M., 1974. *Independence and Deterrence: Britain and Atomic Energy, 1945–1952*, 2 Vols. (London: Macmillan).

[120] Hartcup, G., 1988/1970. *The War of Invention: Scientific Developments, 1914–1918* (London: Brassey's) and *The Challenge of War; Scientific and Engineering Contributions to World War Two* (Newton Abbot: David & Charles).

[121] Kaldor, M., 1982. *The Baroque Arsenal* (London: Deutsch).

[122] Pearton, M., 1983. *The Knowledgeable State: Diplomacy, War and Technology since 1830* (London: Burnett).

[123] Sumida, J.T., 1989. *In Defence of Naval Supremacy* (London: Unwin Hyman); Travers, T., 1987. *The Killing Ground: The British Army, the Western Front and the Emergence of Modern Warfare, 1900–1918* (London: Unwin Hyman).

[124] Drake, S. (ed), 1957. *Discoveries and Opinions of Galileo* (New York: Anchor).

[125] Young, R.M., 1987. "The Scientist as Guru: The Explainers", *Science as Culture*, Pilot, Issue, 130–140.

[126] Elliot, G., 1972. *Twentieth Century Book of the Dead* (London: Allen Lane), p. 167.

[127] Anderson, P., 1968. "Components of the National Culture", reprinted in: Anderson, P., 1992. *English Questions* (London: Verso), pp. 48–104. "Mannheim proposed a sociology of knowledge: what is called for here is a sociology of ignorance" (p. 56). In this influential paper Anderson neglected the role of scientific intellectuals in British cultural life, but made much of the empirical tradition in British sociology, and the 'White emigration', particularly from the former Austro-Hungarian empire, which included both Mannheim and Popper.

[128] Ravetz, J., 1993. "A Leap into the Unknown", *The Times Higher Education Supplement*, May 28, p. 18.

[129] One important point is that historiographical positions, and political positions do not always align in the same way over time (for example, in the 1950s 'historicism' was clearly an antiradical position, specifically, anti-marxist, while in more recent years it has been associated with leftist positions) (See Novick, P., 1988. *That Noble Dream: The 'Objectivity Question' and the American Historical Profession* (Cambridge: Cambridge University Press).) Proctor, R.N., 1991. *Value-Free Science? Purity and Power in Modern Knowledge* (Cambridge, MA: Harvard Univ. Pr.), shows that believing in the neutrality of science is not necessarily a conservative position. His work also helps make clear that the 'use/abuse' model is not the same as the 'neutrality' model. The neutrality model is indifferent to use or abuse; the 'use/abuse' model implies a good use is preferable to a bad use, and that science has something to say about which is which. Furthermore, as Bloor has pointed out, the mystificatory power of particular theories is typically not determined by the theories themselves: in different circumstances, both the Enlightenment tradition and the Romantic tradition become mystificatory or naturalistic. Bloor's suggestive 'law of mystification' states that actors with low power in society, who are critical of society, and those with high power, but who are complacent and unthreatened, will produce naturalistic accounts of knowledge; those with marginal power, or who are threatened will produce mystificatory accounts of knowledge (Bloor, D., 1976. *Knowledge and Social Imagery* (London: Routledge), p. 69.) This paper suggests this law is very plausible, in that

perceived threats to science did lead to mystification, and were not confined to any one view of science. However, different views of the relations of science prevailed in different historical periods so this hypothesis cannot be fully tested. The main aim of the paper is, however, to outline what scientific intellectuals had to say, rather than to explain it.

[130] It is worth noting that Steven Shapin, in his "History of Science and its Sociological Reconstructions", *History of Science*, 1982, *20*: 157–211, did not deal with "many admirable studies that treat, for example, images of science, the rhetoric of spokesmen of science, views of scientific method not clearly related to practice, and the sociology of scientists as opposed to the sociology of science" (p. 158). In a footnote he added "Nor will I treat the history of the social sciences, although I do not accept that these materials present an 'easier case' for the sociology of knowledge in anything but a persuasive sense" (p. 199, n. 5). Golinski, J., 1990. "The Theory of Practice and the Practice of Theory: Sociological Approaches in the History of Science", *Isis, 81*: 492–505, makes no references whatever to macro studies of science, or to the subjects which Shapin explicitly excluded.

[131] Shapin, S., 1992. "Discipline and Bounding: The History and Sociology of Science as Seen through the Externalism-Internalism Debate", *History of Science, 20*: 333–69, pp. 346–8.

[132] I have in mind a recent textbook, Yearley, S., 1988. *Science, Technology and Social Change* (London: Unwin Hyman), and a case study, MacKenzie, D., 1990. *Inventing Accuracy: A Historical Sociology of Nuclear Missile Guidance* (Cambridge, MA: MIT Press), which I discuss at length in "Tilting at Paper Tigers", *British Journal for the History of Science*, 1993, *26*: 67–75.

[133] Bauman, Z., 1989. *Modernity and the Holocaust* (Cambridge: Polity).

HELGE KRAGH

TELEPHONE TECHNOLOGY AND ITS INTERACTION WITH SCIENCE AND THE MILITARY, CA. 1900–1930

1. INTRODUCTION

"Le secret de la guerre est dans le secret des communications," Napoleon is to have said in one of his maxims. The revolutionary French army was the first to base its operations and intelligence on an extensive use of communication technology, especially the new system of optical telegraphy invented by Claude Chappe in 1794. While optical telegraphy was specifically designed for military purposes and largely remained a military technology, its successor, the electromagnetic telegraph, was developed, and mainly used, for commercial purposes. This civilian association of electrical communication was enhanced by the introduction of the telephone in the late 1870s. Whereas the (electromagnetic) telegraph was used primarily for commercial and financial affairs, the telephone soon became a domestic technology used by all people of means for social purposes. Yet, civil and peaceful as these technologies were, they have also played an important part in every large military operation since the middle of the nineteenth century.[1]

In this paper I examine aspects of the development of telephone science and technology during the first three decades of the twentieth century, paying particular attention to long-distance cable telephony.[2] This field emerged as a technical possibility in the early part of the century, and by 1930 it had developed into an efficient long-distance network which secured easy communication between most cities in Europe and North America. The telephone networks of the 1920s were based on underground cables and were part of the national and international communications infrastructure; they served civil purposes similar to other large technological systems such as the railways, electrical power networks and the road system. However, the predominantly civil and commercial nature of long-distance telephony, and of the research laboratories advancing this technology, does not mean that telephony developed independently of military interests or was without military ramifications. In modern societies it makes little sense to distinguish sharply between military and non-military technologies. Civil technologies will often be used for military purposes, as was the case with the telephone in World War I. More importantly, and less generally appreciated, technologies of military interest are not limited to those applied in situations of war. Ordinary civil technological services such as transportation, distribution and communications systems are, or may be, parts of a military-technological complex

37

P. Forman and J. M. Sánchez-Ron (eds.), National Military Establishments and the Advancement of Science and Technology. Studies in the 20th Century History, 37–67.
© 1996 Kluwer Academic Publishers. Printed in the Netherlands.

whose wartime functions are integral with its peacetime functions. If the development of telephony is seen from this perspective it becomes clear that this technology, often praised for its contributions to peace and international cooperation, interacted closely, if indirectly, with national military interests. How this interaction took place is the principal problem investigated in this paper.

In Section 2, I survey the main scientific and technological advances in long-distance telephony during the period 1900–1930. These were unrelated to military needs and were mostly the result of technological research based on applied science. The feasibility of cable telephony depended initially on scientific insight, but in its later development it relied on engineering science with little connection to the fundamental, university-based research which traditionally has been the focus of historians of science. Conversely, the prosperity of university physics depended on the industrial laboratories as employers, among which the most important were the big telecommunications companies, such as Siemens & Halske and the Bell System. Section 3 examines the development of telephony from a military perspective, especially the considerable role it played in the Great War, 1914–1918. The telephone was not only used for traditional military purposes such as artillery guidance and intelligence, but also changed strategic thinking and attitudes to war. In the 1920s a long-distance cable network was created in Europe, with the telephone ostensibly returning to its old semi-ideological role as an instrument of conciliation and international understanding. However, analysis of this development, which is the subject of Section 4, shows that the telephone network was an integral part of the political and ideological history of the period. It was a key factor in the rebuilding of German political and economic power and a cause for concern to her former enemies. Although the network was built and operated by civil authorities, indirectly it served as part of the preparedness policies and military interests of the European nations.

2. THE SCIENTIFIC ORIGIN OF LONG-DISTANCE TELEPHONY

At the end of the nineteenth century telephone communication was firmly established, but mainly limited to intra- and interurban lines. Long-distance and submarine lines existed both in Europe and in the USA, but they were few and costly in construction, of poor transmission quality, and they carried little traffic. The main technical problem was the attenuation of the speech caused by the cable or wire through which the telephone current propagates. Traditional cable technology's answer to this problem was, by and large, to increase the thickness of the copper cores and the insulating layer surrounding the cores (thereby reducing the resistance and capacitance, respectively). Although these measures were not without effect, they were too uneconomical to be applied on a large scale. Attempts to amplify the current by means of mechanical repeaters were made, but were not successful because the devices increased the distortion of the transmitted speech. About 1900 long-distance telephony seemed to have reached an impasse: In order to compete with the telegraph system and become a universal communication technology, the construction of telephone lines required a fundamentally new approach.

The answer to the problem came from science. It consisted essentially in two separate inventions, the loading coil and the vacuum tube amplifier. Of these the loading coil was a result of mathematical physics while the tube amplifier had its background in experiments with the newly discovered electrons. Together they provided a powerful demonstration of the usefulness of science in technologies which had emerged from, and which had traditionally been developed by, practical considerations.

As far back as 1887 Oliver Heaviside in England and Aimé Vaschy in France had demonstrated theoretically that the attenuation of telephone currents would be reduced in conductors of high self-inductance. This surprising result of mathematical analysis suggested the construction of inductively "loaded" cables, either by inserting regularly spaced coils or coating the copper cores uniformly with soft iron. However, the insight was ignored or rejected by most engineers, who distrusted theory and stuck to their tested empirical methods.[3] It was only with an important extension of the Heaviside-Vaschy theory, produced independently in 1899 by George Campbell at the Bell System and Michael Pupin at Columbia University, that the idea of loaded cables was transformed into a technologically useful theory. Campbell and Pupin deduced the optimum number and spacing of loading coils for a given line and thereby provided telephone engineers with definite prescriptions of how to reduce the attenuation.[4]

The system – subsequently known as "pupinization" – was quickly developed into a practical innovation. In America, this was done by researchers at Western Electric, the manufacturing branch of the Bell System which had acquired Pupin's patents. In Europe, Pupin sold the rights of his invention to Siemens & Halske in 1902, who developed it independently of the Americans. In 1905 August Ebeling and Friedrich Dolezalek, two academic physicists who had "prostituted" themselves by becoming industrial researchers at Siemens & Halske, developed the Pupin coil into a method for inserting coils in submarine cables.[5] Beginning in that year, coil loading after Pupin's system was introduced on a large scale, immediately resulting in improved speech quality over still longer distances. At the same time the alternative method of uniform loading was developed by a Danish telegraph engineer, Emil Krarup, who obtained increased self-inductance by winding the copper cores with fine iron wires.[6]

By 1912 long-distance telephony had experienced a minor revolution through the introduction of loaded cables or wires. Cables of the Krarup type were widely used in Europe for submarine purposes and long-distance communication over land was realized by an increasing number of national and international lines provided with Pupin coils. Shortly before the First World War, traffic was opened on the 1350 km long pupinized line between Berlin and Milan, and still longer lines were in operation in America. These and other lines in Europe were operated by the national post- and telegraph offices, but constructed by private electrotechnical companies dominated by the British Siemens Brothers, the Bell-owned Western Electric, and the two German firms Siemens & Halske and Felten & Guilleaume.

Successful as the loading method was, it was not suitable for lines much longer than 1500 km and could not have secured an extended long-distance network comparable to that of telegraphy had it not been supplemented with another science-based invention, the tube amplifier. Primitive electronic tubes were invented between 1904 and 1906 by John Ambrose Fleming in England (the vacuum diode), Lee de Forest in USA (the grid triode or "audion") and Robert von Lieben in Austria (the mercury-vapour tube). The early tubes were mostly intended as detectors for radio waves and not immediately used for amplifying telephone currents. Their potential as amplifiers was recognized about 1912, and the first crude devices were quickly transformed into reliable and efficient repeaters by American and German engineers. The leading companies in this development, as in most other telephonic innovations, were Siemens & Halske and the Bell System.

The electronic amplifier or repeater was not a rival to or substitute for loading, but a supplementary technology used for further increasing the transmission range. The first commercial application of the combined inventions furnished an impressive demonstration of science-based engineering and, at the same time, American superiority in long-distance communication: early in 1915 the first transcontinental line, a pupinized aerial line supplied with triode repeaters, carried speech over the 3400 miles separating New York and San Fransisco. This much publicised accomplishment was primarily due to the work of John J. Carty, Bell System's dynamic and visionary chief engineer. The new possibilities of telephony, further enhanced by numerous advances in switching systems, automation and apparatus, were results of engineering science rather than applied basic science. The inventive research, whether theoretical or experimental, took place in the research departments of private corporations and not at the universities; it was published in technical journals such as *Elektrotechnische Zeitschrift*, *Bell System Technical Journal* and *Journal Télégraphique*, and not (with a few notable exceptions) in *Annalen der Physik*, *Physical Review* or *Philosophical Magazine*. Yet the engineering research which created the first revolution in long-distance telephony was thoroughly scientific and exemplified an increasingly close integration of pure science with industrial research.[7] At the end of the 1920s loading and amplifying had been absorbed in practical telephony, which now progressed in a more regular manner by gradually improving existent technology and supplementing it with new inventions of a less radical nature. As Frank Jewett, Western Electric's chief engineer, expressed it, telephone technology had become "ordinary engineering – that is, the judicious mixing of physics, chemistry and mathematics with dollars".[8]

This judicious mixing led to a number of important innovations, which, together with improved organization and international cooperation, resulted in a fully developed long-distance telephone network by the late 1920s. At the same time as the amplifier technique was perfected by the Bell System, the company's engineers greatly improved the coil loading method by producing smaller and more efficient Pupin coils. In 1916 J. Buckner Speed found a way to compress iron powder under high pressure into coil cores, which resulted in new coils with less waste of electrical energy and better suited for circuits with repeaters. About the same time another Western Electric engineer, Gustaf Elmen, invented a new nickel-iron alloy with magnetic permeability much higher than that of the best soft iron. Permalloy, as it was called, became widely used for uniformly loaded telegraph cables and telephone Pupin coils. By 1925 the Bell engineers had combined the two inventions into a new generation of permalloy-powder coil cores which superseded earlier types of coils. The ever-inventive Bell engineers also pioneered the system of carrier telephony in which the telephone currents are propagated by a high-frequency wave modulated by the voice-frequency wave. Electric filters for transmission of a high-frequency carrier modulated by low-frequency speech were studied by Campbell and the German engineer Karl Willy Wagner before the war, but only used for practical purposes after 1918 by the Bell System. The mixing of the waves was effected by vacuum tube circuitry and led in the early 1920s to the development of the band filter technique with many high-frequency channels over a common circuit. This method, used extensively from about 1928, not only made possible a great increase in traffic but also the transmission of radio waves directly over the regular telephone network.

As mentioned above, most of the important advances in telephone technology took place in America within the Bell System, and more particularly within Western Electric's Engineering Department, which in 1925 was transformed into the Bell Telephone Laboratories, owned equally by Western Electric and AT&T, the holding company of American Bell Telephone Company. With a technical personnel of 3390 and a research staff of about 600 scientists and engineers in 1925, the Bell Telephone Laboratories was the world's largest and richest institution for industrial research.[9] In Europe advances in telephony were developed and utilized primarily by International Western Electric. Many of them were copied or developed independently by Siemens & Halske, the only European company which was capable of competing with the American communications giant. The innovative contributions of the national post and telegraph offices were modest, the only exception being the Imperial Telegraph Test Department (Kaiserliches Telegraphen-Versuchsamt)

in Berlin where important work on loading and transmission theory was done
by Wagner and Franz Breisig.

3. MILITARY USE OF TELEPHONY

The telephone made its entry in the military in the early 1880s, when most
major armies became equipped with portable telephone sets and wire drums
carried on mule- or horseback, but it was only at the turn of the century that
the new technology proved its value in war. It was first used extensively by the
Americans in the Spanish-American war in 1898. For example, the bombardment of Santiago de Cuba by US warships was directed by telephoning from
the front to the shore, where range and direction were then flagged to the fleet.
The Army Signal Corps, under command by General Augustus Greely, not
only used the telephone tactically, but also strategically, for communication
with the War Department in Washington: "[We were] able to keep in touch
with the President, the Secretary of War, and the Commanding-General of the
Army, so as to receive at critical moments such advice, encouragement, or
assistance as might advance the interests of the campaign."[10] Greely's electrical programme consisted in the laying of submarine telegraph cables between
Cuba and the United States, and interlacing the entire front by telephone lines.
The war was the first in which telephony provided news service for the home
front, and exemplifies the characteristic role of electrical communications in
mobilizing popular sentiment for war. Thus, in April 1898, when war was imminent, The Chicago Telephone Company promised its subscribers that they
would be notified by operators within twenty minutes of any declaration of
war. For, as the general manager of the company said, "We are a quasi-public
corporation and we rather consider it our duty to act in this way."[11]

In the Boer war the telephone was used by the British forces for defense
purposes, and it played an important role in the Russo-Japanese war, where
superior Japanese communication lines contributed to the outcome of the
war.[12] The Japanese divisional headquarters were wired to General Iwao
Oyama who sat safely behind the line and sent his orders. By that time the
military importance of electrical communications, including wireless, was
generally recognized, but in practice the new technologies had not yet become
integrated in the armed forces, which either relied on the tested telegraph or,
more commonly, personally dispatched messages.

Until the outbreak of the First World War, military use of electrical communications was limited and resisted by many officers. The British Commander-
in-Chief during the Crimean War, General Simpson, was assisted by a 545
km submarine cable across the Black Sea from Varna to Balaclava, which
gave direct connection by telegraph to Paris and London. But he was not
enthusiastic about the technological wonder which meant that he could no
longer conduct the warfare without constant interference from bureaucrats

and politicians in Whitehall. "The confounded telegraph has ruined everything," he is to have complained.[13] The uneasiness about telegraphy expressed by General Simpson was shared by many officers in the following decades where the telegraph and telephone only slowly became integrated in the military services. The US Signal Corps, for example, had great trouble in making electrical communications accepted in the army. Officers often refused to accept field telephone systems, which, they felt, threatened the integrity of command authority. Greely later reminisced that he had "to force electrical facilities upon the army."[14]

Accustomed as they were with the telegraph, officers distrusted the telephone because it left no written record. They also saw the informal nature of telephonic communications as a potential threat against the rigid military hierarchy. The telephone was "democratic" in the sense that it permitted truly dialogical communication, contrary to the telegraph, whose sequences of monologues seemed to harmonize better with military traditions of orders flowing down the hierarchy. The telegraph invited orders and messages, the telephone invited discussions, interruptions and contradictions. Marshall McLuhan, the media analyst, has described a similar effect in the business world, where the telephone introduced "a 'seamless web' of interlaced patterns in management and decision-making." According to McLuhan, "It is not feasible to exercise delegated authority by telephone. The pyramidal structure ... cannot withstand the speed of the phone to bypass all hierarchical arrangements."[15]

When the telephone was introduced large-scale during the war, it helped break down the rigid rules of command which characterized European armies built after the Prussian model. The strategical as well as tactical importance of the telephone was often emphasized in early 20th-century visions of "electrical warfare". According to one such visionary, weaving together communication technology and social darwinism, "As the struggle for existence in social life makes the telephone indispensable, so will it be even more valuable in the greatest of all struggles for existence, war."[16] And another military visionary, General Alfred von Schlieffen, emphasized that the new communication technologies would lead to an entirely new concept of warfare. The army commander would no longer be an active participant in the battle, guiding his troops by his experience and personal courage, but rather a chief engineer directing the war operations far from the battle front by means of an extended electrical communications network.[17] This vision was fiction in 1910, but a generation later it became an ever more dominating reality. It reached a provisional climax in 1991 in the war against Iraq.

It is customary to contrast the technological unpreparedness for war of Britain and France in 1914 with a technologically determined and sophisticated Germany. This picture exaggerates German integration of science and technology in its military machine and the significance of Germany's militarily oriented research. Economically, not militarily, oriented research predominated in Wilhelmian Germany's research policy. Thus, in the period 1910–1914 only about 5% of the total research expenditures went for militarily oriented research, compared to 35% for economically oriented research.[18] Moreover, scientific and engineering manpower were only mobilized for military research hesitatingly and to a small extent. Germany's relative failure in integrating technological research with the military is exemplified by the telephone. The rhetoric of von Schlieffen and other visionary officers was followed only slowly and hesitantly by practice. Large-scale introduction of telephones in the European armies took place only from about 1912. Equipment and organization of the communications units was still not in good order in the summer of 1914. When war was declared on August 1, 1914, the Germans, who foresaw a short war, were unprepared for the massive need for electrical communications. So were the French, whose civil and military telephone lines were based on different types of equipment and not coordinated. In the absence of an efficient telephone network, French engineer-officers were forced to use existing telegraph lines for telephony. The results were bad and telephonic communications often failed.[19] The deficiency was only partly compensated by the efficient French system of wireless communications directed by Gustave Ferrié of the Military Telegraphic Service.

If French telephone technology was inferior to that of the invading Germans, as it undoubtedly was, patriotic French engineers did find some consolation in the great accomplishments of their scientists in the past. In the propaganda war which was fought within the real war, it was common to present the history of science, culture and technology in a highly chauvinistic way. One common element in this propaganda was to claim that the enemy's scientific accomplishments were either plagiarized or based on foreign creativity. German success in telephone technology was based on the loading method, but didn't this method owe its origin to French scientific genius? "Modesty," boasted a French technical journal in 1917, "is a quality which French scientists have always held an honour to cultivate." Unfortunately, this noble Gallic quality often resulted in French scientific work being attributed to Germans. It was, therefore, welcome that French historians "use their erudition in the defense of our intellectual inheritance and in demonstrating that in the application of science, as well as in the sciences themselves, French scientific genius has most often been a precursor."[20] As a result of such nationalistic historiography, Vaschy was now celebrated as the real inventor of the loading method and a victim of the plagiarism so characteristic of *la science allemande*. The claim had no basis in reality, but was typical for much of the historiography of science and technology which the war produced.

German army communications units counted only 6350 men in August 1914, while the demand was closer to the 190000 men which served in the communications units four years later. Telegraphy and telephony, so highly praised by von Schlieffen, were given only low priority in his master plan to launch a knockout blow to the French army. Railways were used extensively in the campaign, but communication between headquarters and the front, and between different sections of the front, was inadequate and contributed to the failure of von Schlieffen's plan. When General Alexander von Kluck attacked the French-British forces at Marne, failed communications were an important factor in the German defeat. "It is certain," wrote a French officer after the war, "that if von Kluck on 2 September 1914 had been able to telephone to General [Helmuth] von Moltke, he would not have committed the mistakes which facilitated our victory."[21] The lack of good communication is summarized by a historian of World War I as follows: "[Kluck] had sent three messages to OHL [Supreme Headquarters] announcing his intention to cross the Marne but as wireless communication with Luxembourg was even worse than with Coblenz, they did not get through until the following day. Out of contact with the First Army for two days, OHL did not know Kluck had disobeyed the order of September 2."[22] In fact, in order to find out what was happening in northern France, Moltke had to send a staff-officer to the front.

During the first phase of the war, the German telegraph and communications network experienced an enormous increase in traffic (partly due to the many mobilization orders) which caused tensions between the Reich Post and Telegraph administration and the military authorities, which wanted part of the network for their exclusive use. The most serious immediate effect of the war was the complete disconnection of German overseas telegraph lines, which were taken over or destroyed by the British Navy. In Germany telephony was gradually militarized: lines to foreign countries were closed down or reserved for military and diplomatic use; non-military use of the telephone network in the frontier areas was forbidden; and censors listened to domestic calls. Although the civil post and telegraph administration kept control of most of the telecommunications system, and struggled with the military to maintain its autonomy, in practice the military bureaucracy decided how and when to use the system. As the war went on, the post and telegraphy administration became increasingly integrated in the military system.[23]

Fighting a war on two widely separated fronts, long-distance communications were of vital strategic importance to the Germans, who had to start a crash programme to establish the needed lines. When Russian troops attacked East Prussia there was no telephone line to Berlin, but within one month a military line supplied with repeaters had been established. Other long lines, pupinized and repeatered, secured communications between the Eastern and Western fronts and between the headquarters in Berlin and faraway Constantinople. These lines were successful insofar as they served their strategic purpose and marked the first large-scale application of vacuum tube amplifiers in European telephony. The military long-distance cable system was specified by Wilhelm Ohnesorge from the Imperial Post and Telegraph administration. (Ohnesorge became an early follower of Hitler and in 1934 Secretary of State of the Third Reich's Postal Ministry.) The combination of loading and repeater engineering, including perfection of the Lieben electronic tube, brought about traffic over a line of more than 3000 km, a European long-distance record.[24] The theoretical physicist Walter Schottky, working part-time for Siemens & Halske, developed an improved vacuum tube (the screen tube) by introducing an extra grid between the anode and the normal control grid. The tube was tested as amplifier on military lines, but was made effective only in 1926, when it was modified by the British engineer H.J. Round. Schottky's work included a study of the fundamental physics of vacuum tubes and led to the discovery of discrete noise-effects, which he explained in a now classic contribution to information theory.[25] Another eminent theoretical physicist, Arnold Sommerfeld, did research on listening to enemy telephone communication by detecting the weak "earth currents" propagated through the ground. Tube-amplified listening devices based on Sommerfeld's research were able to detect signals more than 1 km from a line and proved of considerable value to the German army.[26]

The German efforts to maintain an efficient electrical communications system was hampered by lack of trained personnel and shortage of material because of the blockade. With supplies of copper and guttapercha caught low, the Germans invented substitute materials for cable insulation and used the copper of civil lines for the more important military lines; the civil lines were instead supplied with pupinized iron wires. By and large, the German army managed to secure efficient strategical telephone communications during the war, especially after von Moltke had been replaced with generals Falkenhayn, Hindenburg and Ludendorff. The latter, in particular, made full use of telephone technology in his direction of the Eastern divisions. Also for tactical and local purposes the telephone became a key instrument, serving such diverse functions as artillery guidance and air-attack warning. By the end of 1917 German signal and communications units operated a telegraph and telephone network of 920000 km wire, about 23 times the circumference of the earth at the equator.[27] But German efforts to secure military communications only succeeded at the expense of the civil system and was characterized by ad hoc solutions narrowly directed towards pressing military needs. In particular, German engineering research stagnated because of the private companies' immediate duties to the Fatherland and the lack of an independent research department under the Government. The biggest of the German electrotechnical companies, Siemens & Halske, almost stopped whatever fundamental research had been conducted before 1914. Hans Gerdien, in charge of the company's Laboratory of Physics and Chemistry and after the war the first director of its research department, described Siemens & Halske's activities during the war as "a complete transformation towards work directly and indirectly related to war purposes; this was the only imaginable [course] and, in a patriotic sense as well as in the interest of the firm, a must."[28]

The military importance of advanced electrical communications became highly visible when the Americans entered the war in April 1917. Already in January 1916 the Bell System had been asked to assist the US army and navy in the application of the newest developments in communications technology, and the company started mobilizing for war and propagandizing for high-tech warfare. Carty lectured on "The organization, plants, and personnel of the Bell System" before the Army and Navy War Colleges. The officers were impressed and "there were springing up in high places a very definite realization of the military importance of the latest telephone developments, and a belief that the research facilities of the Bell Telephone organization could probably contribute still other new devices of value in the national defense."[29] In his talk to the war colleges, Carty likened the operation of a telephone system with a military maneuvre, and he concluded that "Battles are nowadays conducted with the telephone, and it is an important part of the work of a commanding general to use the telephone correctly." The people of the Bell System, Carty said, operated like military personnel, an analogy which at the same time was also used by Siemens & Halske in Germany. But whereas the Bell engineer-soldier had much to teach the military, Siemens & Halske deplored that their engineers did not act with the independence and dynamics which allegedly characterized the Wilhemian officer.[30] The picture of the scientists as members of an army organized in military fashion was widely accepted as part of the endeavours to mobilize science for the war. In a lecture delivered at University College in 1915, the telephone and radio engineer John A. Fleming thus advised that "the learned societies in Great Britain should act in some capacity like the great general staff of an army towards the subordinate generals and corps commanders."[31]

With the intentions of demonstrating the military significance of the long-distance telephone system, Carty arranged in March 1916 a demonstration in Washington D.C. for politicians, business magnates and high-ranking officers, who were offered telephone calls to San Francisco, El Paso and Seattle (with a distance of 3700 miles from Washington). The first call was to General John Pershing, the later commander-in-chief of the expeditionary force in Europe, who was then stationed on the Mexican border. Considering that the Bell System in the summer of 1915 had successfully transmitted the first radio speech across the Atlantic from Arlington, Virginia, to Paris, the point was nicely brought home.

When war was declared a programme was worked out under the Council of National Defense to utilize the nation's scientific and technological resources optimally.[32] The army commissioned Carty and a select number of other scientists and engineers as officers in the Signal Corps, whose Chief, General George Squier, strongly advocated scientific research. (Squier held a Ph.D.

in physics from Johns Hopkins). Carty organized a Science and Research Division which included scientists and engineers from universities and private corporations. Among the members were Jewett and the leading electron physicist Robert Millikan, who headed the Science and Research Division of the Signal Corps. Contrary to some other military programmes, the collaboration between corporation engineers, university scientists, and Signal Corps officers went smoothly. The Bell System itself supplied within a few months 4500 engineers and operators, organized in 14 battalions, and thereby tripled the staff of the Signal Corps.[33] Throughout the war the Army Signal Corps was as much part of the Bell System as the Bell System was part of the Signal Corps. The Americans had little confidence in the electrical communications systems of the British and French forces, which they considered old-fashioned and inefficient. The Old World might be superior in science and culture, but it was hopelessly inferior to the yankees when it came to technology and organization. After the war, Carty, who had then advanced to the rank of general, summarized the situation as follows:[34]

When the war broke out it was not possible for any of the European nations to provide a communication system adequate for the conduct of war. It remained for the Signal Corps of the U.S. Army in nine months to construct a long distance telegraph and telephone system which the Governments of Europe had failed to do in forty years. ... We were able to talk all over the Continent. That gives you an idea of some of the work done by methods absolutely unknown to any of the nations engaged in the combat.

Although the American communications technology was superior to that of the European nations, this was a gross exaggeration of the technology gap. In fact, none of the methods applied by American engineers – such as vacuum tube amplifiers, phantom circuits, and composite systems for simultaneous telegraphy and telephony – were unknown to European engineers. The telephone system built by the Americans served strategic as well as tactical purposes. It extended throughout France, from Marseilles to Le Havre, and included a new submarine cable across the Channel. The system greatly facilitated military communications between the different headquarters, between the headquarters and the fronts, and between the many camps. "It is believed," reported the Army's Chief Signal Officer in 1919, "that the telephonic communication thus supplied by the Signal Corps was one of the essential factors in the winning of the war."[35] A similar high appraisal of the telephone's military importance was given by a French engineering captain in a retrospective assessment of electrical communications during the war. He described the use of the telephone as "one of the many revelations of the war."[36] Such testimonies should be read with the awareness that officers in the signal and telegraph units had a self-interest in presenting communications technology as a crucial factor in the outcome of the war.

The lines erected by the Americans were used simultaneously for telegraphy and telephony and applied so-called phantom circuits to obtain maximum traffic over the wires. While long-distance telephony was usually realized either by means of loading coils or using copper wires of heavy gauge, the Signal Corps engineers decided to use unloaded wires of a low gauge. This was possible only because of an extensive use of vacuum tubes as telephone repeaters. The efficiency of the long-distance plant of the American army was entirely due to the vacuum tube, which was produced in two versions, the tungsten-filament triode manufactured by General Electric and the Wehnelt cathode tube manufactured by Western Electric. These types of tubes were studied by Harold D. Arnold, who had been hired by the Bell System from the University of Chicago in 1911, and Clinton J. Davisson, who had been recruited from the Carnegie Institute of Technology in 1917 (and twenty years later would become Bell's first Nobel laureate[37]); they were developed for military use by F.J. Clark, a Bell specialist in telephone repeaters and captain in the Signal Corps, and Oliver E. Buckley, a Cornell-trained Bell physicist who was in charge of the Signal Corps' laboratory in Paris. The demand for high-quality standardized tubes both for telephone and radio purposes was satisfied by American companies, who drastically increased their production. When the United States entered the war, the country's weekly production did not exceed 400 tubes; two years later the production was about 80000 per week.[38] The engineering advancement in tube production under the pressure of military necessity is an important example of war-induced technological progress.

Both the US Signal Corps, the French Central Bureau of Telegraphy and the German Imperial Telegraphy Test Department experimented with what was called "wired wireless", that is, telephone signals impressed upon high-frequency radio waves guided by telephone wires.[39] General Squier was a pioneer in multiplexing telephony by means of high frequencies, in which area he held several patents. Under his command, the Signal Corps used the method to communicate over buried bare wires instead of cables. Breisig and Wagner of the German Test Department devised another system of multiplex telephony by superimposing high-frequency currents along telephone lines. In this way the Germans succeeded to reduce the problem of crowded traffic on their lines and spare the valuable supplies of copper.

The importance of electrical communications technologies to the war was considerable, but hardly greater than the importance of other technological advances such as in acoustics, gunnery, airplanes, optical glasses, explosives and chemical warfare. The conclusion of the Signal Corps, quoted above, that telephone technology was "one of the essential factors in the winning of the war," seems somewhat exaggerated. Yet there is no doubt that telephony, especially as a strategic and logistic resource, greatly affected the military operations. The role of long-distance electrical communications has not been properly appreciated by military historians, presumably because strategic electrical communications is not a destructive technology of the same type

as, e.g., poison gas, submarine detection, and aircraft. It belongs rather to
the same category as sanitation and preventive medicine. Like telephony,
these are civil and peaceful branches of applied science that did not directly
influence warfare. But indirectly their influence was great.[40]

If one excludes such developments as field telephones and exchanges,
special telephones for gas-masks, and telephone communication between
pilot and bomb crew in airplanes, the needs of the war did not result in
important new telephone inventions. As mentioned, the perfection of the
vacuum tube depended directly on the war, but tubes were used as much in
radio as in telephony. Important innovations made during the war, such as
permalloy and the iron-powder coil core, were not directly related to the war,
but the fruits of Bell System's new research programme which had started
before 1914.[41] Also experiments with carrier telephony antedated the war,
and the method of "wired wireless" was only developed to a commercial
level about 1920. Perhaps the biggest impact of the war on the development
of telephony was that American communications technology came out of the
war in a much strengthened position relative to that of Germany. Before the
war German companies such as Siemens & Halske and Felten & Guilleaume
were technologically competitive and only slightly behind the Bell System in
telephone research. The war changed that, so that in 1918 Siemens & Halske
was no longer the innovative leader it had once been. The German long-
distance telegraph and telephone network had not been developed during the
war and had to be rebuilt and modernized. Most of the submarine telegraph
lines had been conquered by the enemy during the war, and at the Versailles
Conference Germany was forced to give up almost her entire network, a total
line length of 40508 km corresponding to an economic loss of 87 million
Reichsmark.[42] In general, both for Germany and other European countries,
the Great War had a detrimental effect on the growth of the telephone system
and increased the technology gap between America and Europe in this area.

4. THE POLITICS OF COMMUNICATION TECHNOLOGY

As is well known, academic science functioned in the early Weimar republic
as a kind of *Macht-Ersatz*, a substitute for the lost military, political and
economic power.[43] The doctrine was anti-utilitarian in that it emphasized
the cultural value of science *per se*, and not science as an instrument of
wealth and power. Engineering science and its accomplishments, such as
German communications engineering, was seen in a somewhat similar light,
as *Macht-Ersatz* and a source of national pride. But whereas engineering
science shared with academic science its role as substitution for military
and political power, obviously it was not anti-utilitarian. Success in German
technological research was considered important in itself, but at the same
time a vehicle for economic growth; and although it could not recreate the

Fatherland's military and political strength, in the long run it might provide the conditions for such recreation.

In the attempts to rebuild Germany, threatened by civil war and haunted by inflation and mass-unemployment, the Weimar government gave high priority to the establishment of a long-distance communication network. Such a network would not only tie together the politically, economically and geographically fragmented nation, it was hoped, but also furnish the means for economic cooperation with foreign countries and thus help to develop Germany's economy. It was part of a larger political programme towards centralization which also included a nationalized power network and railway system. In spite of the stagnation of the country's civilian communication lines during the war, Germany remained in 1919 Europe's leader in communications technology and its electrotechnical corporations were largely intact. A determined effort was made by the new post and telegraph administration (the Reichspostministerium, RPM) and communications companies, including Siemens & Halske, Felten & Guilleaume, Telefunken, and Allgemeine Elektrizitätsgesellschaft (AEG), to create a national and international telephone network of the highest quality. In this work the increased emphasis on scientific research in German industry was important. In 1920 Siemens & Halske reorganized and strengthened their research efforts, which included a central laboratory under Fritz Lüschen and a laboratory for telegraphy and telephony under Bruno Pohlmann.[44]

Contrary to the American long-distance lines, most of which were aerial wires on poles, the German network would be based on underground multiwire cables, a technology in which German industry had specialized since the very first Prussian telegraph cables in the 1840s. Germany's traditional predilection for underground cables, whether telegraphic or telephonic, was partly rooted in the obvious military and security advantages of such lines.[45] A large telephone cable network had already been initiated before the war with a cable from Berlin to Magdeburg, the first part of a planned Rhineland cable network. This cable was not, however, prepared for repeaters, and had to be changed after the war. The Rhineland cable project was completed in 1921. This cable system had a special significance to the defeated German nation as it demonstrated the strength of German science and technology under very difficult external circumstances.[46] In the eyes of the Germans it also demonstrated the unreasonable policy of her enemies, who occupied most of the Rhine district: the occupation army denied permission to complete the Western part of the network and prevented its connection with Germany's neighbours.

Equally important for German communications politics and technology in the first difficult years after the war were Felten & Guilleaume's laying of submarine cables in the Baltic Sea. In 1919, and again in 1921, Krarup cables were laid to Sweden over a record distance of 120 km, and three similar cables were laid between Pomerania and East Prussia between 1920 and 1922. These lines were followed by other cables, pupinized and laid by Siemens & Halske, so that by 1929 there was excellent submarine communication between mainland Germany and East Prussia. Since Germany had been cut off from East Prussia as a result of the Versailles Treaty, it was of strategical and political importance to link East Prussia to the rest of the nation without passing through Polish territory. The construction of the cables for the Rhineland project was a cooperative effort between private industry and the technical branch of RPM. The success of the project and the need to provide uniformity in the entire national network led to the establishment in April 1921 of the Deutsche Fernkabelgesellschaft (German Long-Distance Cable Company) as a joint state-private organization. Apart from the Reich Post Ministry, Germany's seven biggest cable manufacturers participated in the organization, with Siemens & Halske, AEG and Felten & Guilleaume being permanent members of the board. The capital outlay for the society was 6 million Mark, divided between RPM with 2 million, Siemens & Halske with 1.5 million, and AEG and Felten & Guilleaume with each 1.25 million. The Cable Company ensured exchange of patents and experience between the participants and supervised all the telephone cable plant with its own staff, which amounted to about 800 engineers, technicians and workers. Under the leadership of the Cable Company German work on an extended cable network accelerated.

In the early 1920s French-German relationships were tense over the question of war reparations, which in January 1923 reached a violent culmination with the French-Belgian occupation of the Ruhr district. In this climate of hatred and suspicion, Germany's ambitious cable plans did not get a friendly reception by all her neighbours. Some of them, and France in particular, felt that the plans were a cover for teutonic revanchism and felt it improper that the loser of the Great War should embark on a technological project, the likes of which the victors could neither afford nor manage. At that date the nationalized French telephone system was hopelessly inferior to the German and greatly in need of modernization. Interurban communication was seldom possible and it was generally recognized that, as the president of the union of French electrical industries expressed it, "in matters of telephony, France belongs in fact to the eighth division among the civilized nations."[47] The country already depended on equipment from Western Electric and in 1922 Sosthenes Behn, president of the International Telegraph and Telephone Company (ITT), offered to take over the entire French telephone network and manufacture of equipment. Behn's offer was declined and the minister of post and telegraphy initiated a long-needed, but slow reform.[48] The embarassing backwardness vis-a-vis the Germans, combined with the tense political situation, made the Interallied Commission object to the German cable plans, which were branded a "luxury" unfit for a defeated nation.[49] Other responses were more direct. A French journal accused the cable plan of being a revanchist plan to score a propaganda victory and prepare for a new war: "For this campaign they [the Germans] will use the same means as before the war, when the aim was to raise new money for armament."[50]

The accusation reflected the realization, made clear by the lessons of the war, that industrial research was a key factor for a nation's ability to succeed in modern warfare. In his lecture of 1915, Fleming not only emphasized this, but also predicted a later "commercial and industrial war" as a continuation of the military war. "After this war is over in a military sense," he said, "we shall immediately commence another war of a different kind, in which the weapons will not be bullets and shells, but our national powers of invention, scientific research, commercial organisation, manufacturing capabilities, and education, and these will be pitted against those of a highly organised Germany determined to win back in commerce by any and every means, fair or foul, that which has been lost in war."[51] To many Frenchmen, this was what Germany was up to with her new telephone cable plans.

With the aim of rebuilding and integrating European telephone communication, a conference was held in Paris in March 1923. Already before the war there had been intereuropean telephone conferences: 1908 in Budapest and 1910 in Paris; a third one was scheduled to take place in Berne in 1914, but cancelled because of the war. The Paris conference of 1923 reflected the need to create a long-distance European network and also the tense political situation after the war. The immediate background was an inaugural address given in November 1922 by Frank Gill, newly elected president of the British Institution of Electrical Engineers and chief representative of Western Electric in Europe.[52] Analyzing the deplorable state of European telephony from the perspective of the Bell System, Gill argued that the problems were not technical, but rooted in mental and political differences between Europe and the United States. In a language rich in military metaphors, he criticized engineers and administrators in the European government-owned telephone companies for lacking the "aggressive attitude" which characterized the Bell System. They saw it as their task merely to satisfy public demands, while the proper attitude should be aimed at "a resolute, purposeful and well-directed campaign of education" of the public. Attitudes apart, the greatest problem, according to Gill, was the lack of unity and cooperation between the various nations. Once again he took recourse to military language and comparisons. Great advantage had been gained in the War by a unitary command, he reminded his audience. If such unity could be established in warfare, why not in telephone business? Gill advocated the creation of a unitary, intereuropean long-distance company and urged the governments and telephone operating companies to discuss the possibilities of cooperation at a conference.

The French vice-secretary for post and telegraphy, Paul Laffont, who feared that Germany might become the centre of a European cable network, took advantage of Gill's well-publicized talk and hastily arranged the conference in Paris. At that time the allied boycott of German science was at its very height. Most international scientific and technological conferences held in allied or associated countries excluded delegates from Germany.[53] The Paris telephone conference was no exception. Participation was limited to England, France, Belgium, Italy, Spain and Switzerland; neither Germany nor the Netherlands and Scandinavian countries – which possessed the most developed telephone systems in Europe – were invited. The wounds of the war, reopened by the Ruhr crisis, were too fresh to include former enemies and nations suspected to be pro-German. Yet the French general-inspector of post and telegraphy, M. Dennery, emphasized in his opening address the peaceful aim of the conference:[54]

Disastrous wars could have been avoided if the governments had had at their disposal the means to obtain rapidly and safely a complete agreement in the hours of world crises. There is no better means than speaking and listening among people. From this point of view international telephony is a mighty assistance to universal peace.

The six countries agreed on basing technical recommendations on the standards and equipment of Western Electric, but rejected Gill's vision of a private intereuropean telephone organisation à la Bell.[55] The Germans objected to the unofficial boycott and Paris as the site of the conference, but they knew, of course, that this was France's way of demonstrating her political power as a compensation for her lack of technological strength. "France claims to have the leading position in the communications area in Europe, and the other countries, having once been parts of the Entente, have bowed to the claim. ... [But] if power decides in politics, industry is governed by technical competence. This realization will sooner or later dawn upon the minds bewildered by the present shift in power."[56]

As a countermove the Reich Post Ministry published in German, French and English an account of the German cable plans and contributions to long-distance telephony. The work had a double purpose, emphasizing the high standards of German cable industry and countering the American influence on European telephony: "We regard it as necessary not merely to copy the American constructions, but also to take advantage of the knowledge of other nations, especially that of German science and technique."[57] The German authors, representing the participants in the Cable Company, and Siemens & Halske in particular, presented their version of the modern history of telephone technology, which largely meant a history of successful German innovations prior to and superior to the innovations of Western Electric. Expectedly the propaganda caused counter-propaganda in the form of a strongly worded "commentary" from the European engineering department of International Western Electric. Far from leading the technological development, the Germans merely plagiarized the innovative work of Western Electric, the Americans claimed: "In reality, the work done in Germany is nothing but imitation and adaption of the methods of long-distance telephony introduced by the American Bell System and Western Electric many years ago."[58] Western Electric's comparison concluded, understandably but unfairly, that German telephone technology was hopelessly inferior and only survived because of government support and trade barriers. The principal reason for the controversy was not questions of priority or national pride, although such factors undoubtedly played a role in Germany's attempt to restore national self-confidence. What was at stake was whether the future European long-distance network, a project of immense industrial scale, should be based on American or German technology. This was not only an important political and economical question, but also one of great potential military importance.

A follow-up conference in May 1924, also held in Paris, demonstrated that the German campaign had been largely successful, and marked the beginning of the European long-distance cable network. The conference took place shortly after the acceptance of the Dawes plan, which offered more realistic terms for Germany's payment of reparations and resulted in a more conciliatory political atmosphere in Europe. With a milder political climate and a growing foreign realization of the central importance of the Weimar Republic, Germany was now invited to participate together with 18 other countries. The conference created a permanent body, CCI (Comité Consultatif International des Communications Téléphoniques à Longue Distance), which was to meet regularly in Paris; to satisfy French feelings, also the official language and the general secretary (Georges Valensi) of the new organization were French.[59] With regard to the standards and equipment to be used in the future cable network a compromise was obtained, so that systems based on both American and German technology could be used. As a result of the work of CCI, a plan for cooperation between the European post and telegraphy administrations was established. The heart of the new long-distance network was to consist of underground cables supplied with loading coils and repeaters satisfying certain standards of uniformity and quality; the cable network was to be connected with repeatered, unloaded aerial circuits.

Although the European telephone cooperation respected national differences and was far from the unifying, transnational scheme suggested by Gill and realized in the Bell System, it was an important element in the cooperative and peaceful spirit which permeated Europe in the years after 1924. It was paralleled by the League of Nations' Committee for Communications and Transit (Commission Consultatif et Technique des Communications et du Transit), which, among other tasks, worked toward establishing efficient telephone communications between the European capitals and the League of Nations in Geneva. Telephone lines were seen as important diplomatic instruments and means to prevent a new European war. In this short era of optimism, the telephone, which had been used so extensively in the Great War, was felt to be technology's promise of a better world. This belief was routinely expressed at ceremonial openings of new cable lines and was part of the rhetoric of leading telephone engineers. "In the telephone we have the most perfect means of communication ... to remove misunderstandings," said Gill in his above-mentioned address of 1922. "If only we will use it, not alone will it benefit the industry of the nation, but we shall be making a definite step towards reducing the international jealousies and fears and increasing the good-will without which there cannot be peace on earth." John Carty – General John Carty, that is – was no less sanguine in his visions of what international telephony might accomplish:[60]

Some day we will build up a world telephone system making necessary to all people the use of a common language, or common understanding of languages, which will join all the people of the earth into one brotherhood. ... [Then] we can all believe there will be heard, throughout the earth, a great voice coming out of the ether, which will proclaim, "Peace on earth, good will towards men."

Needless to say, although an international telephone system had emerged by 1939, its consequence was not 'peace on earth.'

The years following 1924 saw a rapid development of a European telephone network, which was largely completed about 1929, when the economic crisis put a temporary stop to further development. Programmes for national cable networks had already started before 1924, when a number of international submarine telephone cables had also been laid. Siemens Brothers, Siemens & Halske, and Felten & Guilleaume supplied most of these cables, which connected England with France, Belgium and Holland; Germany with Denmark and Sweden; and Denmark with Sweden. Submarine telephony over Krarup and Pupin loaded cables was a European specialty, and the only area of telephone technology where Europe was clearly ahead of the Americans.[61] In March 1926 the first international connection between separate cable networks, those of Germany and Switzerland, took place. The first cable line across the German-French border was established in 1927, and the following year a direct German-British telephone line began service. These technological accomplishments were important events, both economically and politically and as symbols of a new relationship between the former enemies. By 1928 there was regular service between London and Rome, and London and Stockholm; the latter line, extending over 2500 km of which 316 km were submarine, was the longest European telephone line at the time.[62] By connecting the cable network with the transatlantic radio service, it was in 1928 possible to communicate between Chicago and Stockholm. Not only were distances increased as a result of improved vacuum tube amplifiers, traffic was also greatly increased by using cables with up to 1200 cores and, from the late 1920s, systems of carrier telephony. These new generations of cables carried not only telephone currents, but also telegraphy and radio broadcast.

The new European cable network placed Germany in a central position (see Fig. 1). It also allowed the German cable and telephone industry to ex-

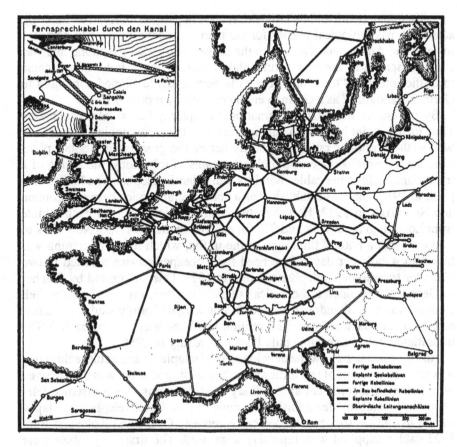

Fig. 1. The state of the European cable network in early 1927. Inserted in the upper left corner are the most important cables across the English Channel. Source: *Europäische Fernsprechdienst* no. 5 (1927) 41.

pand beyond the borders and traditional markets (such as Scandinavia) and to compete successfully with Western Electric, which since 1925 had become Standard Electric. One sign of German success was the construction of the French cable network, the first parts of which had been installed by French companies, but which did not function well. When the important line from Paris to Bordeaux, planned to continue to Spain, had to be built, Siemens & Halske and Felten & Guilleaume were chosen to do the work, which they completed efficiently and by themselves. The German intervention in the French communication structure was allowed in spite of objections from military authorities, who realized the strategic importance of the cable network. But further penetration was obstructed, partly for military reasons and partly to protect the interests of French cable manufacturers.[63] The success of the European cable network and its reliance, at least in part, on German technology, acted as a psychological plaster for German engineers' inferiority-complex towards the Americans. True, even the longest European telephone line was short compared to the transcontinental American lines, but when it came to traffic intensity "the United States is very underdeveloped compared to German standards," as a German engineer joyfully reported.[64]

By the end of the 1920s, international telephony had greatly progressed since its low point in 1918. As an indication of the technological progress as well as the economic optimism shortly before the great crisis, serious plans were made to realize one of the oldest dreams of telephony, the laying of a transatlantic telephone cable. Wireless transatlantic telephony had first taken place in 1915, and was established as a commercial service between London and New York in 1927. However, the costly service – £5 for each minute from London to New York and $75 for each three minutes from New York to London – put a natural limitation to the traffic, which in the beginning was less than 4 calls per day.[65] The need for a cheaper service resulted in two plans for transatlantic telephone cables in 1929. One, suggested by Wagner in Germany, remained a theoretical study; the other, instituted by the Bell System, was only abandoned after a 20-mile long uniformly loaded test cable had been manufactured and $2.2 million had been spent on research.[66] The transatlantic telephone cable only became a reality in 1956.

I have emphasized how the development of telephone and cable technology was not only a matter of scientific and technical progress, but was interweaved with political, ideological and military considerations. Long-distance telephony in Japan provides a further example of this. Cable telephony in Japan started in 1922 with a coil-loaded cable between Osaka and Kobe and from 1925 cables supplied with repeaters were used. The first long cables were produced and installed by Western Electric and Siemens & Halske, who continued to provide loading coils for the cables. In 1928 the main-section of the country's new long-distance network, the Tokyo-Kobe line, was completed; the 600 km aerial cable was manufactured by Japanese companies. As in Europe and USA, all the Japanese cables were loaded and amplified by vacuum tubes, but in the early 1930s a Japanese engineer, Shigeyoshi Matsumae, argued that non-loaded cables (with repeaters and carrier transmission) should be used. Matsumae's arguments were technical, but also political in that he considered loading a result of Western imperialism and the non-loaded system a particular Japanese form of cable engineering. Writing in 1936 in a not entirely correct, nor entirely intelligible, English, Matsumae opined:[67]

The wide field in Orient is waiting our civilized expansion. On this occasion for the completion of its communication network to produce excellent system of having our own individuality by these investigations and introduce the Nippon civilization and the same time make economical expansion is our great mission and is also our special right. The author prays and expects that our own excellent system is to contribute to the Orient and to the world by the researches of such as non-loaded cable elsewhere in the future.

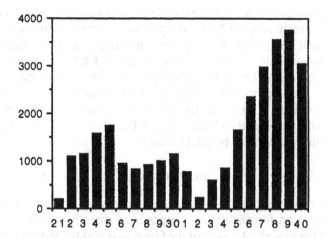

Fig. 2. Development of German long-distance telephone cable network (in km) between 1921 and 1940. Data from Craemer, *Zwei Jahrzehnte*, 121.

The fruit of Nippon civilization was at that time experienced by the Chinese people. Non-loaded cables after Matsumae's design were produced since 1934 by the Fujikura Electric Wire Works and used for the Nippon-Manchukuo line, which was an important part of Japanese imperialist expansion in China and Manchuria. Such military application was fine with Matsumae, who reminded his readers that Japanese communication engineering was all "for the peace of mankind, for the exchange of civilization, for the solemn purpose of creation of universe by God."

Another example, illustrating some other features of the political significance of telephony, is provided by the formation and early development of Soviet Russia. Lenin, always aware of the importance of modern technology, urged in the tumultuous fall of 1917 that "We must mobilize the armed workers ... immediately secure the telephone and the telegraph, install our insurrection headquarters in the central telephone exchange in Moscow, and connect it by telephone to all factories, all regiments, all points of armed uprising."[68] As it happened, the Moscow exchange became a hindrance for the revolution when the telephonists' union joined forces with the provisional government and refused to cooperate with the Reds. After the Bolshevik victory the development of a telephone network received only scant attention. The new leaders recognized the importance of the telephone, but shortage of money, unrest, and blockade slowed down the development. In fact, the number of telephones in operation decreased between 1917 and 1925, and as late as 1924 there were fewer Russian subscribers than in 1910.

Although not the only reason for the backwardness of the Soviet telephone system, the political authorities' suspicion that residential telephones might be used by counter-revolutionary groups was an important factor. The communist leaders realized what McLuhan pointed out many years later, that the telephone might be a politically dangerous instrument because of its unmediated and uncontrollable form of communication. Rather than investing in residential telephones, they gave high priority to "top down communication" such as radio and newspapers which more easily could be controlled and serve as propaganda. As a result, the telephone became an instrument of the state administration to a much higher degree than in the West. It is only in the 1990s that this situation has begun to change.

5. CONCLUSION

The military's interest in science and technology is not limited to those areas which can be used for destructive purposes or directly as part of military operations. One important lesson of the Great War was the strategic necessity of maintaining an efficient system of communication not only in the vicinity of the front, but through the whole extent of the militarily controlled territory. Military logistics had to include the entire nation, not merely its military personnel. The new concept of total war, as it materialized in 1916, made it impossible to distinguish in a clear-cut way between civil and military research. The chemical research on poison gases was clearly military science, but the "civil" research on synthetic nitrogen fertilizers was no less important for Germany's war effort. The point was emphasized by John Desmond Bernal shortly before the outbreak of World War II. "Under modern industrial conditions war is no longer fought only by the men in the field of battle but by the whole national industrial complex," he wrote.[69] One important component of the industrial preparedness of a nation is the electrical communications system, which serves a triple purpose in war: for military communications (tactical and strategic), for propaganda, and as a means of maintaining commercial and social stability. During the first decades of the twentieth century telephone technology proved its military value in the field and military strategists began to realize the potential value of an extended telephone cable network. This network, as it was established in Europe during the years of conciliation and relative prosperity, was strictly under civil political control, but indirectly it was part of the power-struggle in the European theatre and thus also of military significance. For, as Carl von Clausewitz said in his famous dictum, "War is nothing but a continuation of political intercourse with the admixture of different means."[70] The modern concept of total war not only extends to every part of the nation's life, it also transcends the old dichotomy between war and peace by conceiving peace as periods of preparation for war.

The military-political-technological relationships within the field of telephone communications were not easily recognizable in the peaceful 1920s, but became clearly visible in the 1930s. In addition to the Japanese case, mentioned above, a brief reference to the situation in Nazi Germany before the war may illustrate the military significance of the civil cable technology.[71] Shortly after the take-over of power in 1933, the Nazi authorities started expanding and modernizing the cable network at an unprecedented pace (see Fig. 2). The expansion took place in cooperation with the military, who wanted an efficient communication system for a coming war and were eager not to repeat the mistakes of World War I. At the same time the entire communication system was used as a means to control and indoctrinate the German people. Telephone conversations were controlled by Gestapo and radio (and television) broadcasting over the telephone cables provided the propaganda. To speak of telephony as a purely civilian technology under these circumstances makes little sense.

University of Oslo

REFERENCES

Bakis, H., 1986. "Le développement du réseau téléphoniquie dans l'espace français, 1879–1940," *Bulletin d'Histoire de l'Électricité, 7,* 67–79.

Bernal, J.D., 1967. *The social function of science.* Cambridge, Mass.: MIT Press (first published 1939).

Bertho-Lavenir, C. "The telephone in France, 1879–1979: National characteristics and international influences." Pp. 155–177 in Mayntz and Hughes, *Large technical systems.*

Blake, G.C., 1926. *History of radio telegraphy and telephony.* London: Radio Press Ltd.

Blondin, J., 1917. "L'oeuvre de Vaschy en téléphonie à grande distance." *Revue Générale de l'Électricité, 2,* 281–282.

Breisig, F., 1914. "Über Fernsprechkabel großer Reichweite, insbesondere das Kabel Berlin-Rheinland." *Elektrotechnische Zeitschrift, 35,* 646–72.

Brittain, J.E., 1970. "The introduction of the loading coil: George A. Campbell and Michael I. Pupin." *Technology and Culture, 1,* 36–57.

Brown, F.J., 1930. *The cable and wireless communications of the world.* London: Pitman & Sons.

Buchholtz, F.A., 1922. "Das Rheinlandkabel." *Siemens-Zeitschrift, 2,* 45–58.

Campbell, G.A., 1924. "Mathematics in industrial research." *Bell System Technical Journal, 3,* 550–57.

Carty, J.J., 1916. "The relation of pure science to industrial research." *Science, 44,* 511–17.

Clark, P.W., 1976. "Early impacts of communications on military doctrine." *IEEE Proceedings, 64,* 1407–1413.

Clausewitz, C. von, 1968. *On war.* Harmondsworth: Penguin.

Craft, E.B., 1924. "The Bell System Research Laboratories." *Electrical Communication, 2,* 153–63.

Craemer, P., 1923. "Das europäische Fernkabelnetz." *Das Fernkabel,* no. 4, 3–24.

Craemer, P. et al., 1923. *Das Fernsprechen im Weitverkehr.* Berlin: Reichspostministerium.

Craemer, P. (ed), 1941. *Zwei Jahrzehnte Deutsche Fernkabel-Gesellschaft 1921–1941.* Berlin: EF Verlag.

Devereux, T., 1991. *Messenger gods of battle: Radio, radar, sonar; the story of electronics in war.* London: Brassey's.

Fagen, M.D. (ed), 1975. *A history of engineering and science in the Bell System: The early years, 1875–1925*. Murray Hill, N.J.: Bell Telephone Laboratories.

Eckert, M. and Schubert H., 1986. *Kristalle, Elektronen, Transistoren: Von der Gelehrtenstube zur Industrieforschung*. Munich: Rowohlt (Engl. trans., New York: American Institute of Physics, 1989).

Federal Communications Commission, 1938. *Proposed report. Telephone investigation*. Washington D.C.: US Government Printing Office.

Fleming, J.A., 1915. "Science in the war and after the war." *Nature, 96*, 180–85.

Fleming, J.A., 1927. *The interaction between pure scientific research and electrical engineering*. London: Constable & Co.

Forman, P., 1973. "Scientific internationalism and the Weimar physicists: The ideology and its manipulation in Germany after World War I." *Isis, 64*, 151–78.

Genth, R. and Hoppe, J., 1986. *Telephon! Der Draht an dem wir hängen*. Berlin: Transit.

Gerdien, H., 1926. "Das Forschungslaboratorium der Siemens & Halske AG und der Siemens-Schuckertwerke GmbH in Berlin Siemensstadt." *Siemens-Zeitschrift, 6*, 413–19, 469–77, 525–33.

Gifford, W.S., 1937. "The American communication companies and national preparedness." *Bell Telephone Quarterly, 16*, 3–24.

Gill, F., 1922. "The future of long distance telephony in Europe." *Electrical Communication, 1*, 8–26.

Grallert, 1921. "Das Telegraphen- und Fernsprechwesen vor, in und nach dem Kriege." *Archiv für Post und Telegraphie, 49*, 1–31.

Greely, A.W., 1927. *Reminiscences of adventure and service*. New York: Scribner's.

Grundmann, S., 1965. "Zum Boycott der deutschen Wissenschaft nach dem 1. Weltkrieg." *Wissenschaftliche Zeitschrift der Technischen Universität Dresden, 14*, 799–806.

Hartcup, G., 1988. *The war of invention: Scientific developments, 1914–1918*. London: Brassey's Defense Publishers.

Headrick, D.R., 1991. *The invisible weapon: Telecommunications and international politics, 1851–1945*. New York: Oxford University Press.

Hoddeson, L., 1981. "The emergence of basic research in the Bell Telephone System, 1875–1915." *Technology and Culture, 22*, 512–44.

Hoddeson, L., 1981. "The discovery of the point-contact transistor." *Historical Studies in the Physical Sciences, 12*, 41–76.

Höpfner, K. "Development and actual state of the repeater practice in Germany." Pp. 43–56 in Craemer et al., *Fernsprechen im Weitverkehr*.

Höpfner, K. and Dohmen, K., 1924. "Amerikanische und deutsche Technik des Fernsprechweitverkehrs." *Das Fernkabel*, no. 7, 37–49.

Hunt, B.J., 1983. "'Practice vs. theory': The British electrical debate, 1888–1891." *Isis, 74*, 341–55.

Inada, S., 1930. "Telegraph and telephone in Japan." *Electrical Communication, 9*, 114–37.

Jewett, F.B., 1920. "Recent advances in long distance telephony." *Transactions of the American Institute of Electrical Engineers, 39*, 191–227.

Jewett, F.B., 1937. "John J. Carty: Telephone engineer." *Bell Telephone Quarterly, 16*, 160–177.

Jordan, D.W., 1982. "The adoption of self-inductance by telephony, 1886–1889." *Annals of Science, 39*, 433–61.

Kevles, D.J., 1987. *The physicists: The history of a scientific community in modern America*. Cambridge, Mass.: Harvard University Press.

Kieve, J.L., 1973. *The electric telegraph: A social and economic history*. Bewton Abbot: David & Charles.

Kleinert, A., 1978. "Von der *science allemande* zur deutschen Physik." *Francia, 8*, 509–525.

Kline, R.R., 1992. *Steinmetz: Engineer and socialist.* Baltimore: Johns Hopkins University Press.

Kragh, H., 1994. "The Krarup cable: Invention and early development." *Technology and Culture, 35,* 129–57.

Kreger, W., 1937. "Deutschlands Seekabelnetz vor und nach dem Weltkriege." *Archiv für Post und Telegraphie, 65,* 117–31.

Legouez, R., 1923. "Les câbles téléphoniques." *Revue Générale de l'Électricité, 13,* 878–81.

Löser, W., 1969. "Der Bau unterirdischer Telegraphenlinien in Preußen von 1844–1867." *NTM (Schriftenreihe für Geschichte der Naturwissenschaften, Technik und Medizin), 6,* no. 2, 52–67.

McLuhan, M., 1964. *Understanding media.* New York: New American Library.

Martin, G., 1921. "La téléphonie à grande distance en Europe." *Annales des Postes, Télégraphes et Téléphones, 10,* 263–70.

Marvin, C., 1988. *When old technologies were new: Thinking about electrical communication in the late nineteenth century.* New York: Oxford University Press.

Matsumae, S., 1938. *Study on the long distance communication system by non-loaded cable.* Tokyo: Corona-Sha (preface dated 1936).

Mayntz, R. and Hughes, T.P. (eds), 1988. *The development of large technical systems.* Frankfurt am Main: Campus Verlag.

Michaelis, A.R., 1965. *From semaphore to satellite.* Geneva: The International Telecommunication Union.

Müller, E., 1930. "Auf dem Wege zum Ozeanfernsprechkabel." *Europäische Fernsprechdienst,* no. 20, 379–85.

Petzold, H., 1990. "Deutsch-französische Rivalität und Zusammenarbeit bei der Errichtung des europäischen Telefonnetzes nach dem Ersten Weltkrieg." Pp. 263–80 in Yves Cohen and Klaus Manfrass, eds., *Frankreich und Deutschland: Forschung, Technologie, und industrielle Entwicklung im 19. und 20. Jahrhundert.* Munich: C. H. Beck.

Pfetsch, F., 1970. "Scientific organisation and science policy in Imperial Germany, 1871–1914: The foundation of the Imperial Institute of Physics and Technology." *Minerva, 8,* 557–80.

Radkau, J., 1989. *Technik in Deutschland vom 18. Jahrhundert bis zur Gegenwart.* Frankfurt am Main: Suhrkamp.

Reber, S., 1904. "The telegraph, telephone and cable in war." *Transactions of the International Electrical Congress in St. Louis 1904,* 3 vols., vol. 3 (Albany), 449–459.

Reynolds, T.S. (ed), 1991. *The engineer in America.* Chicago: The University of Chicago Press.

Russo, A., 1981. "Fundamental research at Bell Laboratories: The discovery of electron diffraction." *Historical Studies in the Physical Sciences, 12,* 117–160.

Sampson, A., 1973. *The sovereign state: The secret history of ITT.* London: Hodder and Stoughton.

Schottky, W., 1918. "Spontane Stromschwankungen in verschiedenen Elektrizitätsleitern." *Annalen der Physik, 57,* 541–67.

Schroeder–Gudehus, B., 1978. *Les scientifiques et la paix: La communauté scientifique internationale au cours des années 20.* Montreal: Presses de l'Université de Montréal.

Schubert, H., 1987. "Industrielaboratorien für Wissenschaftstransfer. Aufbau und Entwicklung der Siemensforschung bis zum Ende des Zweiten Weltkrieges anhand von Beispielen aus der Halbleiterforschung." *Centaurus, 30,* 245–292.

Schwarte, M. (ed), 1920. *Technik im Weltkriege.* Berlin: E. S. Mittler.

Siemens, G., 1951. *Geschichte des Hauses Siemens.* 3 vols. Munich: Karl Alber.

Siemens & Halske, 1929. *Fernsprech-Fernkabel Paris-Tours-Bordeaux.* Berlin: Siemens & Halske.

Sola Pool, I. de (ed), 1981. *The social impact of the telephone.* Cambridge, Mass.: MIT Press.

Sola Pool, I. de, 1983. *Forecasting the telephone: A retrospective technology assessment.* Norwood, N.J.: Ablex.

Solnick, S.L., 1991. "Revolution, reform, and the Soviet telephone system." *Soviet Studies, 43,* 157–76.

Squier, G.O., 1919. *Annual Report of the Chief Signal Officer, 1919.* Washington D.C.: Government Printing Office.

Thomas, F., 1988. "The politics of growth: The German telephone system." Pp. 179–213 in Mayntz and Hughes, *Large technical systems.*

Thomas, F., 1989. "Korporative Akteure und die Entwicklung des Telefonsystems in Deutschland, 1877 bis 1945." *Technikgeschichte, 56,* 39–65.

Toché, G., 1921. "Les enseignements de la guerre au point de vue de la télégraphie militaire." *Annales des Postes, Télégraphes et Téléphones, 10,* 246–62.

Trendelenburg, F., 1975. *Aus der Geschichte der Forschung im Hause Siemens.* Düsseldorf: VDI-Verlag.

Tuchman, B.W., 1962. *The guns of August.* New York: Macmillan.

Tyne, G., 1977. *Saga of the vacuum tube.* Indianapolis: H.W. Sams & Co.

Vincenti, W.G., 1992. *What engineers know and how they know it.* Baltimore: Johns Hopkins University Press.

Wagner, K.W., 1929. "Plan einer Fernsprechkabelverbindung zwischen Europa und Amerika." *Sitzungsberichte der preussischen Akademie der Wissenschaften, Phys.-Math. Klasse, 7,* 1–15.

Wasserman, N., 1985. *From invention to innovation: Long-distance telephone transmission at the turn of the century.* Baltimore: Johns Hopkins University Press.

Weart, S.R., 1976. "The rise of 'prostituted' physics." *Nature, 262,* 14–17.

Weart, S.R., 1979. "The physics business in America, 1919–1940: A statistical reconnaissance." Pp. 295–358 in: Reingold, N. (ed), *The sciences in the American context.* Washington D.C.: Smithsonian Institution Press.

Weiher, S. von and Goetzeler, H., 1981. *Weg und Wirken der Siemens-Werke im Fortschritt der Elektrotechnik 1847–1980.* Wiesbaden: Franz Steiner.

Wessel, H.A., 1983. *Die Entwicklung des Elektrischen Nachrichtenwesens in Deutschland und die Rheinische Industrie.* Wiesbaden: Franz Steiner.

Western Electric, 1924. *Commentaires de l'International Western Electric Co. Inc. sur "Das Fernsprechen im Weitverkehr."* Paris.

Wilson, R. W., 1979. "The cosmic microwave background radiation." *Science, 205,* 866–74.

NOTES

[1] Michaelis, *Semaphore to satellite*, 175. Devereux, *Messenger gods*, 7–11.

[2] A version of this paper was presented at the 4S-EASST Conference in Gothenburg in August 1992. Paul Forman's criticism and helpful comments of an earlier version of the manuscript are gratefully acknowledged.

[3] Hunt, 1983. *Isis, 74,* 341–55; Jordan, 1982. *Annals of Science, 39,* 433–61.

[4] Wasserman, 1970. *Invention to innovation*; Brittain, *Technology and Culture, 1,* 36–57, reprinted in Reynolds, *The engineer*, 261–82.

[5] For the "prostitution" of physicists in early twentieth century, see Weart, 1976 and 1979. Although the entrance of academic physicists in industrial research was most marked in the United States, it was also common in Germany, especially in the electrical industry. See Eckert and Schubert, *Kristalle*, 36–44.

[6] Kragh, 1994. *Technology and Culture, 35,* 1–22.

[7] Carty, 1916. *Science, 44,* 511–17. See also Fleming, 1924. *Interaction,* and Campbell, *Bell*

System Technical Journal, 3, 550–57.

[8] Jewett, 1920. *Transactions AIEE, 39,* 195. For critical analysis of the concept of engineering science and the relationship between science and technology, see, e.g., Vincenti, *Engineers,* and Kline, *Steinmetz,* 105–20.

[9] Craft, 1924. *Electrical Communication, 2,* 153–63; Hoddeson, 1981. *Technology and Culture, 11,* 512–44; Russo, 1981. *Historical Studies, 12,* 117–160; Fagen, *Engineering and science.*

[10] Report of the Chief Signal Officer, 1898, as quoted in Reber, *Transactions,* 456.

[11] Quoted in Marvin, *Old technologies,* 222.

[12] Reber, *Transactions,* 542–53. Devereux, *Messenger gods,* 51–58.

[13] Kieve, *Electric telegraph,* 240.

[14] Greely, *Reminiscences,* 186. See also Clark, 1976. *IEEE Proceedings, 64,* 1407–1413.

[15] McLuhan, *Understanding media,* 234. See also de Sola Pool, *Forecasting the telephone,* 60–61, 89–91.

[16] Schott, 1903. "Die Telegraphie im Kriege," *Kriegstechnische Zeitschrift,* 599–614. Here quoted from Genth and Hoppe, *Telephon!,* 73.

[17] Radkau, *Technik in Deutschland,* 244.

[18] Pfetsch, 1970. *Minerva, 8,* 557–80. The figures include expenditures of both the imperial and state governments, of which the latter predominantly went to academic and medical research. In the same period the imperial government allocated 50% of its research money to economically oriented research and 29% to militarily oriented research (ibid.). For a survey of the contributions of German technological research to the First World War, see Schwarte, *Technik im Weltkriege.*

[19] Genth and Hoppe, 1986. *Telephon!,* 75. Bakis, *Bulletin d'Histoire de l'Électricité, 7,* 67–79.

[20] Blondin, 1917. *Revue Générale de l'Électricité, 2,* 281–82. For the ideological use of history of science during and after World War I, see Kleinert, 1978. *Francia, 8,* 509–525.

[21] Toché, 1921. *Annales des Postes, 10,* 253.

[22] Tuchman, *The guns of August,* 414.

[23] Thomas, *Large technical systems,* 192. See also Thomas, 1989. *Technikgeschichte, 56,* 39–65.

[24] Grallert, 1921. *Archiv für Post und Telegraphie, 49,* 1–31; Höpfner, *Fernsprechen im Weitverkehr;* Genth and Hoppe, *Telephon!,* 74–84.

[25] Schottky, 1918. *Annalen der Physik, 57,* 541–67. See also Weiher and Goetzeler, *Weg und Wirken,* 68–69.

[26] Hartcup, *War of invention,* 77.

[27] *Elektrotechnische Zeitschrift, 39* (1918), 94.

[28] Quoted from Schubert, 1987. *Centaurus, 30,* 251.

[29] Jewett, 1937. *Bell Telephone Quarterly, 16,* 170.

[30] Carty, as quoted in Genth and Hoppe, *Telephon!,* 70; Schubert, 1987. *Centaurus, 30,* 250.

[31] Fleming, 1915. *Nature, 96,* 185.

[32] Kevles, *The physicists,* 117–38.

[33] *American Telephone and Telegraph Company, Annual Report 1918,* 25.

[34] Testimony of Carty before the Committee on Military Affairs of the House of Representatives. Quoted from Gifford, 1937. *Bell Telephone Quarterly, 16,* 8.

[35] Squier, 1919. *Annual Report,* 109.

[36] Toché, 1921. *Annales des Postes, 10,* 253.

[37] Davisson's discovery of electron diffraction, which provided the first experimental proof of quantum mechanics, was the unplanned outcome of his work with vacuum tubes. See Russo, 1981. *Historical Studies, 12,* 117–60. It was not the last time that Bell telephone research would result in a Nobel prize. Also the invention of the transistor by Bell Laboratories' William Schockley, John Bardeen and Walter Brattain in 1948 (for which they were awarded the Nobel prize in 1956) had its basis in research related to telephony, and so had the discovery

of the cosmic background radiation for which Robert W. Wilson and Arno A. Penzias were awarded the prize in 1978. See Hoddeson, 1981, *Historical Studies, 12*, 41–76, and Wilson, 1979, *Science, 205*, 866–74.

[38] Squier, 1919. *Annual Report*, 116. For details on the early manufacture of vacuum tubes, see Tyne, *Saga of the vacuum tube.*

[39] See Blake, *Radio telegraphy and telephony*, 328–34.

[40] Hartcup's *War of invention* gives some attention to the use of wireless, but totally ignores the role of wire or cable telephony. Headrick, *The invisible weapon*, provides a detailed account of the political and military role of radio and cable telegraphy, but has almost nothing to say about telephony. And in Devereux, *Messenger gods*, the military role of telephony is dealt with on half a page (p. 28). The effect of preventive medicine on warfare is illustrated by the fact that during the Boer war more British soldiers died by diseases than by enemy action (Hartcup, *War of invention*, 166–180). Medicine is clearly not a military science, but like communication technology it decides the life and death of thousands of soldiers.

[41] Indirectly, the development of powder cores and the quick transformation of the invention into a practical innovation was influenced by the war: Early during the war importation from Europe of diamond dies, which were necessary for the production of the hard-drawn iron wires of which Pupin coils were manufactured, had become impossible. The import stop might have threatened the entire production of loading coils had the new method not been invented. The feared scarcity of loading coils may have contributed to the Signal Corps' decision to use repeaters instead of coils on their European lines. See Fagen, *Engineering and science*, 806.

[42] Kreger, 1937. *Archiv für Post und Telegraphie, 65*, 117–31.

[43] Schroeder-Gudehus, 1973. *Les scientifiques et la paix*, 223–29; Forman, *Isis, 64*, 151–78.

[44] Siemens, 1926, *Siemens*, vol. 3, 33–35. See also Gerdien, *Siemens-Zeitschrift, 6*, 413–19, 469–77, 525–33, and Trendelenburg, 1975. *Siemens*.

[45] In 1910 German telephone cables extended over 21000 km, more than ten times the length of the British cable circuits. Wessel, *Elektrisch. Nachrichtenwesens*, 566. For the Prussian telegraph cable system and its bearing on military considerations, see Löser, 1969. *NTM, 6*, no. 2, 52–67.

[46] Breisig, 1914. *Elektrotechnische Zeitschrift, 35*, 646–72; Buchholtz, 1922. *Siemens-Zeitschrift, 2*, 45–58.

[47] Legouez, 1923. *Revue Générale de l'Électricité, 13*, 881.

[48] Bertho-Lavenir, *Large technical systems*, 155–77. Behn, an American businessman who had served in the Army Signal Corps during the war, founded ITT in 1920 and started an aggressive expansion in Latin America and Europe. The expansion accelerated after 1925, when ITT took over International Western Electric and transformed it into International Standard Electric Corporation. Sampson, *The sovereign state*.

[49] *The Telegraph and Telephone Journal, 8* (1922), 89; *Vossische Zeitung*, 21 July 1922.

[50] *Revue des Téléphones, Télégraphes et T.S.F.*, 1923, no. 3, 201–02.

[51] Fleming, 1915. *Nature, 96*, 184.

[52] Gill, 1922. *Electrical Communication, 1*, 8–26. Similar ideas had earlier been proposed, for example in Martin, 1921. *Annales des Postes, 10*, 263–70, but it was only with Gill's address the problem received wide attention. See, e.g., "The unification of European lines," *The Times*, 3 November 1922.

[53] Between 1922 and 1924, 86 out of 106 international conferences hosted by allied or associated countries boycotted Germany; of the 20 conferences on technological subjects, only three included German participants. See Grundmann, 1965. *Wissenschaftliche Zeitschrift, 14*, 800.

[54] *Comité Technique Préliminaire pour la Téléphonie à Grande Distance en Europe* (Paris: Libraire de l'Enseignement Technique, 1923), 56. Compare with the similar views of Gill and Carty, quoted below.

[55] "Comité technique pour la téléphonie à grande distance en Europe," *Annales des Postes*,

Télégraphes et Téléphones, *12* (1923), 1016–1061. The British, less anti-German than the French and Belgians, deplored the fact that the Germans were not invited to Paris and realized that the exclusion of the former enemies and most of the neutral countries made the conference of little practical use. See *The Post Office Electrical Engineers Journal*, *15* (1923), 178.

[56] Craemer, 1923. *Das Fernkabel*, 5. *Das Fernkabel* was the journal of the German Cable Company, from 1926 published under the name *Europäische Fernsprechdienst*.

[57] Craemer, *Fernsprechen im Weitverkehr*, 6.

[58] Western Electric, *Commentaires*, 1. With a reply in Höpfner and Dohmen, 1924. *Das Fernkabel*, 37–49.

[59] "Le comité consultatif international des communications téléphoniques à longue distance," *Revue Générale de l'Électricité*, *17* (1925), 421–28. CCI's permanent committee consisted of representatives from France, Great Britain, Germany, Austria, Belgium, Italy, Holland, Sweden, Switzerland, Czechoslovakia and Yugoslavia (Serbia-Croatia-Slovenia); the other members of CCI were Poland, Norway, Denmark, Luxemburg, Latvia, Hungary, Finland, and Spain. Estonia and Lithuania entered CCI in 1925, Portugal, Rumania and USSR in 1926, Albania in 1928, and Danzig in 1929.

[60] Gill, 1922. *Electrical Communication*, *1*, 26. Carty, as quoted in de Sola Pool, *Social impact of the telephone*, 129.

[61] Craemer, *Fernsprechen im Weitverkehr*, 39–42. Brown, *Cable and wireless communications*.

[62] *Europäische Fernsprechdienst*, 1927, no. 6, 44. See also the May 1931 issue of the same journal, a special issue devoted to the European long-distance cable network.

[63] Siemens & Halske, *Paris-Tours-Bordeaux*. Bertho-Lavenir, *Large technical systems*, 166, and Petzold, *Frankreich und Deutschland*.

[64] *Europäische Fernsprechdienst*, 1927, no. 3, 65. In 1925, the average daily number of calls on the New York-San Francisco line was only 14, and even the important line beween New York and Boston carried less than 2300 daily calls. In comparison, the Berlin-Hamburg line had a traffic of 2880 calls per day and the Berlin-London line, with increased fares for the submarine part, 124 calls per day.

[65] *The Telegraph and Telephone Journal*, 1927, no. 149, 217.

[66] Wagner, 1929. *Sitzungsberichte*, 7, 1–15; Müller, 1930. *Europäische Fernsprechdienst*, no. 20, 379–85; Federal Communications Commission, 1938. *Proposed report*, 254–56; Fagen, *Engineering and science*, 813–14.

[67] Matsumae, *Long distance communication*, 116. For earlier Japanese telephone cables, see Inada, 1930, *Electrical Communication*, 9, 114–37.

[68] Solnick, 1991. *Soviet Studies*, *43*, 161. The information concerning telephony in Soviet Russia is based on Solnick's article.

[69] Bernal, *Social function of science*, 174.

[70] Clausewitz, *On war*, 402.

[71] Craemer, *Zwei Jahrzehnte*; Thomas, *Large technical systems*.

MICHAEL ECKERT

THEORETICAL PHYSICISTS AT WAR: SOMMERFELD STUDENTS IN GERMANY AND AS EMIGRANTS

In the years immediately after World War II, theoretical physicists in different countries gave quite different appraisals of their own and their enemies' war work. In particular, theoretical physics in Germany, once an important aspect of German scientific superiority, became a subject of controversy between German and Allied physicists. Werner Heisenberg, for example, Germany's outstanding theorist of this period, had confidently believed that German nuclear research during the war excelled that of the Allies – until the news of Hiroshima shattered this over-confidence. But even so he remained proud of the German achievements and of his own.[1] Arnold Sommerfeld, the grey eminence of the discipline in Germany, emphasized the great amount of fundamental research which was done during the war in his country. He singled out the S-matrix-theory, by which his favourite pupil Heisenberg opened "a new era of quantum theory".[2] Another former Sommerfeld pupil, Max von Laue, felt himself carried away by "the roaring stream of research", as he euphorically introduced a post-war report on German nuclear research during the war.[3] Quite an opposing appraisal was given by Samuel Goudsmit, the scientific leader of the Allied intelligence mission to uncover the German nuclear effort. He portrayed his German colleagues as complacent and even observed a "deterioration of interest in pure science". His book *ALSOS* gave a grim account of what appeared to him as the "errors" of German science under the Nazi regime.[4]

Obviously, the "physicists' war" – as World War II came to be called, confusing a part with the whole – was won by the Allied scientists as much as the war on the battlefields was won by the Allied soldiers. The Allied victory in physics clearly demonstrated that German scientific superiority was a matter of the past. The Manhattan Project provided the most conspicuous evidence of the Anglo-American rise to scientific primacy. German physicists indeed were doomed to be "all second-raters", as Otto Hahn declared them upon hearing the news of Hiroshima.[5] The bold mobilization of physicists for military research in Great Britain and in the USA during the war, and the resulting new weapons such as the atomic bomb and microwave radar, make the equivalent German efforts appear rather bungling. There were and still are many disputes among physicists as well as historians about the causes for the unequal accomplishments in this scientific war; Goudsmit's view of the German "errors" was countered by a "conspiracy" thesis, according to which

P. Forman and J. M. Sánchez-Ron (eds.), National Military Establishments and the Advancement of Science and Technology. Studies in the 20th Century History, 69–86.

German physicists abhorred employing their scientific skill for Hitler's war; other interpretations followed, giving rise to a never ending dispute.[6] But on what basis can the German and the Allied scientific war efforts be compared? Is it correct to call the German nuclear war effort, the Uranium Club, the counterpart of the Manhattan Project? The same problem arises with radar. Although there was some German research on microwave radar, there was certainly no equivalent to the MIT Radiation Laboratory, the Allied centre for this particular research. How then may we arrive at some general conclusions, if there is no common ground for comparisons?

As indicated by the subtitle, I restrict my study to a few theoretical physicists with a common background, namely their descent from the Sommerfeld school: on the one side those forced from Germany and finally joined together in the Anglo-American war effort; on the other side, those engaged in war work in Germany. But just to mention their common scientific heritage is not enough. It is necessary to study how they became absorbed in their different environments, in order to understand the divergent tendencies on both sides. I call such a study "eco-biographical"[7] because it is primarily concerned with the various environments of an individual's life and work. This eco-biographical perspective is derived from the assumption that science and scientists are shaped by the various environments in which they become absorbed. So we have to start some time before World War II and focus our attention on those environments, in which theoretical physics traditions began to diverge, resulting in distinct local, national, or continental characteristics.

THE "HAPPY THIRTIES": DIVERGENT TRENDS IN AMERICAN AND EUROPEAN PHYSICS

From the early 1920s to the 1930s, theoretical physics passed through a decade of tremendous growth. With the older quantum theory followed in the latter half of this decade by the new quantum mechanics, the whole discipline acquired its modern identity. As pursued in the European centers of the quantum revolution, theoretical physics became a subject for intense international collaboration. Niels Bohr's and Max Born's institutes in Copenhagen and Göttingen, respectively, the Sommerfeld school in Munich and its colonies in Zurich and Leipzig under Wolfgang Pauli and Werner Heisenberg, together with a few other sites, were marketplaces for ambitious theorists from all over the world, particularly from America, where the import of quantum mechanics would become the vehicle to modernize and expand many physics departments.[8] Already before World War II, some of these American departments were no longer subordinate to the European centers and ranked among the premier places of international theoretical physics.

However, the coming of age of American theoretical physics was not just a duplicating of the European model. Quantitatively, American physics soon surpassed European scales; it was an ever expanding enterprise throughout the 1920s and 1930s, with only minor setbacks by the Great Depression.[9] Qualitatively, too, there was a growing divergence: theoretical physics in America was pursued with "the empiricist temper regnant".[10] Slater's school at MIT, for example, was quite different from the German centers of early solid state theory in Munich or Leipzig. Slater considered quantum mechanics first of all as an instrument to calculate real properties of solids,[11] while many of his German colleagues considered solids primarily as an opportunity to demonstrate how quantum mechanics works – without a proper interest in real solids. As a consequence of his more practical orientation, Slater was keen to develop useful methods and to reconcile theory with empirical evidence, not merely in principle but also in practice. Slater's colleagues shared this same spirit. Philip M. Morse, for example, who had joined Slater's group after spending some time as a research fellow in Sommerfeld's institute in Munich, profited from Vannevar Bush's "stimulating electrical engineering acquaintance", as he recalled in his memoirs. Bush invited him to use his differential analyzer, the first computing machine capable of solving differential equations, to perform numerical calculations on the scattering of slow electrons by atoms.[12]

It is interesting to note how much Slater's school had imported from the European centers of quantum mechanics, without losing sight of its own priorities. Slater himself had spent some time in Heisenberg's institute in Leipzig. Several of Slater's colleagues had spent considerable periods at their beginning careers as travelling fellows in Sommerfeld's institute. Others knew the Munich professor from his visits in America. Jacob Millman, who would later do pioneering work in band structure calculations, wrote to Sommerfeld in 1933, after he had returned from a one-year stay in Munich, "Your many friends here, president Compton, and the professors Frank, Guillemin, Morse, Allis and others send their greetings."[13] No doubt, the Sommerfeld school represented a style of theoretical physics which was very close to the American empiricist temper. Instead of bothering with the philosophical foundations of quantum mechanics, students were introduced to some tangible applications such as the electron theory of metals.[14] (This research field had been initiated by Sommerfeld's own semi-classical free-electron gas model in 1927, a precursor of the subsequent quantum mechanical theories.) This was just the right environment for future solid state theorists, who would later combine this basic know-how with the empirical approaches in order to calculate the properties of real materials. Slater and his colleagues, therefore, did not copy the Sommerfeld tradition but used it as a basis upon which they could found their own research programs.

Migration between institutes and continents was an important element of the dynamical growth of American physics. Just as the travelling fellows and guest lecturers between Europe and America in the 1920s contributed to the modernization programs in many physics departments, so after 1933 did a number of outstanding European emigré theorists who were forced to leave Germany by the Nazi regime. But the European traditions were reshaped in the different American environment. As S.S. Schweber has observed, "these refugees did not initiate the theoretical physics tradition in the United States; they resonated with it".[15] But it should be added that only those who were willing and able to adapt themselves to their new environment, and to overcome antisemitism and xenophobia, had the opportunity to resonate. Only an elite was allowed to enter permanently into the American academic system and contribute to basic physical research.[16] Thus only a few distinguished and fortunate immigrants could reminisce about the "happy thirties", as Hans Bethe titled a review of this period.[17]

Bethe's case well illustrates the diverging traditions of German and American physics in the 1930s. Bethe was one of Sommerfeld's favourite pupils. He had written the major part of a famous handbook article on electrons in metals, co-authored jointly with his teacher, which work became a most valuable job recommendation when Bethe was forced to emigrate. In 1935, on Sommerfeld's recommendation, Bethe was offered a permanent professorship at Cornell University in the USA.[18] He quickly adjusted to the prevailing boom of nuclear physics. "What I have done myself here," Bethe reported back to his teacher in 1936, "you will see essentially from the *Physical Review* and *Reviews of Modern Physics*. It is all about nuclei".[19] Indeed, the series of long articles in the latter journal soon became the "Bethe Bible," as famous among nuclear physicists as his previous reviews on metal electrons had become among solid state physicists. Later Bethe recalled how he came to commit himself with this task: "I found great enthusiasm and interest in nuclear physics. In many universities, including Cornell, new laboratories were being started for experimental research in nuclear physics, with very competent experimenters in charge. However, the physicists concerned knew relatively little about nuclear theory and about the advances in nuclear physics since the discovery of the neutron. After talking individually to many nuclear physicists in this country, I decided that it would be more effective if I would write a comprehensive article on the status of nuclear physics".[20]

Both the subject – nuclear physics – and the way Bethe was involved in it reveal to what extent the conditions for his work were set by his new environment. As Bethe confessed to Sommerfeld in 1936, he had come to America with very mixed feelings, like "a missionary coming to the darkest parts of Africa in order to spread the true faith". But "already half a year later I no longer held this opinion and today I hardly would return to Europe even if I would be offered the same amount of dollars as at Cornell." Then he reported on those features of the American system which most contrasted with his German experience. "What is characteristic of physics in America is team work. Working together within the large institutes, in each of which everything that exists in physics is being done, the experimentalist constantly discusses his problems with the theorist, the nuclear physicist with the spectroscopist... More team work: The frequent conferences of the American Physical Society... One discusses what one is just then interested in, which, of course, is nuclear physics. With the result that 90% of all work in this field is done in America". Furthermore, Bethe cited the Ann Arbor summer sessions as well as the smaller Washington spring symposia and the Cornell summer sessions, in order to illustrate to his teacher the vivid cooperative spirit in American physics. As far as his personal working conditions at Cornell were concerned, he especially valued the enthusiasm of the students and colleagues. In particular, the collaboration with the experimentalist M. Stanley Livingston, Cornell's cyclotron specialist, with whom he co-authored one part of the nuclear physics review, helped to change his view of scientific life. Another new experience concerned the entrepreneurial role of the chairman of the department, who – though not a very good physicist – was the perfect manager. "He is busy the whole day," Bethe explained, "to get money for the institute and to take care that everybody has available what he needs for his work."[21]

Such was the first-hand report from this American enviroment of the midthirties, which confronted Bethe with a different scientific way of life than he had experienced in Germany. Meanwhile, back in National Socialist Germany, physics was no longer an expanding science. In contrast with the continuing expansion of American physics in the 1930s, German physics entered a phase of stagnation. The chairman of the Deutsche Gesellschaft für Technische Physik, for example, wrote to Sommerfeld late in 1930 that "for the coming years we cannot expect the same capacity in industry for young physicists as previously, and therefore one should slow down somewhat the output of physicists."[22] The situation of academic physicists was even worse. Born, writing to Sommerfeld, remarked at just this same time on the "personnel politics" of their profession:

In Germany for the foreseeable future all posts for theoretical physics are occupied, the creation of new ones is excluded. On the other side, there is a number of very good people; I count what just comes to my mind: Wigner, London, Heitler, Bethe, Nordheim, Peierls etc. In your institute new ones are growing up, and I myself have some very good people: Weisskopf, Stobbe, Delbrück, Maria Göppert. Certainly there are even more in Berlin, Leipzig, and Zürich. What should become of all of these?[23]

After several decades of growth, physics in Germany stagnated. The overproduction of PhDs in the 1920s – especially in theoretical physics – was followed in the 1930s by a decline both in the number of university personnel and in the number of students. This trend is found not only in physics but to a greater or lesser extent in all disciplines in German universities.[24]

Furthermore, there was a divergence between Germany and the United States in the institutional practice of physics. In Germany, theoretical physics had become a rather autonomous activity; theoretical and experimental physicists worked in separate institutes, and there was usually little collaboration between them, even if they had similar research programs. This institutional autonomy was the result of the academic organization of physics established in Germany in the 19th century.[25] At Munich University, for example, although solid state physics was a major research field in both institutes, the theoretical solid state research of the Sommerfeld school and the experimental research in magnetism and metal physics in Gerlach's institute proceeded without interaction. Similarly, there was not much collaboration between Heisenberg's and Debye's institutes at the University of Leipzig, although both professors were Sommerfeld pupils, and although Debye was as much a theorist as an experimentalist. The growth of German theoretical physics in the period 1880–1930 was a consequence of this institutional autonomy, because it made possible theory's emancipation from the experimental physicists. But by the 1930s this autonomy seemed rather anachronistic. The more capable theoretical physics became of guiding experiments, for example in nuclear and solid state physics, the more this autonomy became inefficient.

This long-term divergence between German and American institutionalizations of physics was aggravated by the divergent political development of the two nations. As Bethe reported euphoric news about physics in America, Sommerfeld's institute was becoming the target of an onslaught of the socalled "German physics" movement, an extreme faction of national-socialist physicists who considered relativity and quantum theory a Jewish bluff, and who had by this time started a campaign in order to destroy the tradition of Sommerfeld's institute as a centre of modern theoretical physics. The timing of this attack was a consequence of Sommerfeld's imminent retirement, and its principal target was Heisenberg, Sommerfeld's favorite candidate for his succession. This incident has been well described in the literature, so that I merely summarize its outcome: Although the "German physics" movement won the battle of the Sommerfeld succession, it lost its war against modern theoretical physics in Germany.[26] Nevertheless, the effect of this particular war was further to deepen the gulf between physics in the "Third Reich" and elsewhere.

Thus, even before the outbreak of the Second World War the divergent trends in America and in Germany had created quite different environments for "the physics business" (Weart) in these countries. Theoretical physics in America was a flourishing activity, increasing in numbers and in effective interaction with experiment, while in Germany it was stagnant and even forced to defend itself against ideological attacks.

MOBILIZING FOR WAR: DIVERGENT MOTIVATIONS FOR MILITARY-SCIENTIFIC ALLIANCES

Obviously these divergent trends implied different involvements of scientists in the war efforts on both sides. But instead of these differences being accounted for as the result of diverging historical traditions, they were interpreted after the war by both German and Allied scientists as a consequence of errors made by "Nazis" on the one side versus virtues possessed on the other side. In particular, the "German physics" movement was singled out as a major cause for Germany's defeat in the physicists' war. For quite different reasons, this myth was held to by both Allied and German physicists. For the former, this myth provided welcome evidence of the inferiority of science under a dictatorial regime. For the latter it was used as a whitewash: opposition to "German physics" was implicitly equated with resistance to National Socialism, thereby hiding the rather complex pattern of the physicists' acquiescence and collaboration with this regime.[27]

The foregoing observation is not intended to play down the ideological perversion that was the "German physics" movement. As shown in the case of the Sommerfeld succession, the adherents to this movement were aggressive and destructive. But this same incident also shows that the orthodox physicists knew how to defend themselves and how to prevent further damage to their profession. Among such defensive measures was alliance with more powerful factions in the National-Socialist regime than those supporting the "German physicists". Consequently, successful opposition to this movement contributed to the alignment of the orthodox physicists with the leading Nazi circles – and not to resistance against them.

Let us look more closely at the growth of this alliance. One of the key men was Ludwig Prandtl. He had grown up in a scientific environment similar to Sommerfeld's. Both Sommerfeld and Prandtl had extended the tradition of Felix Klein's Göttingen school, Sommerfeld in theoretical physics, Prandtl in technical mechanics and aerodynamics, and both had become grey eminences in their respective fields. Aerodynamics flourished with the building up of the Nazi air force, and Prandtl was in touch with, among others, Göring, the most powerful authority of the Nazi regime, not only for the air force but for the whole armament sector of the Third Reich. Prandtl used this connection when informed by Sommerfeld about the deterioration of his institute under his infamous successor, Wilhelm Müller: "You perhaps have not yet heard that Müller has thrown me out of the institute", Sommerfeld wrote to Prandtl in March 1941, suggesting to his influential friend that "something should be done in this affair".[28] Prandtl followed Sommerfeld's advice and wrote a letter to Göring, explaining that theoretical physics "is indispensable for the training of new leaders in physics", and complaining that "a group of physicists, who unfortunately have the Führer's ear, rages against theoretical physics and insults the most deserving theoretical physicists".[29]

To support his initiative, Prandtl mobilized a devoted admirer of Sommerfeld, the chief physicist of the Zeiss company, Georg Joos, and also Carl Ramsauer, the chairman of the German Physical Society.[30] Thus, the counter-attack against the "German physics" movement was made the official concern of the German Physical Society. It won the support of several high-ranking Nazi representatives and ministries, with the result that the "German physics" movement lost its influence completely. Indeed, by 1943, Ramsauer found this movement not worthy of notice when reviewing the status of physics in Germany before the German Academy of Aeronautical Research, a body of Göring's ministry which had become a sympathetic forum for the physicists' affairs. Although Ramsauer confronted his audience with a sober comparison of German and Anglo-Saxon physics, in which the German backwardness became conspicuous, he did not blame the "German physics" movement as a scapegoat. When he contrasted the 37 American cyclotrons with the single German one, he rather blamed the widespread mentality of physicists in Germany, who despised such "Mammut-Physik" as something which depends "more on money than on intellect." He criticized this attitude, but his own proposals to incorporate mammoth physics in Germany were made only half-heartedly: one could expand the Physikalisch-Technische Reichsanstalt or found a German "Zentral-Forschungsinstitut für Physik", he suggested, but preferably one should strengthen Germany's traditional assets, "our physical institutes at the universities and technical universities, upon which was based our past superiority in physics and which, in accord with their unique characteristics, have grown out of the spirit of our race."[31]

Clearly, the spokesman of the German Physical Society was exploiting the new alliance with the military in order to enhance the traditional autonomy of his profession. Or, in Prandtl's formulation, the physical institutes should become again what they had been in Helmholtz' or Röntgen's days: autonomous enterprises under the leadership of charismatic "physics leaders". The spokesmen saw themselves as such "physics leaders", accustomed to be the sovereign directors of autonomous institutes. They guarded their mutual independence from each other and usually did not concert their efforts beyond the borders of their sovereign enterprises. Even in joint projects such as the Uranium Club they did not refrain from exerting their traditional sovereignity. Scattered over many separate institutes, this project never had a coherent research strategy, nor was Heisenberg, its scientific leader, in total control of it. There was no teamwork between various institutes, and within a single institute hierarchical order and traditional barriers precluded useful collaboration. The only German cyclotron, completed in Walter Bothe's laboratory in Heidelberg in autumn 1943, was run by experimental physicists only, in conspicuous contrast to the teamwork of engineers, experimentalists, and theorists, characteristic of cyclotron operations in America. In retrospect, one of Bothe's pupils observed how badly the group needed a theorist.[32] This restrospective observation is the more noteworthy because at that time Heisenberg judged that there were "too many theorists" in the project, and rebuffed an inquiry by Heinrich Welker, another capable Sommerfeld pupil, who had asked in 1941 whether he could participate in this effort.[33] Beyond the mere quantitative difference in scope between the Manhattan Project and the Uranium Club, this spirit of institutional autonomy is another fundamental characteristic of physics in Germany which renders a direct comparison between the two efforts futile.

Moreover, the procedures followed in mobilization for war work reveal another important difference between the German and Allied attitudes. During the early phase of the war, physicists usually were drafted for military service irrespective of their particular knowledges, except when they were considered indispensable for some special war research. Even within research projects such as the Uranium Club, which was considered "kriegswichtig" (important for the war), the directors of the various institutes were in a constant quarrel with the military over exemption of their young researchers from service at the front. For this, Ramsauer, in his April 1943 address to the German Academy for Aeronautical Research, criticised the military. Although one "should not go as far as the Anglo-Saxons and exempt the physicists in principle from service at the front", nonetheless "3000 soldiers less" would not greatly damage the army, while "3000 physicists more" could perhaps decide the war.[34] Only late in the war and as a result of constant criticism was this rigid system of drafting revised and many physicists sent home to the various physics institutes or to industrial firms, where war work was assigned to them.

There was also a difference in motives in the particular case of the mobilization for the Uranium Club in contrast to the Manhattan Project. Although there was certainly no conspiracy among German nuclear scientists to direct their effort into a blind alley, one must concede that their zeal to make the bomb was less than that of their colleagues in the Manhattan Project – at least as far as Heisenberg and his circle are concerned. Beyond all disputes about Heisenberg's attitude[35] it is evident that his dedication to this effort cannot be compared with that of Manhattan Project theorists like Rudolf Peierls, Hans Bethe, Edward Teller and Viktor Weisskopf, to name only those with whom he had shared around 1930 the same environment; as emigrés in Great Britain and in the USA they were "desperate to do something – to make some contribution to the war effort", as Bethe recalled.[36] In view of the German aggression during the "Blitzkrieg" they felt a double obligation toward their host-countries, which obviously sparked their motivation to place their special skills at the disposal of the Allied military.

Here again, Bethe's case is illuminating. His first involvement in war work began in summer 1940, after the fall of France, with an unsolicited contribution, a theory about armor penetration. When he was cleared for classified military projects in December 1941, several months after he had obtained United States citizenship, he started to work for the radar project. He established a group of theorists at Cornell, including Frank Carlson, Julian Schwinger, Robert Marshak and others. In summer 1942, Bethe's group joined the Theory Group at the MIT Radiation Laboratory and moved to Cambridge; their work ranged from the theory of wave guides to the problems of rectification of high frequency currents by crystals.[37] Bethe later recalled the stimulating atmosphere: "I was encouraged to go around the lab and listen to people and see what might be interesting, and then I would do it; for example, somebody would suggest to me to think about the junction of semiconductors to a little wire, a so-called cat's whisker, as they called it then. I tried to understand that theory and got maybe a quarter of the way to a transistor in this connection...".[38] While others from Bethe's group like Julian Schwinger remained in this environment throughout the war, Bethe himself was lured away somewhat later by Robert Oppenheimer to head the theoretical division of the American atomic bomb project at Los Alamos, another environment which was described by Bethe as "very enjoyable."[39]

Needless to say, Bethe's excitement about work in the exceptional environment of Los Alamos was shared by most of his colleagues. Weisskopf's memoirs, as well as those of Peierls and Teller, provide ample evidence for the devotion with which they participated in this effort.[40] Although the Los Alamos environment was without precedent, it was typical for the operational style of American theoretical physics. For those who had already resonated with this style before the war, Los Alamos became a climax in their career. Compared with the German environments in which these theorists had begun their careers more than a decade earlier, Los Alamos represented the widest gap between both traditions.

THE PHYSICISTS' WARTIME ACHIEVEMENTS IN DIFFERENT PERSPECTIVES

Many of the postwar misconceptions about the physicists' war work may be explained by their limited perspectives on their respective environments. The physicists did not trace their mutual achievements back to some common starting point, but, disregarding how far the traditions on both sides had diverged, they interpreted the others' results from the perspective of their own present frame of reference.

Although it is tempting to present the postwar disputes between German and Allied scientists about their respective nuclear projects as an illustration of the extent of such misunderstanding, I leave this to the excellent literature which addresses this topic in great detail.[41] Here it may suffice to recall how Heisenberg and his colleagues viewed their achievements when they were first interviewed by Goudsmit, i.e., before they learned of those of the Allied Manhattan Project: they were convinced that they had done a very good job and that they had achieved as much as was possible during the war. Underlying this conviction was their tacit assumption that a bomb could not have been developed during the war, which in turn was based on the limitations of the environment in which their project was embedded. Goudsmit, however, although not knowing the details of the Allied nuclear project, had experienced during the preceding decade a totally different environment; in addition, his parents had been killed by the Nazis in a concentration camp. From this perspective, Heisenberg's reaction looked complacent if not cynical. What followed was a year-long dispute, extending to other scientists in both camps, which never was resolved, and which recently has been resumed with new vigour.[42]

Instead of further arguing about the widely publicized case of the atomic bomb, I rather want to choose the less known achievements of German and Allied scientists in radar research, in order to analyse in more detail how different perspectives provide different historical assessments. Here, too, Sommerfeld's pupils were active on both sides: Bethe, as I mentioned above, for the Allied side, and Welker, as I will demonstrate now, for the German side. From the Allied perspective, German radar, in particular microwave radar, was a bungled effort. By the end of the war, the MIT Radiation Laboratory employed about 4000 people; it had developed microwave radar into an operational weapon, and its achievements have been compared with that of the Manhattan Project with the phrase, "the bomb ended the war, but radar won it". As far as the German counterpart was concerned, the official historian of the MIT Radiation Laboratory, Henry Guerlac, laconically concluded: "The Germans themselves did not develop any useful microwave techniques." By the appositional intensifier "themselves" he indicated that while there was some German microwave technology, it was derivative from the Allied. In Germany, Guerlac asserted, "microwaves received no attention until the capture in the spring of 1943 of the British 10-cm H_2S-set."[43]

Although it is true that there was no German counterpart to the MIT Radiation Laboratory, it is not true that German scientists paid no attention to microwave radar before they were confronted with it in the form of captured Allied aircraft radar. Well before 1943, for example, in the course of German radar research the same discoveries of the rectifying properties of silicon and germanium detectors were made as in Allied radar research. But here again, the importance we grant to this achievement depends on the perspective from which we view it. This will become evident with the case of Welker's war work.

Welker, too, had received his initiation as a theoretical physicist through Sommerfeld, who also stimulated his particular interest for electrons in solids, a pet topic in the Sommerfeld school since the late 1920s.[44] He was Sommerfeld's assistant, while the onslaught of the "German physics" movement against modern theory was in full swing. Sommerfeld suggested to him the riddle of superconductivity, and Welker – like so many other prominent theorists before him – produced a paper about it which subsequently turned out to be just another futile effort. At the time, however, Welker's theory was received with great interest among the German solid state theorists, including Friedrich Hund, Walter Schottky and Heisenberg. After studying Welker's ideas, Heisenberg remarked to Sommerfeld that he was now "more optimistic as far as superconductivity is concerned"[45] – quite a compliment in view of the persistent failure over a decade of previous attempts by Bohr, Bloch and other prominent theorists. But perhaps more interesting than the theory itself is its timing in the years 1938 and 1939: This shows that even now, immediately before Sommerfeld's infamous successor arrived in Munich, the theorists performed their research almost as usual, apparently undeflected by the attacks of the "German physics" movement.

As far as Welker's professional future was concerned, however, things were different. There was no possibility of an academic career for him under Sommerfeld's successor. A career as an industrial physicist seemed to Welker a more promising alternative. And indeed, he managed to get a job in a "research station for wireless telegraphy and air-electricity", where radar research was performed under contract with the Air Force. Welker's research was considered important enough for the war that he was exempted from military service. His particular task was to contribute in the development of microwave detectors. The problem immediately caught his interest. Electronic tubes could not rectify microwaves, because the distance between the electrodes in a vacuum tube could not be made small enough (i.e. the electronic transit time short enough) to respond to the high microwave frequencies. In this situation, in Germany as in Britain and in America, crystal rectifiers were considered as an alternative, even though the rectifying mechanism was still unclear. Welker, with his background in solid state theory, was aware of a recent theory of rectification by Schottky, with whom he had recently discussed his ideas about superconductivity. Schottky's theory related the rectifying properties of semiconductor-metal contacts to the different electron mobilities within a small boundary layer between metal and semiconductor.[46] Welker immediately grasped its pertinence to the microwave problem: the distance between the two surfaces of the boundary layer could be made small enough to provide the desired capabilities of microwave detectors. In 1941, he outlined the basic mechanism in a report to his research unit.[47]

Welker's colleagues, however, electronic engineers who still recalled the unreliabilty of crystal detectors from the infancy of radio, were more than sceptical. Welker felt out of place in this uncongenial environment and looked for another job that would preserve his exemption from service at the fighting front. His attempt to join Heisenberg's project failed, but eventually he persuaded his bosses that his research should be farmed out to the physical-chemical institute of the university in Munich, where also other "war-important" research was going on, for example the study of isotope-separation for the Uranium Club.[48]

Still employed by the same unit under the umbrella of the Air Force, Welker now was back in an academic environment. Here he found colleagues who shared his enthusiasm and did the physical-chemical investigations of semiconductor-metal contacts which were necessary in order to prove Welker's theory. First, he concluded that the traditional semiconducting materials such as pyrite were unsuitable because the heat in the vicinity of the point contact caused chemical and structural instability. Pure elementary semiconductors with a high melting point, such as silicon and germanium, therefore were given preference. In a patent granted in October 1942, Welker and his coworkers described their invention which finally resulted from these studies: an "electrical rectifying device with germanium as semiconductor, and a method for the production of germanium for such a rectifying device".[49] Although silicon, too, was mentioned in this patent as equally suitable because of its even higher melting point, germanium was finally chosen because it could be manufactured with higher purity.

Without much delay the invention was taken over by industry. As Welker wrote to Schottky in May 1943, "our work on detectors in the physical-chemical institute has now achieved practical importance. Our detectors have been transferred to the Siemens radio-tube laboratory, Dr. Jacobi, for production."[50] Based on Welker's achievement, Siemens produced several types of microwave radar detectors, suitable for wavelengths down to 3 cm. Long after the war, the Siemens company counted this effort among the first steps of its semiconductor activities.[51] The military use of Siemens' germanium detectors was less efficient. Detectors based on silicon, which were manufactured by the Telefunken company after an investigation of captured Allied radar sets, had a better operational performance. Karl Seiler, who was charged with this job at Telefunken, is an example of a physicist who was called back from the military front in 1943. He had been a student of Sommerfeld's pupils Peter Paul Ewald and Erwin Fues, and he, like Welker, contributed after the war to the entry of solid state physics into industry.[52]

From an immediate postwar perspective, Welker's and Seiler's contributions seemed rather unimportant. Of course, this research was far surpassed by the host of semiconductor researches which were performed under the umbrella of the MIT Radiation Laboratory.[53] Even if we restrict ourselves to the German radar story, it was only marginal. Detectors were just one topic out of a host of scientific and technological problems, and these in turn were embedded in a complex system of political, military and industrial developments.[54] Nevertheless, this same effort had been a stimulus for pioneering developments in German semiconductor electronics. In Welker's case, it was not only

the starting point for germanium research at the Siemens company, but also
the background for a new era of semiconductor electronics. Siemens hired
Welker after the war as director of a new solid state physics division, where
Welker initiated the investigation of a new class of semiconductors, the III-V
compounds. In 1975, the American Physical Society acknowledged this research by granting Welker its first "International Prize for New Materials".[55]
From the long-term perspective of solid state electronics, therefore, Welker's
case is not at all marginal; it provides yet another illustration for the birth of
this discipline as a result of the marriage between science and the military,
however variously this marriage was motivated and performed, and however
late its offspring entered the scene.

CONCLUSION

What is the lesson from all these examples? I have tried to underscore with this
ecobiographical study that one has to analyze each case of scientific advance
under military auspices within its own context. Although this seems to be
evident when stated in an abstract manner, many accounts of the physicists'
work in World War II are written as if there was a unique frame of reference for
it. This does not mean that a comparison between different cases is impossible.
On the contrary, my examples from the Sommerfeld school should provide
a simple model showing how such comparisons may be obtained without
distorting one case by judgments taken from the other's frame of reference.
By tracing back the paths of Sommerfeld's pupils through these different
environments to their common point of departure, it is possible to keep
track of the diverging traditions. Bethe's and Welker's cases were chosen in
order to illustrate this approach. More generally, a study along these lines of
Sommerfeld's circle is not restricted to the war period. Sommerfeld's central
position in 20th century theoretical physics provides a unique opportunity to
take the case of his school as a red thread through the history of the whole
discipline. This is the more general subject of my research, from which this
study originated.[56]

Regarding the theorists' war work we may conclude from this study that
the apparent divergence of opinions about its quality and character is itself
part of the story, because it reveals to what extent the divergent research
traditions had prevented both camps from a sober assessment of their mutual
achievements. The German scientists' self-assurance, so astonishing to their
Allied colleagues, was not unwarranted from their own point of view. German
scientific performance continued largely along its traditional avenues, while
Allied traditions diverged from those avenues. Underrating such divergence
is, in effect, ignoring the social history of science, as scientists themselves
usually tend to do. Their silent assumption of a universal science, based on a

universal frame of reference, gave rise, in this case as in others, to misunderstanding. The challenge for historians of science is to do better.

Universität München

NOTES

[1] Walker, M., 1989. *German National Socialism and the Quest for Nuclear Power, 1939–1949* (Cambridge, Engl.: Cambridge University Press; German translation 1990, here pp. 194–200. D.C. Cassidy, 1992. *Uncertainty: The Life and Science of Werner Heisenberg* (New York: Freeman), pp. 501–522.

[2] Sommerfeld, A. "Bericht über theoretische Physik," Unpublished manuscript. Sommerfeld-Nachlaß, Deutsches Museum München (in the following abbreviated as SN).

[3] Introduction in Bothe, W., and Flügge, S. (ed), 1947. *Kernphysik und kosmische Strahlung.* Naturforschung und Medizin in Deutschland, 1939–1946, Vol 13. FIAT (=Field Intelligence Agency Technical)-report. ("... mitten im brausenden Strome der Forschung...").

[4] Goudsmit, S., 1947. *ALSOS* (New York: Schuman; reprinted 1986), pp. 242–243.

[5] *Operation Epsilon: The Farm Hall Transcripts.* Introduced by Frank, C., 1993. (Institute of Physics Publishing: Bristol and Philadelphia), p. 71.

[6] The history of this dispute was summarized in Walker, M., 1990. "Heisenberg, Goudsmit, and the German atomic bomb," *Physics Today*, Jan., 52–60, which itself became subject of a new controversy in *Physics Today*, May 1991, 13–15 and 90–96. For a more comprehensive account see Walker, *German National Socialism* (n. 1). A recent revival of the conspiracy thesis was presented by Powers, T., 1993. *Heisenberg's War: The Secret History of the German Bomb* (New York: Knopf/Cape), and vigorously criticized in a book review by Cassidy, D., 1993. "Atomic Conspiracies," *Nature*, 363, 311–312.

[7] In analogy with David Cassidy's appreciation of Lewis Pyenson's *The Young Einstein*, where the various environments of Einstein's early life are analysed in great detail in order to understand the advent of relativity. See: Cassidy, D., 1986. "Understanding the history of special relativity: Bibliographical essay," *Historical Studies in the Physical Sciences, 16*: 177–195, here p. 182.

[8] Sopka, K.R., 1980. *Quantum Physics in America, 1920–1935* (New York: Arno).

[9] Weart, S.R., 1979. "The physics business in America, 1919–1940: A statistical reconnaissance," in Nathan Reingold (ed), *The Sciences in the American Context: New Perspectives* (Washington D.C.: Smithsonian Institution), pp. 295–358.

[10] Schweber, S.S., 1986. "The empiricist temper regnant: Theoretical physics in the United States 1920–1950," *Historical Studies in the Physical Sciences, 17*: 55–98.

[11] Schweber, S.S., 1990. "The young John Clark Slater and the development of quantum chemistry," *Historical Studies in the Physical and Biological Sciences, 20*: 339–406. See also J.C. Slater, 1975. *Solid State and Molecular Theory: A Scientific Biography* (New York: Wiley).

[12] Morse, P.M., 1977. *In at the Beginning: A Physicist's Life.* (Cambridge, Mass.: MIT Press), p. 121.

[13] Millman to Sommerfeld, 21 Oct. 1933. SN.

[14] Eckert, M., 1987. "Propaganda in science: Sommerfeld and the spread of the electron theory of metals," *Historical Studies in the Physical Sciences, 17*: 191–233.

[15] Schweber, S.S., "The empiricist temper regnant" (n. 10), p. 81. On the role of migrations in scientific development see P.K. Hoch, 1987. "Migration and the generation of new scientific ideas," *Minerva, 25*: 209–237.

[16] Hoch, P.K., 1983. "The reception of Central European refugee physicists of the 1930s: USSR, UK, USA," *Annals of Science, 40*: 217–246. S. Wolff. "Immigration of physicists from Germany into USA: Immigration of physics?" Paper presented at the XIX International Congress of History of Science, 29 August 1993, Zaragoza, Spain.

[17] Bethe, H.A., "The happy thirties," in R.H. Stuewer (ed), 1977. *Nuclear Physics in Retrospect: Proceedings of a Symposium on the 1930s.* (Minneapolis: University of Minnesota Press), pp. 11–31.

[18] Sommerfeld to Gibbs, 14 Jun 1934. SN.

[19] Bethe to Sommerfeld, 1 Aug 1936. SN. ("Was ich selbst getan habe, sehen Sie im wesentlichen in der Physical Review und Reviews of Modern Physics. Es ist alles über Kerne...")

[20] Preface in *Basic Bethe: Seminal Articles on Nuclear Physics, 1936–37.* (New York: American Institute of Physics, 1986). Reprint of the three *Reviews of Modern Physics* articles, published in 1936–1937 by Bethe and his co-authors R.F. Bacher and M.S. Livingston.

[21] Bethe to Sommerfeld, 1 Aug 1936. SN. ("...wie ein Missionar, der in die schwärzesten Teile Afrikas geht, um dort den wahren Glauben zu verbreiten... Das Charakteristische der amerikanischen Physik ist team work. Zusammenarbeit in den großen Instituten, wo in jedem eigentlich alles gemacht wird, was es in der Physik gibt, und wo dauernd der Experimentator mit dem Theoretiker, der Kernphysiker mit dem Spektroskopiker seine Probleme diskutiert. Durch diese Zusammenarbeit werden viele Schwierigkeiten sofort erledigt, die in einem spezialisierten Institut Monate kosten. More team work: Die häufigen Kongresse der American Physical Society... Was einen gerade interessiert, ist natürlich Kernphysik. Mit dem Resultat, daß 90% aller Arbeiten in diesem Gebiet in Amerika gemacht sind... Das meiste Verdienst an der Entwicklung von Cornell hat Gibbs, der zwar kein sehr guter Physiker ist, aber ein fabelhafter Institutsdirektor... Er beschäftigt sich den ganzen Tag damit, Geld für das Institut zu kriegen und dafür zu sorgen, daß jeder alles hat, was er zum Arbeiten braucht".)

[22] Gehlhoff to Sommerfeld, 3 Dec 1930. SN. ("... für die nächsten Jahre nicht mit der Aufnahmefähigkeit der Industrie für junge Physiker zu rechnen ist wie bisher, und daß man daher die Zahl der auszubildenden Physiker vorerst etwas verlangsamen sollte...")

[23] Born to Sommerfeld, 13 Nov 1930. SN. ("...Personal-Politik: In Deutschland sind für absehbare Zeit alle Stellen für theoretische Physik besetzt, die Gründung neuer ausgeschlossen. Andererseits gibt es eine Schar besonders tüchtiger Leute; ich zähle auf, was mir gerade einfällt: Wigner, London, Heitler, Bethe, Nordheim, Peierls usw. Bei Ihnen wachsen sicher neue heran, auch ich habe einige sehr gute: Weisskopf, Stobbe, Delbrück, Maria Göppert. Sicher gibt's in Berlin, Leipzig, Zürich noch mehr. Was soll aus all denen werden?...")

[24] Ferber, C. von, 1956. *Die Entwicklung des Lehrkörpers der deutschen Universitäten, 1864–1954.* (Göttingen: Vandenhoek), Tab. I, p. 197. Quetsch, C., 1960. *Die zahlenmäßige Entwicklung des Hochschulbesuchs in den letzten fünfzig Jahren* (Berlin: Springer), pp. 4–7, 13, 42.

[25] Stichweh, R., 1984. *Zur Entstehung des modernen Systems wissenschaftlicher Disziplinen: Physik in Deutschland, 1740–1890* (Frankfurt a.M.: Suhrkamp). Jungnickel, C. and McCormmach, R., 1986. *Intellectual Mastery of Nature: Theoretical Physics From Ohm to Einstein.* (2 Vol.). (Chicago: University of Chicago Press). Eckert, M., 1993. *Die Atomphysiker: Eine Geschichte der theoretischen Physik am Beispiel der Sommerfeldschule.* (Braunschweig/Wiesbaden: Vieweg Verlag).

[26] Beyerchen, A.D., 1977. *Scientists under Hitler* (New Haven: Yale University Press). Walker, M., 1989. "National Socialism and German physics," *Journal of Contemporary History*, 24:63–89. Cassidy, *Uncertainty* (n. 1), pp. 346–364.

[27] A detailed account on the emergence of this myth and its functions for both sides is given by Walker, M., *German National Socialism* (n. 1); see also: Walker, M., 1990. "Legenden um

die deutsche Atombombe,", *Vierteljahreshefte für Zeitgeschichte, 38*: 45–74.

[28] Sommerfeld to Prandtl, 1 March 1941. Prandtl papers, Max-Planck-Institut für Strömungsforschung, Göttingen. I am grateful to Dr. Willi Pricha and Dr. J. Rotta for providing copies of this correspondence. ("Sie wissen vielleicht noch nicht, daß Müller mich aus meinem Institut hinausgeschmissen hat... irgendetwas in der Sache zu unternehmen...").

[29] Prandtl to Göring, 28 April 1941. Prandtl papers. ("Es dreht sich kurz gesagt darum, daß eine Gruppe von Physikern, die leider das Ohr des Führers besitzt, gegen die theoretische Physik wütet und die verdientesten theoretischen Physiker verunglimpft... Unstreitig ist jedenfalls, daß die theoretische Physik ein gerade für die Ausbildung des Führernachwuchses in der Physik unentbehrliches Fach ist...").

[30] Prandtl to Ramsauer and Joos, 28 April 1941. Prandtl papers.

[31] Ramsauer, C. "Über Leistung und Organisation der angelsächsischen Physik mit Ausblicken auf die deutsche Physik," Vortrag vor der Akademie der Deutschen Luftfahrtforschung, 2 April 1943. Peenemünde-Archiv, Deutsches Museum. ("... bei der es mehr auf Geld als auf Geist ankommt... Man könnte z.B. daran denken, die vorhandenen großen Forschungsstätten, wie die Physikalisch-Technische Reichsanstalt, die Kaiser-Wilhelm-Gesellschaft und die großen Forschungsinstitute der Industrie wesentlich zu erweitern oder ein Deutsches Zentral-Forschungsinstitut für Physik zu begründen. Tatsächlich liegt die optimale Lösung aber an einer ganz anderen Stelle. Wir haben ein großes Aktivum, das wir nur als solches erkennen und weiter ausbauen müssen. Das sind die physikalischen Institute unserer Universitäten und technischen Hochschulen, welche unseren früheren Vorrang in der Physik begründet haben und welche ihrer ganzen Eigenart nach aus dem Geist unserer Rasse entstanden sind...").

[32] Maier-Leibnitz, H., in: Edingshaus, A.-L., 1986. *Heinz Maier-Leibnitz: Ein halbes Jahrhundert experimentelle Physik* (München: Piper), p. 61. See also: Osietzki, M., 1988. "Kernphysikalische Großgeräte zwischen naturwissenschaftlicher Forschung, Industrie und Politik: Zur Entwicklung der ersten deutschen Teilchenbeschleuniger bei Siemens 1933–1945," *Technikgeschichte, 55*: 25–46.

[33] Heisenberg to Welker, 17 July 1941. Welker papers, Deutsches Museum.

[34] Ramsauer, *"Über Leistung ..."* (n. 31). ("... sollten wir nicht so weit gehen wie die Angelsachsen und die Physiker grundsätzlich vom Frontdienst ausnehmen... '3000 Soldaten weniger' muß die Wehrmacht ertragen können, '3000 Physiker mehr' kann vielleicht den Krieg entscheiden...").

[35] See n. 1.

[36] In: Bernstein, J., 1981. *Prophet of Energy: Hans Bethe* (New York: Elsevier-Dutton), p. 61.

[37] Guerlac, H.E., 1987. *Radar in World War II* (New York: American Institute of Physics), pp. 625–632.

[38] Interview with Bethe by Charles Weiner, 17 November 1967. American Institute of Physics, Transcript p. 156.

[39] Ibid., p. 157.

[40] Blumberg, S.A. and Owens, G., 1976. *Energy and Conflict: The Life and Times of Edward Teller* (New York: Putnam's Sons); Peierls, R., 1985. *Bird of Passage: Recollections of a Physicist* (Princeton: Princeton University Press); Weisskopf, V., 1990. *The Joy of Insight* (New York: Basic Books; German translation 1991).

[41] See n. 1.

[42] Walker, Heisenberg (n. 6).

[43] Guerlac, *Radar in World War II* (n. 37), p. 1073.

[44] Interview with Welker, 4 December 1981. Sources of the International Project for the History of Solid State Physics, Deutsches Museum. The role of Sommerfeld's school for the emergence of solid state theory is described in Hoddeson, L., et al., 1987. "The development of the quantum-mechanical electron theory of metals: 1928–1933," *Reviews of Modern Physics,*

59: 287–327. Eckert, M., "Propaganda in science" (n. 14).

[45] Heisenberg to Sommerfeld, 15 Feb 1939. SN. ("... hinsichtlich der Supraleitung viel optimistischer...")

[46] Schubert, H., 1986. "Walter Schottky und die Halbleiterphysik," *Kultur und Technik, 4*: 250–258.

[47] Welker, H., 1941. "Über den Spitzendetektor und seine Anwendung zum Nachweis von cm-Wellen," *Jahrbuch der deutschen Luftfahrtforschung, 3*: 63–68.

[48] The work of Klaus Clusius, who directed this institute, on isotope separation is described in Walker, *German National Socialism and the Quest for Nuclear Power* (n. 1).

[49] Patent no. 966387, 3 October 1942, Deutsches Patentamt. Welker papers, Deutsches Museum.

[50] Welker to Schottky, 27 May 1943. Welker papers. (... daß jetzt unsere Arbeiten über Detektoren im Physikalisch-Chemischen Institut eine praktische Bedeutung bekommen haben. Unsere Detektoren sind jetzt der Firma Siemens, Funk-Röhrenlabor, Dr. Jacobi zur Fertigung übergeben worden...").

[51] Schubert, H., 1987. "Industrielaboratorien für Wissenschaftstransfer: Aufbau und Entwicklung der Siemensforschung bis zum Ende des Zweiten Weltkriegs anhand von Beispielen aus der Halbleiterforschung," *Centaurus, 30*: 245–292.

[52] Interview with Karl Seiler, 2 June 1982. Sources of the International Project for the History of Solid State Physics, Deutsches Museum. A technical report on German detector research was authored by Seiler for the FIAT reviews: Seiler, K., 1953. "Detektoren," *Naturforschung und Medizin in Deutschland, 1939–1946* (Weinheim: Verlag Chemie) (=FIAT-report, vol. 15:272–292).

[53] For a summary of the semiconductor research during World War II, including references for the respective activities at the various places, see Braun, E. "Selected topics from the history of semiconductor physics and its applications," chapter 7 in: Hoddeson, L., et al. (eds), 1992. *Out of the Crystal Maze: Chapters from the History of Solid State Physics* (New York: Oxford University Press).

[54] Kern, U., 1984. *Die Entstehung des Radarverfahrens: Zur Geschichte der Radartechnik bis 1945* (Doctoral dissertation at the University of Stuttgart). Frank Reuter, 1971. *Funkmeß: Die Entwicklung und der Einsatz des RADAR-Verfahrens in Deutschland bis zum Ende des Zweiten Weltkrieges* (Opladen: Westdeutscher Verlag).

[55] Weiß, H., 1975. "Steuerung von Elektronenströmen im Festkörper," *Physikalische Blätter, 31*: 156–165, 208–212, here p. 156. Welker, H., 1976. "Discovery and development of III-V compounds," *IEEE Transactions on Electron Devices, 23*: 664–674.

[56] Eckert, M., *Atomphysiker* (n. 25).

HERBERT MEHRTENS

MATHEMATICS AND WAR: GERMANY, 1900–1945

All Wars are wars of technology. Twentieth century wars are distinguished – and this has been remarked upon since the beginning of this century – by being wars of science as well. The First World War was called "the chemist's war"; the Second World War "the physicist's war". Biological warfare has been a constant, though as yet hardly realized, possibility. A "mathematical war", however, is hard to imagine. Even with computer science becoming central to military technology, mathematics still remains in the background, being "just a tool". The history of the relation of mathematics to the military and to war has therefore to be analyzed as a mediated connection.

The first section of this paper presents a conceptual frame for this analysis, considering the sociology, ideology and epistemology of "mathematics and war". The following sections then make use of and supplement the interpretational frame in surveying the German situation in successive phases demarcated by the years 1900–1914–1933–1939–1945. The larger part of this paper is devoted to National Socialism and World War II. This last period is treated in three sections; the first presents a survey of mathematical war work, while the second describes the "self-mobilization" of German mathematics since early 1942. The historical presentation then ends with an account of the establishment in 1944 of a mathematical department in a concentration camp. Here, for a brief period, scientific slave labor was exploited for calculational tasks assigned by the army. This slave-labor institute illustrates the special situation of science in National-Socialist Germany at war, and raises a further problem of historical interpretation: in that context "mathematics and war" extends far beyond "mathematics and the armed forces" to include every aspect of the pursuit, promotion, and propagation of mathematics in a militant, destructive, and self-destructive social order, even touching on the genocide of the Jews.[1]

1. VALUES, PRACTICES, AND EPISTEMOLOGY

Shortly after Germany's utter defeat in the Second World War, Alwin Walther, editor of the official report on applied mathematics in Germany during the war, wrote: "Comparing today [accomplishments in Germany with those in] other countries, we find duplications of research-work miraculously bearing witness to the autonomous life and power of mathematical ideas across all borders."[2] Shortly before the end of the First World War while the Germans still imagined themselves on the verge of a great victory, Felix Klein, eminent organizer and

P. Forman and J. M. Sánchez-Ron (eds.), National Military Establishments and the Advancement of Science and Technology. Studies in the 20th Century History, 87–134.
© 1996 Kluwer Academic Publishers. Printed in the Netherlands.

politician in mathematics and natural sciences, said in a ceremonial adress at
the twentieth anniversary of the Göttingen Association for the Advancement
of Applied Physics and Mathematics:

And now, during the war, we realize what has been effective ..., namely that which we like
to call, following Leibniz, pre-established harmony. After all, precisely those directions of
physical and mathematical research gained immediate importance during the war which the
Association brought to bear in Göttingen.[3]

What is common to these two declarations is that both relate mathematics,
pure and applied, to some higher, transcendental order: the miraculous life of
mathematical ideas transcending all borders and military confrontations, and
the Leibnizian pre-established harmony reconciling all differences of purposes and power. In twentieth century science such rhetoric of transcendental
legitimation is ubiquitous.[4] Here it serves to sanctify the mathematician and
his scientific and political activities confronting the sober realities of scientific
research and science politics tightly integrated into militarized, belligerent societies. War as a reality is suppressed from the image of science; belligerent
science and the rising involvement with military R&D is, in some sense, a
taboo.[5]

Felix Klein had been the architect of the Göttingen Association and he had
taken great care to bind the military into the technoscientific complex he was
constructing in Göttingen and beyond. Klein's successor as the "manager" of
Göttingen mathematics was Richard Courant, and Alwin Walther had been
his assistant. After 1928 Walther had concentrated work at his institute in
Darmstadt on mathematics for engineering; he was well prepared to take
up military assignments during the war years. Klein and Walther's phrases,
"pre-established harmony" and the "miraculous life of ideas", are paroles that
veil the individual and collective efforts to establish and expand the scientific
enterprise and to market its products. Science's marketing strategy uses a
double construction of legitimacy. On the one hand, there is straightforward
utility; on the other hand, the appeal to a transcendent reality, or to the higher
ethical value of the quest for knowledge.

Since the late nineteenth century the world powers each have developed
a large and intricate technoscientific complex, including "pure" research and
mathematics as a base of political, economic, and military power.[6] The
"miraculous" parallels in R&D obviously result from growing homogeneity of such work caused by cooperative competition, military and otherwise,
between nation states. During wars, cooperation ceases but communication
still goes on, with the weapons and "intelligence" crossing the battle lines and
national borders serving as messages. The clearest example of indirect communication despite all secrecy is perhaps the "radar war", in which technical
development on each side was closely locked to that on the other side in an al-
ternation of measures and counter-measures. Another notable example is the
not so miraculous life of the idea of using nuclear fission to create supremely

powerful explosives.[7] The "self-mobilization" of nuclear physicists, mathematicians, and others in Germany for such purposes since 1941/42 was an unconditional offer: "Use me!" Their availability to the military was an effort to regain official acknowledgement and material resources for science as useful but autonomous work. In this effort they did not fail to point at the technoscientific potential of the USA and its military relevance (below, Section 6).

In the construction and cultivation of the public image of science, utilitarian legitimation is extremely prominent in times of war. At other times it is used as well but in a flexible and cautious manner. In the absence of war's overriding utilitarian imperative, transcendental constructions of meaning move in, claiming legitimacy in a way that ensures scientists independence from the aims and interests of the users of the useful. By constructing these exalted values and aims as equally binding upon the various elites in politics, economy, and military, transcendence helps to negotiate conflicts of interest between scientists and those other social elites, and to defend demarcations. A war, as the all-embracing task of the nation, is of similar binding force, but it calls predominantly for practical engagement. The shift is best expressed in Fritz Haber's words: "In peace for mankind, in war for the fatherland!"[8] The very terms of this motto imply the immediate convertibility of scientific institutions from exalting all to subjugating others. Thus in World War II Alwin Walther's institute became the central "computing factory" for military R&D. After the war he praised the autonomous life of mathematical ideas, confessed to pacifism, and tried to retain and exploit the recognition, resources, and results gained during the war.[9]

In the negotiation of autonomy, transcendentality also functions to emphasize science as power *in potentia* only, to provide assurance that, per se, the enterprise of science does not make money, nor aspire to positions of political decision making, nor build and use weapons of destruction. Thus in this structure of science-in-society the boundaries with other social systems are carefully respected. Difficulties do arise when the respect is not mutual, as happened in the Third Reich. By and large, however, with the help of transcendental values, this social contract between the enterprise of science and the other powers ensures the innocence and impotence of science. This contract, with its problematic features, is then mirrored, if distorted, in the popular image of the mad scientist associating with the forces of evil. And the following examination of German mathematics before and during World War II will show that this contract did not need much revision to integrate science in a deeply destructive system.[10]

To study the interaction of mathematics, the military, and war in the first half of this century we have to look at the mediations of the systematic integration of mathematics in the technoscientific complex. One aspect is the social structure and integration by practices, roles, and institutions, underlying the "pre-established harmony" of professional politics and organization. In the following section (Section 2) I shall point to "finalized" fields of research as mediating instances. A second aspect is rhetoric and ideology, flagrantly militaristic at times, at others serving to veil utilitarian orientations. This is discussed in Sections 3 and 4 in connection with the use of school mathematics as a carrier of ideology. The third aspect considered is the specificity of mathematics' role in the technoscientific complex as a type of knowledge that is always "just a tool".

Here a further theoretical consideration is necessary. Epistemologically seen, mathematics is the production of complex systems of symbols signifying nothing but rules and options for handling these and other symbols. There is at most a highly indirect reference to the world outside that of the signs on paper. Modern mathematics is basically self-referential. Linguistically speaking, mathematical systems are systems of imperatives. The prescriptive complexes of rules-and-signs need an operator, usually the mathematician himself, to perform what is prescribed. The operator is, to be precise, the semiotic agent to test the prescriptions. This agent is to check that there are no contradicting orders and no ambiguities or unclear spots where the operator could himself have a choice. He is the ideal slave, willing and able to do whatever he is told to do. His operations are to be completely controlled by the rules. Mathematics is the art and technique of creating completely mastered, prescriptive signification systems.[11]

If we accept such a semiotic epistemology, the worldly impotence of mathematics is obvious. The real world does not provide ideal slaves. There is no *pre-established* harmony between mathematics and reality. To get a mathematical sign complex to function along with some signified physical complex, further means of control, physical or social, are necessary. Mathematics proper remains in the pure world of signs. Crassly said, the "rigour" it claims is much like that of an ideal military regime where all orders are rationally coordinated and disobedience cannot occur. Thus the mathematical field-marshals insisting on the purity of their field are correct to this extent: mathematics *can* be a game, the playful exploration of a "paradise", of a beautifully ordered "imaginary universe".[12] It can, however, also become the very opposite of a game: when adequate means of physical or social control are brought to bear, the otherwise merely formal symbols of mathematics provide the rules for mechanisms through which physical or social power may act with heightened effectiveness. In the modern technoscientific enterprise, mathematics is deeply involved in the business of control and domination.

The next question is how, specifically and concretely, these masterfully constructed, highly complex systems of signifiers come to control physical objects and procedures. As a first approximation to an answer we can take the word "applied mathematics" literally. Thus the signs are applied to objects and actions, and both sides have to be adapted to function together. However, "applied mathematics" is not the most apt term for this process of mutual adaptation through which the mathematical system is attached to the regular world. Getting the signs and the objects to function together should rather be called "attached mathematics". In a regular world, constructed as such by signification, this not so much of a miracle, and I shall not enter upon any further epistemological discussion.

It is, however, necessary to make some further distinctions. We find a spectrum of mathematical work, starting on one side with "pure", or rather "autonomous", mathematics like algebraic number-theory, uninterested in the real world and constructed by the discipline as the center of its self-definition.[13] Then there is "applicable" mathematics, like the theory of partial differential equations or probability theory. It is concerned with areas the external usefulness of which is proven and taken into consideration, but it aims at the development of generalized knowledge without regard for any specific field of application. "Applied" mathematics – which is still a step short of "attached" mathematics – takes into consideration the objective of application and, to some extent, the physical objects involved. So, for example, in the theory of propagation of errors in numerical analysis, the objective is the control of the process of calculation on given types of machines. "Attached" mathematics finally ceases to be mathematics in the disciplinary sense of the word, i.e., it is the actual process of complex calculations or constructions for some extra-mathematical purpose.

I propose to use these terms although the borderlines between them are diffuse. Especially "applied mathematics" is a traditional term laden with historically variant meaning. The concepts designate work on and with mathematics, not any substance or characteristic inherent in the sign systems. Every mathematical sign system is applicable, be it merely to other sign systems. Of this the use of number theory in cryptology is the now notorious example; there is no mathematics wholly without potential for external (and in this case, military) application.[14]

Lastly, a consideration concerning attached and incorporated mathematics, including the very elementary parts, like counting. The shape of an artillery shell to be delivered at supersonic speed poses a mathematical problem. This problem having been solved theoretically, as it was in World War II, the mathematical characteristics of the conditions of stable flight and optimal range have to be realized in material projectiles. The sign system acts as a prescription for the procedures of manufacturing the projectile. Here problems may arise: either the requirements of the prescription will have to be modified to such as personnel, materials, and machinery can perform, or technological innovation is needed to adapt the physical to the mathematical. In testing the combined system of gun, projectile, and atmosphere, further steps of mutual adaptation compromising the prescriptive rigour might be taken. In the end, if there is success, symbolic prescription, technological production, and the physical process have combined into an artefact and its use.

It is important to recognize that, notwithstanding all those modifications and compromises, the mathematics involved – the prescriptive symbolic system – is present right down into the artefacts. Successfully applied and attached mathematics is partially still visible in the prescriptions for production of the artefact, in blueprints, in the organization of work, and in the numerical control of machines, materials, and shapes. In the artefact itself the mathematics may become invisible to the untrained eye. It is incorporated into the design: the mathematical orders have become corporeal order. The expert, however, still can read the orders from the order. The process of recovering the mathematics from the artefact is one of the types of indirect 'communication' alluded to above. A captured weapon or device is analyzed, the symbolic prescriptions are 'read' from the artefact, and back up the line mathematicians might be set to work to reconstruct the theoretical background.

2. BEFORE WORLD WAR I: SOCIAL INTEGRATION AND FINALIZED NEIGHBOURHOODS

The institutional patterns characteristic of German mathematics in the first half of the twentieth century were created at the turn of the century, mainly in Göttingen. Felix Klein as politician and organizer was the predominant figure although there were many collaborators.[15] In Göttingen a small but paradigmatic technoscientific complex was created with mathematics and mathematicians in a central position. With the backing of the Prussian ministry, Klein and his collaborators managed to institutionalize a strong mathematics department with four chairs. Two of these were oriented towards pure mathematics with inclinations towards theoretical physics and philosophy. The other two were Klein's own position, with tendencies towards applied work, experimental physics and technology, and Carl Runge's chair, the first institutionalization of applied mathematics in a full professorship in Germany. With an institute for technical physics and another for actuarial studies, hybrid fields adjoining natural sciences and social sciences were established. Further, through teaching assignments and summer schools, the pedagogy of mathematics was made part of the system.

The military was involved in three of Klein's plans, two of which were realized. A ballistics laboratory that Klein had proposed for Göttingen was, as the military wanted, established in Berlin at the *Militärakademie*. But in connection with the *Institut für angewandte Eletrizität* (Institute for Applied Electricity) an experimental station for wireless telegraphy was established at Göttingen by the army and the navy.[16] The military also had supported the establishment of the *Motorluftschiff-Studiengesellschaft* (Society for the Study of Motor-Powered Flight) in which Klein and Ludwig Prandtl engaged to establish aeronautical research at Göttingen. Behind all these efforts to create institutions for the 'attachment' of mathematics was the *Göttinger Vereinigung*, founded in 1898, which brought private money into the university and provided a large part of the financing of the new institutions. In creating the *Göttinger Vereinigung* Klein was following the American example, but he emphasized that in Germany it was necessary to work "constantly hand-in-hand with the government" thus ensuring a "higher degree of stability".[17] The military connection was included as a matter of course in the construction of a stable social system.

Ludwig Prandtl, whom Klein had brought to Göttingen as professor for technical physics, became the "father of aerodynamics" and Göttingen the "cradle of aviation research". The *Motorluftschiff-Studiengesellschaft* had been planned by the military, but with the engagement of the *Göttinger Vereinigung* it took a strong civil, scientific turn. Prandtl followed a dual approach: practical, with model research in the wind tunnel of his design; and theoretical, with his theory of the boundary layer. The financial base was meagre, however, and for further institutionalization the new *Kaiser-Wilhelm-Gesellschaft* (KWG) was approached. Only with the war, however, was a commitment obtained from the KWG, and in 1917 the *Aerodynamische Versuchsanstalt* (AVA) of the KWG was founded in Göttingen. Later, in 1925, a department for foundational research, the *Kaiser-Wilhelm-Institut für Strömungsforschung*, was added.[18]

Aeronautical research with aerodynamics as its theoretical branch was established as a new field on the eve of World War I. In 1912 the *Wissenschaftliche Gesellschaft für Luftfahrt* (Scientific Society for Aeronautics) was founded. The board of this society again incorporated science, industry, the state, and the military. In the whole movement the military appears to have taken a sober and quite rational position, rejecting uncertain engagements in technological innovations. The new field and its institutions were built mainly by academics, backed by practitioners and industry, all making use of the popularity of aviation and appealing to the accepted goal of preparing Germany to win a war that was to come. Scientists and engineers were building their civil system in interaction *with* the military, but clearly not *for* the military. However, in looking for the "stability" that the military connection might grant, they prepared for a situation where military needs would take command: in war. One could say that the potential conversion of research to more immediate military aims was immanent in the system.

Aerodynamics posed mathematical problems and needed mathematical competence. Its institutionalization at Göttingen raised the status of mathematics. This was clearly a motive for the applied mathematician Runge and the politician of mathematics Klein to cooperate with Prandtl in establishing the academic status of aerodynamical research. And generally, in Göttingen scholarly cooperation and common interest policies went hand in hand to create an exceptionally rich system of mathematical fields, from pure to attached. The young scholar could choose from a wide variety of topics, ranging from the most esoteric, autonomous mathematics down to experimental work related to immediate technological practice. Part of Göttingen's success was due to its density as a social-intellectual system, the close juxtaposition of so many areas and types of mathematical work. Socialization of the individual mathematician at Göttingen meant learning that all this was serious, and maybe even exciting scientific work. In any case, in the small university town it was thoroughly academic work, removed from shop floors or proving grounds.

The center of all this work was research, but there were other options available; one could, if inclined, concern oneself with teaching, politics, finance, or organization – all with the understanding of a German university professor as a modernized, techno-scientific mandarin.[19] In the tight little yarn ball of Göttingen we find nearly all the strands of the complex network in which mathematics is embedded. Göttingen became a world center of mathematics with unprecedentedly large numbers of regular students, foreign students, and visitors. The rising generations in pure and applied mathematics, in statistics and insurance mathematics, in mechanics and aerodynamics, and, to a lesser degree, in physics and other fields, were largely trained in Göttingen.[20] And all along it had been Klein's plan that at the end of this detour through neighbouring fields of more obvious utility, and thus easier to sell, the mathematicians would obtain an Institute building of their very own. Ironically, it was attained only in the late twenties, only after Klein's death, and only with funds donated by the Rockefeller Foundation in complete disregard of these utilitarian connections.[21]

In general, there was no mathematical research or training directly related to the military. What had been established, however, was a range of applicable and applied mathematics connecting the practice of mathematics to the development of technologies of military import. The single most important example of a new, largely theoretical but technology-oriented field was aeronautical research. Such specialties have been treated under the much criticized label "finalized science". A finalized science is defined by a specific openness to political, economic, or social determination of its possibilities of development. In finalized fields, scholarly autonomy, in the sense of a self-regulated community oriented toward internally determined goals, is given up to some extent in favor of an integration into a partly or largely external dynamic. Although the theoretical concept of "finalized science" as it was presented in the 1970s may deserve criticism, the concept is useful for the discussion of the integration of scientific work in modern societies, and it should be developed further.[22] Aerodynamics is such a finalized field. Here model research is a form of scientific practice connecting theoretical, academic work with testing and measuring under contracts with the military and with industry. The neighbouring fields of attached mathematics Klein found easiest to sell were the finalized fields, where it was clear to those who granted money and recognition that their own aims were sufficiently incorporated in the researchers'.

Mathematics as a scholarly discipline, even applied mathematics, is by (self-) definition not finalized. Wherever research practices in or close to mathematics grow into stable, finalized specialties, they are delimited from mathematics proper. Aerodynamics, insurance mathematics, biometrics, mathematical economics, all developing with a certain autonomy since the turn of the century, are all defined to be "not mathematics, properly speaking". The disciplinary center of mathematics remains autonomous and "pure", untainted by ends to which it would be just a means. It is socially integrated through the neighbouring fields, where trained mathematicians work and where mathematical competence is of highest importance. Thus mathematics, as it was socially and ideologically constructed at the time of Felix Klein, remained far apart from military aims and activities. At the same time, however, the web of practices and specialties connecting it to the military, as also to other goal-defining agents, was made more dense and more stable. Here it is a matter not only of research and knowledge, but also of practices and competence. In the weaving of this web, training plays an extremely important role. To be sure, a graduated mathematician going into aerodynamics, geodetics, or insurance is no longer a mathematician in the disciplinary sense of the word. But he is working with his competence, attaching mathematics to the real world. Between 1890 and 1914 Felix Klein and many others actively produced the web of practices and institutions socially integrating mathematics, adding new, strong threads connected or connectable to the military.

3. FROM WORLD WAR I INTO THE 1930S: IDEOLOGY AND PRODUCTIVITY

No study exists of military work in mathematics during World War I and a rough survey of the published mathematical research and of biographies of mathematicians indicates that there was hardly any such work at all. Needless to say, mathematical knowledge was in demand, e.g., in surveying or ballistics, but hardly on the level of actual research. Mathematicians in expert military services were concerned with attached elementary mathematics as, e.g., Richard von Mises as instructor to the aviation corps (below). The relative insignificance of mathematics for the German military during the First World War is indicated by a contemporary survey publication, for whose chapters the standard title is "X in the war". For 'X' we find physics, meteorology, aeronautics, chemistry, botany, zoology, but not mathematics. Instead we find: "School mathematics and the war".[23] The author, Heinrich Timerding, who also included military examples in a book on probability of 1915,[24] discusses in broad terms the movement for the reform of mathematics education in elementary and secondary schools. Rather than reporting on military activities, he takes the war as an argument for the necessity of a thorough mathematical education. School mathematics, Timerding contends, has to be rigorous mathematics, but at the same time oriented toward practical tasks in real life. This had been the line of argument of the reform movement led by Felix Klein in the two decades before the war, and the war is used only to give it a stronger, nationalistic turn. Among the mathematical tasks of the war, geodesy and cartography, ballistics, and aerodynamics are mentioned. "Aeronautics rests on aerodynamics", Timerding writes, which was not yet quite the case, because early planes were rather experimental designs, and only later did aerodynamics play an effective role in airplane construction.

The whole article, like many others of the time, uses the war to legitimate mathematical training in the context of science and technology.[25] Getting military themes into school mathematics was a matter of professional interest politics. The main task of mathematicians at the universities was the training of school-teachers. Thus the labor market for academic mathematicians depended on the share of mathematics in the curriculum of secondary schools. The interest politics of teachers of mathematics coincided on this particular issue with the interest of the mathematicians, and Felix Klein had used this issue to fuse the two interest groups into a solid professional-political bloc.[26] This combination remained stable through the first decades of this century; in 1922 it was reinforced by the formation of a Union of Mathematical Associations, the *Reichsverband*, that worked exactly at the interface between academic and school mathematics. It lobbied for the preservation and extension of mathematical education, with a strong tendency towards practical orientiations along the same lines advocated by the author quoted above.[27]

Teaching also involves the public image of a discipline, for each is affected by the other. Mathematics, as somehow necessary but aloof, inaccessible and meaningless, obviously has difficulties achieving a favourable public image. The complaints from mathematicians about insufficient recognition are endemic and endless. The military importance of mathematics was quickly used to associate mathematics with the heroic tasks of the nation. But this line was pursued almost exclusively for school mathematics. Ideological salesmanship was restricted to this sphere where the autonomy of academic mathematics was not endangered. Thus, in the web around mathematics a further division of labor was established, namely that between the scholarly association of the mathematicians, the *Deutsche Mathematiker-Vereinigung*, and a separate agency for interest politics, the *Reichsverband*. With respect to the military connection, after 1933 this division of labor was activated again, when the *Reichsverband* re-militarized the pedagogy of mathematics (below, Section 6).

For aerodynamics and the related mathematics the war acted as an accelerator. Prandtl's institute, the AVA, was established by an agreement combining partial funding from the *Göttinger Vereinigung*, with the KWG promising to take financial responsibility for the institution after the war, and the *Kriegsministerium* (Ministry of War) and the *Reichsmarineamt* (Reich Naval Office) taking over the lion's share of the cost of buildings and installations. Prandtl, in turn, promised to work predominantly for the military with the qualification that he would be "maintaining fully its [i.e., the AVA's] scientific nature." The military pressed for quick completion of the installations and, facing rising numbers and speed of war-planes, for a wind tunnel of highest capacity. Thus the Göttingen facilities became the most advanced aerodynamic research installation in the world.[28]

In this case, we indeed find the war and the military as an accelerating agent for finalized science. The most immediate impact was on the practical side of aerodynamics, but in the long run the theoretical side too was influenced. The Göttingen center existed as an exemplary institution and produced a considerable number of highly qualified personnel. In the decade before the war Theodore von Kármán, later a highly important figure in aeronautical research in the USA and world-wide, had come from Hungary to study with Prandtl,[29] and then became professor for aerodynamics at the *Technische Hochschule* at Aachen, in northwestern Germany. In the war he had to serve in the Austro-Hungarian Army. After the German military authorities had requested his services for aeronautical R&D, in 1915 the Austro-Hungarian Army placed him at the head of the experimental division of its aviation corps. Many years later he observed, that "war makes scientists" and that his "lifelong association with military aviation" had started in 1915.[30] After the war Kármán, again at Aachen, stood after Prandtl as the second central figure in German aerodynamics.

Kármán was not a mathematician but an aerodynamicist working with mathematics. In contrast, his compatriot Richard von Mises, although trained as an engineer, became a mathematician and in 1900 was appointed to an associate professorship for applied mathematics at Strassburg University. Like Kármán, von Mises served in the Austro-Hungarian Army. As technical advisor to the aviation corps and instructor of the technical staff he gave lectures on the "theory and calculation of airplanes", which he published in 1918.[31] After briefly holding a chair for hydro- and aerodynamics at the Technische Hochschule Dresden, in 1920 he accepted the invitation of Berlin University to become head of a new institute for applied mathematics. In 1918 the university had argued for the chair in a typically long German sentence:

The progressive pervasion of the practice of technological development with mathematical methods in the last years before the war and, most of all, the unexpected demand for theoretically and practically educated mathematicians which became obvious during the war – in the general staff for problems of cartography, topography, theory of mapping, in the artillery commission for ballistics, in the aviation troops for statistical and aerodynamical calculation etc. – make it an absolute necessity that applied mathematics be represented in full scope at the largest university of the country [i.e. Prussia]...[32]

Von Mises' most important mathematical work concerned the theory of probability, where he combined his mathematical competence with his philosophical inclinations. He had already begun to lecture on this topic before the war, and from 1919 it occupied him constantly. In 1922 von Mises published a paper on the "crisis of mechanics", that has been used by Paul Forman as evidence for the "capitulation to Spenglerism" and the "craving for crises" by German mathematicians and physicists of that time.[33] Only then, Forman argued, were statistics and probability theory construed as an alternative to deterministic causality.[34] This is not the place to discuss the influence of Weimar culture in regard to a continuity of interest in statistical methods. Certainly, the war had brought about a massive cultural change, but it also had widened technical experience and thus set new and more complicated problems.

With the establishment of Prandtl's 'boundary layer' and Kármán's 'vortex street' as the basic theoretical paradigms for aerodynamics, the most prominent problem emerging after the war was turbulence, a ubiquitously occuring phenomenon. In his 1921 introduction to the new journal he edited, the *Zeitschrift für angewandte Mathematik und Mechanik* (ZAMM), von Mises posed the question whether the solution to the problem of turbulence might lie in a new statistical approach that "breaks the boundaries of classical mechanics." He himself did work only on probability, its techniques, foundations, and philosophy, while others attempted to solve the problem of turbulence.[35]

Von Mises' work on probability, stemming more from a philosophical interest than from problems of fluid dynamics, may be seen as interwoven with the technoscientific problematic of aeronautics. Finalized fields of work mark nodes in the otherwise loose networks of theoretical research and practical R&D, allowing for the possible synergy of rather loosely related lines of inquiry. This node was institutionally reinforced by the ZAMM and by the international congresses for mechanics that were held at frequent intervals from 1922 onward.[36] Further, and more important for the present topic, parallel to the journal ZAMM, a German society for applied mathematics and mechanics (GAMM) was founded in 1922, with Ludwig Prandtl and Richard von Mises on the board. Despite some early tensions between the mathematician (von Mises) and the experimental aerodynamicist (Prandtl), the society managed a successful integration of both sides, at least as far as a type of mathematics was concerned which was clearly attached to engineering problems.[37] Felix Klein commented in 1921 on the foundation of the journal that what counted was practical problems; science and mathematics had to serve the country's resurrection, and, as he stated in italics, "The aim of theoretical natural science shall not be mere passive understanding but active domination of nature."[38]

An important base for the type of engineering mathematics in ZAMM and GAMM were the *Technische Hochschulen*, where applied mathematics, closely connected with technical mechanics, had in the preceding decades become well established. Finalized specialties like fluid dynamics or insurance mathematics and biometrics are important from the disciplinary perspective, and it was in this perspective that the concept of "finalization" was developed. But such specialties go along with, legitimate, and stabilize an institutional system where, in training as well as in research, practice-oriented mathematics is established in a wide variety of ways. The topics treated in the ZAMM, as well as its institutional interrelations, evince this.[39] As made clear in both Felix Klein's comment on the journal and von Mises' introduction to it, "applied mathematics" was institutionalized not as a "discipline" or "specialty" in the usual sense but as a conglomerate of practices in mathematics with strong bonds to engineering and finalized fields of theoretical work.

Both, ZAMM and GAMM on the one hand, and the *Reichsverband* on the other, show the more differentiated and strongly institutionalized position that mathematics acquired during the postwar years in the network of technoscience and the educational system. The military is almost completely absent during the twenties, but the experience of the war played its role, and the potential for military orientation of the work was strengthened. Mathematics remained in the background: the technology of symbolic procedures is incorporated but made invisible in the technical artifacts. "Active domination of nature" was inherent in this practice. This does not necessarily yield military uses, but the possibility of such remains wide open.

One field in which this possibility was clearly recognized, and reckoned upon, was aeronautics. As the Versailles Treaty severely restricted the aviation industry in Germany, and especially military aviation, practical aeronautical research had to turn to gliders. But on the theoretical side it could be and was strongly developed, explicitly with the idea of science not only as a substitute for more concrete national power but as the basis of technological and military potential.[40] When National Socialism came to power this potential was intensively used.

Another story, spanning the period from the turn of the century into the Third Reich, illustrates the role of mathematics as a potential means of military power even where, initially, the thought of a military role was completely absent. Around the turn of the century in Göttingen, David Hilbert and his disciples became involved in questions of the foundations of mathematics. In Hilbert's mathematical modernism, existence of mathematical concepts and truth of mathematical theories were both equated with freedom from logical contradictions, a stance clearly distancing mathematics from the real world and from practical tasks. Opposition arose, and the Dutch mathematician L.E.J. Brouwer, Hilbert's main antagonist, claimed that all mathematics ought to rest on the "primary intuition" of the "one-after-the-other". Neither mathematician was thinking of applications or real machines. Brouwer located the truth of mathematics in the human spirit and its intuition, while Hilbert postulated a "pre-established harmony" between thinking and the world, allowing for the use and meaning of the "ideal mathematics" of the infinite.

Pressed by the intuitionist opposition's criticism, in the 1920s Hilbert and others developed a "metamathematics", based on formal logic and on the elementary, finitary use of simple signs, to mathematically prove the logical consistency of theories in "ideal mathematics". In this framework the problem of decidability of propositions and of effective computability of numbers arose and in the early 1930s came to occupy a central position in the field of metamathematics. To solve the problem Alan Turing, a young English logician, conceived the "ideal machine", an automaton that came to be known as the "Turing machine", a concept central to the theory of computers. A few years later Turing applied his competence and conceptions in military work at Bletchley Park, where the German Enigma code was broken with the help of computing machinery. Similarly John von Neumann, a central figure in the history of automatic computing, was a brilliant, abstract mathematician who had earlier worked within Hilbert's formalist program. Finally, Konrad Zuse, the German computer pioneer, learned of formal logic only in 1938, but immediately recognized it as the perfect tool; the duality principle was an "inspiration" for his work on systems of switches.[41] Thus, purely mathematical lines of development had led to theories and conceptions that quickly and effectively merged with technological lines under the accelerating run-up to World War II. Metamathematics, as the theory of symbolic procedures of the prescriptive, mathematical kind, turned out, in part at least, to be the theory of symbolic procedures in machines.

4. 1933–1939: DECLINE, ADAPTION, AND DEMAND

In the election of July 1932 the National Socialist German Workers' Party received the largest share of the vote, 37.8%; in March 1933, after Hitler had become chancellor, 44%. The National Socialist "revolution" was welcomed by a large part of the population. The people, academics included, hoped and expected that a powerful, harmonious, and self-assured Germany would be created by the new leadership. What developed, however, was a state and society full of contradictions, competition, aggression, with an accelerating dynamics of destruction.

Up through 1932 German military had been regulated by the strict limits of the Versailles Treaty. But from early in 1933 these restrictions were disregarded. The air force especially was rapidly developed, and with it the field of aviation research. At the same time applied mathematics in the universities and THs was left to stagnate, at best, due to the racial and political expulsions, and to other factors as well. While Ludwig Prandtl's institute for aerodynamics expanded massively, the purges in the mathematical institute destroyed the special qualities of the Göttingen system. Prandtl tried to re-establish applied mathematics at the university, but without success.[42] Richard von Mises had built a lively research group at Berlin University but he had to leave Germany, and applied mathematics at this university deteriorated rapidly.[43] Further, student numbers declined enormously. Between 1932 and 1937 the number of students of physics and mathematics in Göttingen fell by 90 percent. The THs were in general not as strongly affected by the racial purges, but numbers of students dropped there too. The lack of manpower for R&D was soon noted and publicly discussed. But practically nothing was done to redress the situation.[44] There is little evidence for any systematic policy except in the field of aeronautics. Mathematicians at the THs did their work, and I shall return to Alwin Walther's institute below, but they received as little political backing as the pure mathematicans.

The politics of technoscientific research and development is never a matter of one central political agent. It is made by a cooperative network of representatives of the interested parties, including scientists, politicians, industrialists, and the military. A political elite of technoscience is in charge. For it to function, a certain homogeneity in aims and values is necessary as the basis for compromise in situations of conflicting interests and for the stability of demarcations of the fields of power and competence. Felix Klein had established a functioning network of this kind for the mathematical sciences. When National Socialism came to power it destroyed the homogeneity of this elite, in Göttingen and elsewhere, without establishing a new stable order. The usual catchwords for the Nazist political structure are "polycracy" and the "destruction of politics".[45] This is reflected in historiography: to tell a comprehensive story seems almost impossible; it is bound to become fragmented, a patchwork falling apart.[46] The disconnectedness of what follows here is, thus, a consequence of the historical situation itself.

In this section I shall briefly discuss the pre-war years, first ideological and rhetorical adaptation and, second, education and research.[47] As to ideology and rhetoric, three central elements have to be noted: the ideal of a heroic will tempered by a soldier-like character; the praise of *Anschauung*; and the utilitarian move. In the writings on mathematics with political implications – and these range from outright racist ideology to the more moderate versions of rhetorical adaptation – there is a central motive: the semantics of character and will-power, discipline and order. The comparison of the mathematician to the soldier was drawn; and if it was not to the soldier, still the ascribed tone was of heroic masculinity:

Those of us who were born soldiers saw mathematics and mathematical training as a means for discipline and order [*Zucht und Ordnung*].[48]

Much more important is the educational value flowing from the kinship of the spirit of mathematics and the Third Reich. The basic attitude of both is heroic.[49]

Whether behind the text there was revolutionary conviction, as was probably the case with the first quotation, or, with the second, opportunism and sales-manship – for its author, Georg Hamel, was chairman of the *Reichsverband* – in both quotations the rhetorical figure connects mathematical rigour with manly discipline, and the researcher's heroic labor, his "struggle for truth", with the national struggle for dominance. The higher values in the self-image of mathematics proved to be adaptable to the new ideological order. Militance and heroism had never been distant from the mathematician's self-image,[50] and even the aesthetics of mathematics now found heroic tones as in "the beauty and sublimity of mathematics, its crystalline clarity and unimpeach-able rigour."[51] Even if no outright Nazism is present, the textual structure fits the discourse on mathematics into the rhetorical, textual web of Nazism, showing mathematics and the mathematician to be of the right spirit: "Along-side the fight with weapons, there is the spiritual fight (*geistiges Ringen*) in which our *Volk* is engaged."[52]

The second element of adaptive rhetoric is *Anschauung*, that is "intuition", associating the Kantian notion with that of simple visual imagery.[53] The con-necting element between the traditional motive of *Anschauung* in discourse on mathematics and the political ideology of Nazism is the rhetorical, sym-bolic simplification of the world. Political power was enacted in architecture and mass demonstrations; friend and foe could be identified by the shape of their noses and skulls;[54] the German mathematician had a clear inner image of the problem or theory and he was working only to bring it to "crystalline clarity." Such simplification is possible in the sphere of the symbolic. Reality is more complicated. And thus the symbols had to be forced upon the real world by being re-enacted time and again. Character, will-power, the capa-bility for *Anschauung* are constant qualifications of normality of the German individual in Nazism. The individual is ideologically defined by the capacity to will and to "intuit" what he is supposed to will, in the "spirit of the Third Reich."[55] Implicit is the antithesis: the non-German, non-normal, which has to be *ausgemerzt* (weeded out). This is where the inherent militancy of such rhetoric is located. Viewed against the backdrop of the coming war, with its racist-imperialist rationality, such rhetoric was also militaristic.

For mathematics, and science, such rhetoric had two sides. First, it was adaptive to a destructive political system and ideology, defending the auton-omy of science by subscribing to a new clause of the social contract with the leading powers, namely, that science had to be part of and in the ser-vice of the Third Reich. The definition of "part" and "service" were open to negotiation. Autonomy was defended, but loyalty and kinship of spirit were affirmed. The two contributions on mathematics in the volume that "German Science" dedicated to Hitler's fiftieth birthday illustrate the style. Both praise the soundness of the development of mathematics embedded in the new spirit, and the applied mathematician starts his examples with the "doctrine of the shot", i.e. ballistics, and with aerodynamics, but there is no sign of compromising scientific autonomy other than in rhetoric.[56] Haug, in his analysis of fascist ideology, quotes the physician Sauerbruch in a speech of 1937, who promised that science would fulfill its task, under the precondition that "along with trust and belief in science its freedom and autonomy is granted. Both are preconditions for ultimate achievement. The great successes of natural science have not been achieved by order of the state, but they came in free creation from the heart of scientists (*aus dem Inneren von Forschern*), who did their hard work without regard for practical utility".[57] Between the lines, this untranslatable *Innere*, a combination of spirit, heart, and soul, is promised to be in accord with the spirit of the *Reich*, as is also the other promise: achievement. It is but a small twist to the standard rhetoric of higher values that fits it into the contemporary spirit. Achievement is promised, and, coming from the heart, it goes without saying that, this being a German heart, the achievement will be of the required sort.

Rhetorical adaption into the textual webs of Nazism includes more or less implicit acceptance of the aggressive and militaristic spirit. Mathematics, through such rhetoric of higher values, was seeking official and popular recognition and thus took a first, anticipatory step toward militarization. It found, however, little recognition. Frustrated efforts of this sort are background to the "self-mobilization" of 1942.

A second implication of using of the rhetoric of qualities of character is the practical service that mathematics can offer for the realization of the definition of the normal fascist individual. In racist discourse, qualities are inborn. Therefore they can, as constants, be treated with the elementary mathematics of bureaucracy – the bureaucracy of demarcation, registration, sterilization, internment, and *Ausmerze* (eradication). More scientifically the mathematics of heredity comes in, for it is the hereditary qualities that have to be fostered or eradicated, and mathematical statistics serves as a means to determine the necessary actions.[58] The ideology of qualities in Nazism went along with a social and medical technology of quality control of the population. Mathematical statistics was part of the theoretical basis to that practice. Like rhetoric in the textual web of Nazism, fields of practical mathematics, such as statistics, were part of the socio-technical enterprise, in which physical extermination was not so far off.

The most straightforward rhetorical move effecting adaption to Nazism was assertion of the practical utility of mathematics, including the statistical tasks just discussed. Mathematics, as any other science, was said to serve the nation in tackling its great tasks. *Wehrwissenschaft* (defence science) was established, more rhetorically than in institutional reality, and mathematics was related to it.[59] The militaristic turn was there, from early 1933 on, in almost every text concerning mathematics in relation to Nazism. Praising heroism, *Anschauung*, and militaristic utility, the spokesmen for mathematics had their small share in completing this web of simplification and militance.

In mathematics, however, such spokesmen were relatively few. The great majority of mathematicians kept quiet. Moreover, most texts addressed mathematical education in general or, more specifically, secondary school mathematics in the tradition of lobbying for the curricular status of mathematics. The division of labor installed with the foundation of the *Reichsverband* functioned perfectly. The academic association, the *Deutsche Mathematiker-Vereinigung* (DMV), attempted to defend itself against politicization and kept in the shadow of the *Reichsverband*, which practiced adaptation to Nazism consistently and shamelessly.[60] Especially secondary-school mathematics was loaded with ideological and militaristic messages. Such, for example, was the exemplary collection of mathematical exercises, produced in 1935 by the *Reichsverband*, as a combined effort of mathematicians from universities and schools. Among these exercises, a wide variety were intended for training in militant calculation, and a very large part treating directly military topics.[61]

A similar adaptive move took place at the universities. The mathematicians created the figure of the "industrial mathematician" as a new objective for mathematical training. Prior to 1935 the universities had educated mathematics teachers almost exclusively. To show that mathematics was a practical and useful subject, this new image and curriculum were born. Military mathematics was an obvious part, but by no means central.[62] The mathematicians insisted on the theoretical status of their field, but pushed applicable and applied mathematics and the relation to finalized fields into the foreground of university education. The traditional unity of research and teaching was thus compromised in a legitimatory move to the utilitarian side.

The rhetorical moves aimed at adaptation to Nazism, but how to adapt to a rule and ideology that were contradictory in themselves? The party and the "movement" were antagonistic to the state bureaucracy and regular government; enthusiasm for technology stood in opposition to the ideology of blood and soil. True, utility and heroism appealed to one side without conflicting with the other. But "Anschauung" as an inborn competence characteristic of a particular race, led, in the argumentation of Ludwig Bieberbach and others, to a racially specific "German mathematics", and thus to a clear conflict with the prevailing professional standards of the discipline. The difference in adaptive moves reinforced the conflicts and divisions within the discipline. And the fact that it was impossible to find the one safe rhetorical route to the political authorities added to the sense of frustration. Adaptation happened, but at the cost of frustration and fragmentation of the discipline.

The new type of final exam, the *Diplom*, qualifying for positions in industry and administration, discussed since 1936, was officially established only in 1942. It was part of a general move towards further professionalization in academic fields.[63] There are no statistics on industrial employment of mathematicians, but the sample of biographies I can draw on suggests that the role of the 'mathematician in industry' was created by a combination of political pressure and shortage of qualified personnel, rather than by the educational planning of mathematicians and politicians. Very few mathematicians found jobs in industry, and the few who did were usually hired in lieu of some other sort of specialist. Wilhelm Magnus, for example, reports that he was employed as a physicist in electrical industry; to employ a mathematician would have appeared absurd to his boss.[64] But with the preparation for war, industrial work intensified and graduated engineers and physicists were rare; thus mathematicians could slip in. On the other side industry and the military were a welcome refuge from the ideologically and politically highly controlled civil service positions in schools and universities. Karl Storck, for example, reports that for political reasons he saw no chance for a scholarly career. With the help of his professor, he found a job in the aircraft industry, came to head a development department, and started to hire more mathematicians.[65]

In general, the demand for mathematical applications in engineering and science, and thus for individuals with mathematical competence, continued to rise through the whole of the Nazi period. The single most important field was aeronautics. Early in 1935 the *Luftwaffe* was officially established, thus emerging from its clandestine existence.[66] Already in 1933 aeronautical R&D found its golden opportunities: "Infinite possibilities seemed to present themselves. Research installations, previously unthinkable due to the high costs involved, were suddenly approved without question. The financing problem, which had always been the limiting factor of research, seemed no longer to figure."[67] In the *Deutsche Versuchsanstalt für Luftfahrt* (DVL, experimental institute for aeronautics) centers for scientific information and for the recruitment of engineering staff were immediately established. In Göttingen Ludwig Prandtl's institute expanded enormously, a new wind tunnel was immediately projected and over the next six years the staff rose from 80 in 1933 to about 700 in 1939.[68] Early in 1935 Göring ordered a crash program to catch up with the Americans, then leading in the field. Among the many measures an Academy for Aeronautical Research (*Akademie für Luftfahrtforschung*) was established with Göring as president, – a real "academy" for a technical field. So official an acknowledgement of the 'ideal' value of technology was unthinkable before 1933. Finally, the most important new research institution in aeronautics, the *Deutsche Forschungsanstalt für Luftfahrt* (DFL/LFA) was founded in Braunschweig in 1936, where in the course of time a considerable number of mathematicians were employed and during the war a working group for "industrial mathematics" was installed.[69]

Aeronautics, in industry as well as in research institutes, was the main consumer of applied mathematics, and especially for complex calculations. Mathematical theorists were engaged, but also hosts of female computers. The work was done on desk-calculators, but the calculational methods had to be refined and the large tasks required programming. This programming of the team of human computers was in principle not much different from later programming of machinery. The need to automate procedures and machines used in calculation was obvious. Konrad Zuse reacted to that need, transferring the program of the calculating bureau to the idea of a programmable machine which he thought to build using electro-mechanical switches (relays). But however obvious the need, Zuse's project was not taken seriously until the end of the war. Similarly, his friend and collaborator, Helmut Schreyer, who attempted to develop an electronic computer with vacuum tubes, found no support.[70]

Through these same years work on instrumental mathematics, although less innovative than that of Zuse and Schreyer, was under way in various institutes in THs. Leading was Alwin Walther's institute for practical mathematics, the IPM, at TH Darmstadt. Walther's mathematical work was unusually close to engineering problems. The example most often cited is Walther's slide rule, "Darmstadt", with a systematic instead of the historically accidental organization of scales. Indeed, models, graphical and numerical methods, tables, and nomograms were typical examples of his work. Demands for such tools rose during the war. Thus, for example, special slide rules for the radio location of rockets were developed at the IPM for the Army's rocket research establishment, the *Heeresversuchsanstalt* at Peenemünde (HVP).[71]

Another more advanced line of instrumental techniques was analogue computers, mechanical or electro-mechanical devices for the solution of complex mathematical tasks, especially for differentiation and integration. The typical military use of such devices was gun-laying, computing the direction in which a gun of given ballistic characteristics, mounted on a ship, should be aimed to hit a moving target. It was more of a technical than a mathematical problem to combine a target location device and an aiming device with a computer that could be fed information on the changing velocity and course of the vessel.[72] More interesting to the mathematician were machines for more general, scientific tasks. In 1934 the report on the newly developed "differential analyzer" of Vannevar Bush started a series of attempts to duplicate the machine.[73] Walther's IPM and Robert Sauer of TH Aachen were involved in this project during the war. Already in the summer of 1939, the IPM appears to have been engaged in computing assignments for the rocket project of the HVP, because some of Walther's students were engaged there. Immediately after the outbreak of the war an assignment from HVP was extended to the development of mathematical instruments, and Walther received the means to advance work that he had started before on an integrating machine and on the duplication and improvement of the differential analyzer.[74]

With the exception of the HVP assignments to the IPM, immediate involvement of mathematicians in military R&D before the war appears to be restricted to the obvious situations: military academies and research on ballistics and aerodynamics. Demand for mathematical services was rising, but hardly directly from the military. At the same time the effectiveness of mathematical work was seriously restricted by the racial purges, the decline of leading institutions, and the dramatic drop in the number of students. Moreover, constant uncertainties, political pressures, and the pervasive petty competition for power, for positions, and for survival added up to a climate adverse to innovation and to innovative cooperation.

5. 1939–1945: THE USE OF MATHEMATICS FOR THE WAR

Mobilization in 1939 was directed at increasing armed service and military production; military R&D did not play a role in the conception of the *Blitzkrieg*. Consequently, mathematicians, engineers, and scientists were indiscriminately drafted to unprofessional services. Very few scientists were immediately ordered to R&D services; even Alwin Walther himself was drafted for a week.[75] Requests for the release of experts were by no means always successful, nor were the attempts to protect oneself and the collaborators from conscription by projects and assignments officially stamped as *kriegswichtig*, important to the war.

A systematic effort to get the experts back to the institutes and laboratories came about through the initiative of Werner Osenberg, an engineering scientist who led a torpedo research laboratory at the TH Hannover. His first official assignment to register and employ technoscientific personnel from the universities and the armed services came from the Navy in 1941 and concerned R&D on torpedoes. His untiring efforts to extend the program were eventually successful, and in 1943 he became head of a "planning office" of the *Reichsforschungsrat* (RFR, Reich Research Council). The first report from his office, August 1943, listed 234 physicists, 478 chemists, and 102 mathematicians in non-professional service. Undated statistics, probably in mid-1944, tell of 25 mathematicians called back and of 329 engaged in military research. The general numbers of November 1944 in Osenberg's files include 3430 persons successfully called back and 1660 applications pending.[76] Those files are incomplete, and the number of mathematicians employed and called back for military R&D, including teachers, insurance mathematicians, and students, is certainly much higher. Further, many mathematicians were engaged only very late in the war. For example, in July 1944 a six-months training of fifteen mathematics graduates started at the ballistics department of the *Militärakademie* in Berlin.[77]

Osenberg's success was possible only when the *Blitzkrieg* phase was definitely past. During the winter 1941/42, after the campaign in the East had come to a stop and the United States had entered the war, there was a clear change in policy. It was brought about by the initiatives of representatives of the technoscientific complex like Osenberg, combined with the willingness of political leaders to turn to technical solutions and to give more heed to technical experts. This change coincided with a renewed "dedication to 'self-administration of the economy'" on the part of the politicians.[78] Although in 1942 the steering bodies for general R&D and for that of the *Luftwaffe* were reformed, cooperation was lacking and policies remained rather chaotic, determined largely by competition. The greater part of civil R&D was in the hands of the *Ministerium für Erziehung, Wissenschaft und Volksbildung*, (REM, Ministry for Education and Research) and its RFR, combined with the *Deutsche Forschungsgemeinschaft* (DFG, the German national science foundation). Both of the latter institutions, as well as the research department of the ministry, were headed by Rudolf Mentzel, who, moreover, had a high SS rank. Mentzel's area of authority was limited by the *Forschungsführung* (Steering Committee of Research) of the *Luftwaffe* on one side, and, on the other side, by the *Reichsamt für Wirtschaftsausbau* (RWA, Office for Economic Development) under Carl Krauch from the IG Farben, that controlled chemical R&D. The 1942 reform put Göring in charge of all three agencies, but the effect was rather the opposite: the struggle for authority intensified, with the SS continually expanding its influence into military production and R&D, while the armed services each had their R&D departments and commissioned work at the universities. Mutual sharing of information on R&D was started in July 1943, but a serious attempt at coordination was made only with the *Wehrforschungsgemeinschaft* (Military Research Association) set up in the summer of 1944 and headed by Osenberg. Although the effort was serious, this organization appears to have been effective mainly in producing file cards; more was hardly possible given the conditions of that time. What was achieved resulted largely from the intense organizing activities by technoscientists themselves, what Karl-Heinz Ludwig has dubbed the "self-mobilization" of science.[79]

The efforts toward mobilization of mathematics, and the resulting institutions, will be treated in Section 6. In this section I survey the individual efforts of mathematicians and the types of research done in military contexts. Most useful in this connection is the *FIAT Review of German Science*, the official collection of reports, written by German scientists after the war, to which reference was made above. Due to the immediate post-war situation they are incomplete, but still encompass all important topics. The five volumes on applied mathematics were edited by Alwin Walther.[80] The first volume is on various mathematical methods, instrumental and otherwise. The two chapters of immediate military character treat pursuit curves (for target-seeking projectiles) and cryptology. Volume 2 is dedicated to theoretical and technical mechanics, including the wing flutter. The third volume is on fluid mechanics, mainly oriented towards aeronautics, and the fourth is on geodesy. The fifth volume presents a less systematic collection, including the gearing of calculators, the mathematics of electric technologies, ballistics, with special chapters on the theory of the effects of fragmentation of shells and on the ballistics of the V–2-rocket, and finally two chapters on gyroscopes for various purposes.

The first two volumes of the FIAT series are devoted to pure mathematics and were edited by Wilhelm Süss, who will appear below as the main organizer of mathematics during the war. Here military orientation is never explicit, but it is the obvious background to much of the theoretical work. Wilhelm Magnus, for example, who had worked for the Telefunken company and then for the Navy, wrote on special functions of mathematical physics.[81] To compile and re-edit tables of functions and improve calculation and application had been a major part of mathematician's work at R&D institutions and elsewhere, thus providing tools for engineers and physicists in a variety of fields. Similarly, work on series-expansions of these and other functions, on conformal mapping, on differential equations, etc., was background-work for technology, done partly in or for the military service.

H. Wielandt's chapter on the theory of eigenvalues in volume 2 of the *FIAT Review* deserves special mention.[82] The author himself, and others he quotes, worked on aerodynamical problems. Although there are no cross-references, the corresponding chapters of the applied mathematics volumes where the calculation of eigenvalues is treated are those on the flutter of wings and on numerical methods. The names alone of the authors quoted in the chapters show the close interaction of mathematical theory, calculational practice, and technical design. Wieland had worked on algebra before the war and returned to it later, but his work on the theory and numerics of eigenvalues was then well known. Wielandt, by the way, is a typical case: He was an assistant in Tübingen when he was drafted in September 1939. In the fall of 1941, as policies began to change with the shift from *Blitzkrieg* to sustained warfare, various aeronautical research institutes tried to secure his services. It took a whole year, however, until he was ordered to a "working vacation" at the AVA in Göttingen, and eventually, in February 1945, he was discharged from service for "reserved occupation."[83]

Lothar Collatz, who made the numerical treatment of eigenvalue-problems his specialty and wrote the report on numerical and graphical methods, was a student of von Mises and worked at Walther's IPM during the war, commuting between Hannover, where he was professor, and Darmstadt. He later became one of the leading figures in the field of numerical analysis. The basis for his reputation was laid with his war-work. In general, the Darmstadt IPM produced a considerable fraction of the leading specialists in applied mathematics after the war. The institute had started its military work in 1939 with the assignments for the V–2 and continued to expand. By mid-1942 it occupied 14 researchers, 30 (female) calculators, 2 mechanics, and 3 secretaries.[84] In an interrogation just after the war Walther reported on 45 assignments from the armed forces, industry, and the RFR.[85] For the HVP (Peenemünde) Walther mentioned tables of air density, temperature, etc. as functions of altitude, oscillations of the V–2, stability of projectiles and missiles, computation of orbits for missiles and anti-aircraft rockets, and so forth. Combined funds, partly military, partly RFR, went into the construction of analogue computing machinery.[86] Most of IPM's work for the *Luftwaffe* and the aircraft industry concerned numerical methods for the solution of special problems, including the long-standing problem of wing-flutter, and the recently arisen problem of orbit calculations for guided bombs. The interrogation report named seven *Ausweichstellen* (dispersion points) of the institute after its bombing in September 1944 with eleven responsible scientists and about 70 collaborators.

According to Walther's report, the collaboration with the HVP came about because many of his students worked there. Peenemünde, however, had its own mathematicians, and, as Walther said, collaboration was not as close after 1941, when "Peenemünde became jealous of outside efforts and made itself more and more independent."[87] Among the expressions of this independence was the first electronic analogue computer, developed at Penemünde by Helmut Hoelzer, a student of Walther.[88] That diminished collaboration between the IPM and the HVP might have been expected, as the IPM produced results that were problematic for the HVP enterprise. Heinz Unger recalled that, being in charge of orbit-calculations for the V–2, he determined the radius of impact to be 35 km. This was the reason, Unger said, that at the IPM nobody believed in the operational use of the rocket. "For fun," he says, they calculated the possibilities, or rather impossibilities, of a flight to the moon.[89] The main motive for the HVP's mathematical self-sufficiency, however, was certainly Hitler's decision in 1942 to produce and employ the new weapon. Part of the resulting expansion of the HVP was the establishment of its own "mathematical bureau", which commissioned work mainly from Walther but also from Robert Sauer at Aachen, the LFA Braunschweig, Göttingen University, and the TH Dresden.[90]

The development of calculating devices was the IPM's speciality for which funds were raised and cooperations established wherever possible.[91] Besides the work on analogue computers, digital calculation was done on mechanically or electrically linked desk-calculators, and on punch-card and punch-tape machines modified to be programmable.[92] Walther's *Rechenfabrik* (calculating-factory) is evidence of the heavy demand for complex calculations. Computing was done on a large scale, but the standard "computer" was the girl graduated from secondary school. The outsiders, Zuse and Schreyer, did not find the support they needed, and the insiders, like Walther and Sauer, worked on traditional lines of instrumental mathematics. Walther, referring to his meeting with Zuse, confessed that at the IPM nobody had thought of relay-technology, binary arithmetics, or propositional logic.[93] This is but one example of the German situation, where tradition, competition for funds, positions, and jurisdiction, uncalculable political interventions, time-pressure, and poor coordination were adverse to seriously innovative work.

Political and military authorities realized technical necessities and opportunities only very slowly. The aggravating military situation finally made effective work almost impossible. Nevertheless, work was done practically to the last day of the war. Walther's collaborator, De Beauclair, reports that an application for support of basic research in automatic computing was rejected in 1943 as unrealistic and unnecessary. After news of the American Mark I computer had reached Germany, Walther was interrogated: why had such work not been done in Germany? He pointed to the rejected application, and immediately received an assignment of high priority. A few days later the IPM was destroyed by bombs, and work continued at the dispersion points.[94] Konrad Zuse was seen by some authorities as a dreamer, his machines as mere gadgets. He was first supported on credit basis by the DVL to build a machine for calculations on the problem of wing-flutter. Interest in Zuse's machine grew and the funding also, but "the moment for which I had waited about ten years," when he was able to show a working machine, the "Z4," came only weeks before the end of the war.[95]

Another example of poorly coordinated mathematical war-work can be found in the field of cryptology. Measured against the British success in breaking the Enigma code with computing machinery, the German story is not one of great achievement. But also in Germany the business of deciphering secret messages expanded immensely compared to earlier times, and mathematicians were employed in this endeavour. The High Command, as well as every branch of the services, had its cipher department. Further the *Auswärtiges Amt* (AA, Foreign Office) and the *Reichsicherheitshauptamt* (RSHA, SS Central Security Office) had theirs. At least two of these groups had a considerable number of mathematicians at the end of the war.[96]

The AA's department "PersZ" seems to have been quite efficient. One of the two heads of PersZ held a doctor's degree in mathematics and had since the early twenties combined linguistic and mathematical methods at the department. With the outbreak of the war the AA enlarged the department and employed a number of mathematicians, among them Hans Rohrbach, who wrote the chapter on cryptology for the *FIAT Reviews*.[97] This group also worked, like the IPM, with modified punch-card machines and employed statistical methods. Similar work was done at the group "Chi" of the High Command of the armed forces. Wolfgang Franz reports that he had been ordered into service by the High Command a few months after the outbreak of the war to build a group for the development of cryptoanalytical methods. He started alone and ended with a staff of sixty, including a large number of mathematicians, most of whom were called back from the front, usually after waiting many months. Cooperation between the groups was, in Franz' own words, "miserable." Theoretical cryptology had to start practically from zero, but the analysts had their successes and raised the field to a new level. Rohrbach reports that after the war the Indian government attempted to employ him to build up a school for cryptology, and later he received a similar request from South-Africa.[98]

Documentation on military assignments to mathematicians is fragmented and scattered. My general impression is that by 1944 only very few mathematicians were left who were not involved, and quite a number had multiple projects and multiple engagements. Like Collatz, who commuted between Darmstadt and Hannover, Rohrbach was also professor at Prague, from 1941, while working part time for the AA. In Prague he had a small group doing mathematical work for various industrial firms and also for the HVP.

Listings of commissioned research tasks from July 1943 to the end of 1944 show the *Forschungsführung* of the *Luftwaffe* in the lead with about twice as many entries as the Navy and four times that of the Army.[99] Besides Walther and Sauer, who appear most frequently for mathematical assignments, the ballistics specialist Wolfgang Haack, a professor in Karlsruhe, appears frequently with projects from all the services and from the RFR. Haack had turned from pure mathematics to ballistics at the beginning of the war. His first assignment was for aeronautical research, attained by personal connections. Subsequently he obtained his own research institute, funded by the High Command of the Army, which was soon supplemented with a development and testing laboratory of the *Rheinmetall* company. According to his own report he was extremely successful in the theoretical determination and practical development of light and heavy artillery shells. While Walther used the war to enlarge his institute and largely to work on old lines of interest, Haack had managed to mobilize himself for the war just in time, ensuring a safe position, a considerable career, and also a good income.[100]

The development of guided and self-guided projectiles opened a new field involving mathematics. All service-branches had such projects, but especially active mathematically was the department FEP III of the *Marinewaffenamt* (Navy Ordnance Office) with the mathematics professor Helmuth Hasse from Göttingen in the rank of lieutenant commander in charge. Also involved were Wilhelm Magnus, mentioned above, and the physicist Pasqual Jordan. Hasse worked on the mathematics of pursuit curves himself and also commissioned such work by others. Projects which Magnus had brought in from Telefunken were commissioned to various mathematical institutes. These concerned the theory of propagation of electromagnetic waves and the theory of cavity resonators, contributing to the development of radio location and of radar.[101]

Further topics assigned to university mathematicians from FEP III were sonic location at sea and meteorological problems. Moreover, Magnus also reported about extrasensory perception experiments by Lieutenant Commander Roeder of FEP IV, such as the location of enemy vessels using a pendulum over naval maps. This is not in the Navy's project list, which does, however, include one project of Roeder's on "influencing psychic states by electromagnetic fields" commisioned to the Psychological Institute of Vienna University.[102] Stories like this, told by witnesses, abound with examples of incompetence and absurdities on the part of military and political authorities. Such stories arise not only from ordinary professional arrogance; the German situation appears to have been especially chaotic, and gross absurdities, made possible by the Nazi regime and ideology, had established themselves in many places.

Mathematical research commissioned by the RFR concerned, for the most part, applicable general mathematics.[103] A card file prepared by Osenberg, probably of 1944, contains cards on about 2200 researchers, including about 120 mathematicians. Research assignments, if noted, range over the fields discussed above, and include, further, the dreariest type of work, the compiling of tables for astronomical navigation – mostly simple interpolation work on desk calculators done at the astronomical calculating institute at Babelsberg and also commisioned to universities.[104]

Summing up, the amount of military research in mathematics rose rapidly after 1941/42. Part of it was traditional engineering mathematics and scientific calculation like astronomical tables or standard ballistics. But part of it needed high-level mathematical competence. Such were new technologies, the ever-rising demand for performance of existing military technologies, and the new tasks of acquiring, handling, and safeguarding information, as in cryptology. A wide range of highly sophisticated expertise had to be combined in R&D on devices like the V–2, or on high altitude and high speed aircraft. Performance had to be calculated, that is, theoretically anticipated and controlled. Mathematics as the tool of theoretical control reached a new level of importance with the military race of technoscientific invention. Germany had clearly not prepared itself for that race; finally it was the experts themselves who pressed for Germany to enter it, and who did their best for Germany as soon as they were allowed to engage themselves in it.

6. 1942–1945: THE SELF-MOBILIZATION OF MATHEMATICS

Karl-Heinz Ludwig illustrated the "self-mobilization of science" exclusively with Werner Osenberg's activities.[105] The phrase has since found wider use, and Helmuth Trischler has shown in detail the zeal of aeronautical research experts to serve the military aims of the *Reich*.[106] The physicists made *their* move in January 1942. Carl Ramsauer, industrial physicist and chairman of the *Deutsche Physikalische Gesellschaft* (DPG, German Physical Society), submitted a memorandum on the state of physics to the Education Minister, Bernhard Rust, pointing out Germany's decline relative to her Anglo-Saxon enemies. The Education Ministry did not react, but the memorandum played its part in the shift of political climate.[107] Combined efforts of various scientists for self-mobilization of their fields brought about the change. How much coordination was behind it and how the network functioned, with central individuals like Prandtl, Krauch, Osenberg, or Ramsauer in positions of varying influence, remains open to further analysis. Bonding these activists was a common attitude summarized by Ludwig:

[German scientists] were asking ... for the right to co-operate as experts, which had, in their view, been unjustly withheld from them. The catastrophic course of World War II enabled critical scientists again to present status-generated resentments 'objectively', thus preserving a seemingly unpolitical self-understanding. They compared only the state of technology with that of the enemies, while in general it did not occur to them to consider the political order. Quite 'value-free' the opinion was held that the 'neglect' of research had contributed to the catastrophe or was its symptom.[108]

Ludwig is still a little too cautious. Most experts playing important roles by 1942 had got into their position because they had, initially, assented to some extent to the regime and accepted a series of compromises binding them into the Nazi order. They lived with the ideological web, and wanted to combine their professional pride with national pride. They certainly made political comparisons, but rather contrasted Germany with the Soviet Union. It was also the fear of Bolshevism that led them to put Germany and the USA in the same political category.[109]

The mathematician who became the representative of the discipline in the mobilizing network, Wilhelm Süss, also based his arguments on the "objective" comparison. Süss was chairman of the association of 'pure' mathematicians, the *Deutsche Mathematiker-Vereinigung* (DMV), and rector of Freiburg University. In 1943 he spoke at the conference of the rectors of German universities, comparing German academic capacities with those of the USA, quoting Ramsauer, and apparently also using information from Osenberg. His conclusion:

We have to make use of Germany's scientific potential in an optimal way and mobilize it completely for warfare. ... Further it is necessary immediately to start the training of younger collaborators, since we have to prepare for a long war. And we have to take into account that the longer the war lasts the more heavily scientific discoveries will weigh in its outcome.

In more emotional tones he called for public recognition, for gratitude, emoluments, and decoration:

A scholar of eminence ought to be given the same public recognition as a military leader of great merit or other leading men. As a sign of the gratitude of the people he should be granted, in my opinion, for example, like an outstanding general, a Reich endowment. At his death he should be given a state funeral.[110]

In combination with the rest of the speech and with Süss' career and activities, this can clearly not be read as irony or as shrewd politics in the interest of survival. Such appeals, containing open criticism of political leadership, began to be publicly expressed and circulated only in 1942. In the same vein but less outspoken they were present, at times, in 1941, before anybody could know how the war would end. And they were often heard from academic spokesmen attempting to retain their elite status.[111]

Süss represented mathematics from 1937 onward. Because of his political abilities, the statutes of the DMV were changed so that he could be permanent chairman – until 1945.[112] His official charge from the RFR came about in 1942, initiated by two memoranda of October and November 1941 sent to the RFR by Dr. Johannes Rasch of Siemens AG. Rasch called for mathematical reference-works to be used in industry and drew in his argumentation heavily on the American example. Although there is no archival evidence, it seems likely that Rasch's action was the result of discussions between various interested parties. The memoranda became the subject of a conference arranged by the chairman of the physics department of the RFR, who was also responsible for mathematics. Süss made good use of the opportunity: he was appointed head of an *Arbeitsgemeinschaft Mathematik* (working group) and eventually, in January 1945, of a separate department for mathematics of the RFR. He had to work hard to make things happen, and he produced a long-lasting and historically important result, the mathematical research institute at Oberwolfach in the Black Forest, today of world-wide reputation.[113]

With American (the Institute for Advanced Study) and Italian (the *Istituto per le Applicazioni del Calcolo*) models in mind, German mathematicians had long been hoping for their own national institute. Gustav Doetsch, a colleague and long-standing adversary of Süss at Freiburg University, had visited the Italian Institute for Applied Mathematics and in 1941 proposed to the Education Ministry the foundation of a German equivalent. The plan was, however, rejected in favour of extension of university institutes.[114] The proposal for an institute was advanced again at a March 1942 meeting arranged by the REM to discuss the new international organization of mathematics under German auspices, a meeting that had been planned at a time before Germany's future domination of Europe was yet in doubt.[115] Also in March 1942 the foundation of a *Reich*-institute for applied mathematics was proposed by Süss himself at the conference on mathematical handbooks referred to above. In July he presented his conceptions to Rust, with whom he was well acquainted, and found a positive reaction.

The RFR, however, being reorganized at the time and put under Göring's authority, was unable to act. Süss turned to another potential source of funds, Ludwig Prandtl of the *Forschungsführung*. Prandtl reacted immediately, circulated the proposal, and arranged a meeting including two mathematicians from aeronautical research institutions as experts (C. Schmieden and F. Lösch), who both approved the plan. Gustav Doetsch, however, who, although a member of the staff of the *Forschungsführung*, had not been at this meeting, intervened. He had plans of his own, namely to expand the just established small mathematical division at the *Luftfahrtforschungsanstalt* Braunschweig. To it Doetsch proposed to allocate Süss' mathematical books project, presenting a rather degrading characterization of Süss to Prandtl. From Heidelberg Udo Wegner, who also had his own ambitions (*vide infra*), wrote to Prandtl backing Doetsch's plan. Alwin Walther, in a long letter to Prandtl, took a relatively moderate position. The conclusion of his extensive discussion of Doetsch's plans was that Walther believed his own IPM at Darmstadt was the one and only place for mathematics oriented towards engineering, and a request to the *Forschungsführung* for moral and financial support for the IPM.[116]

While the decision in the *Forschungsführung* took its time, Süss returned to the REM. In September 1942 the RFR via the DFG granted 70 000 Mark for Süss' mathematical books project. At that time twelve handbooks were planned. The project ran until the end of the war, the last transfer of funds being made in April 1945. A report of March 1945 named nineteen titles, eleven of which had been printed or were ready for printing. From the beginning the list included not only new projects, but also books that were already in the making, and German as well as pirated English reprints. Erich Kamke, who had lost his professorship in 1937 because his wife was Jewish, also became involved through his important handbook on differential equations. The first volume had been completed by 1942, funded by the Air Ministry and by the DFG. The second volume on partial differential equations was included in Süss' program; the first half of it appeared in early 1945. Another important piece, ready before the end of the war, was Lothar Collatz' book on the numerical treatment of eigenvalue problems. Both works are mathematically quite sophisticated and not at all of immediate military import. In general, Süss' books program approached applicable mathematics in a way no different than in any other, more peaceful, time.

Already in September 1942 Süss wrote to the ministry that the number of parties interested in a *Reich*-institute for applied mathematics appeared to be steadily growing. Doetsch in the Air Ministry was clearly one. Another was Udo Wegner who also attempted to tap two sources. In Heidelberg an aeronautical research institution was in the making, and Wegner, dean of the department for natural sciences, was to be one of the heads of the new establishment. A rather typical conflict developed. The representatives of the university claimed "freedom of research" to escape the tight control of the Air Ministry. Rudolf Mentzel intervened in their favor, defending his sphere of authority, the universities, against the *Forschungsführung*. Within the Air Ministry Doetsch also intervened on behalf of Wegner and the university, for he had his personal quarrels with Adolf Baeumker, the representative of the Air Ministry in the *Forschungsführung* in charge of the negotiations with the university. The institute came into existence, as did its building, but appears to have done little more than what could have been accomplished in the university with some standard research contracts.[117] Wegner himself received an assignment from the Air Ministry for the study of conformal mapping related to the theory of wings, and he appears to have also been included in Süss' program, collaborating with Ludwig Bieberbach on a book on conformal mapping.[118]

The mathematical department at the LFA Braunschweig, favoured by Doetsch, appears to have come into existence, but only on paper, in mid-1942. It got under way early in 1943 as "working group for industrial mathematics". Wolfgang Gröbner, a well-known specialist in algebraic geometry who, however, had worked at the Italian institute for applied mathematics mentioned above, was recruited in July 1942 as the head of the group. Four other mathematicians joined him between February and May 1943. The first report names seven projects, an aerodynamical calculation, a calculation for the Zeiß company concerning a "curved body" for an aiming device, a table of integrals, a survey of existing literature and tables on integrals, differential equations, and special functions, a survey of industrial needs for tables, a problem of statistics, and collaboration with Wegner on his book on the practice of linear equations and matrices. Later reports added lectures for the staff of the LFA and one new task of obvious importance at the time: calculation of the probability of destruction of planes by fragmenting anti-aircraft shells. Quite typically the problem of lack of calculating personnel is lamented in the reports, and as a remedy cooperation with Rohrbach's group in Prague is mentioned.[119] Doetsch appears to have arranged his own program for textbooks at that time.[120] After difficulties with Doetsch, Gröbner left the group in 1944 and joined Rohrbach. Doetsch himself took over an institute for ballistics at the LFA in September 1944, into which the group was then integrated.[121]

Wegner's ambitions for mathematics could not be realized with the Air Ministry; Doetsch could not and probably also did not want to be very helpful. By mid-1943 Wegner had turned to Krauch's *Reichsamt*. Krauch was willing to establish one of his "four year plan institutes" for him. The funding envisaged is unknown. Such institutes ranged from large establishments to mere door-plates serving as protection against induction orders.[122] The proposal had to be coordinated with the RFR and thus came into Süss' hands. He used it to pursue his old line. In 1942 he had already pointed out that everybody else was planning "special institutes" and not the necessary, general-purpose *Reichsinstitut*. Quite in accord with Rudolf Mentzel's ambitions and playing on the competition between Mentzel, Krauch, and the *Forschungsführung*, Süss argued that the Ministry had to remain the responsible authority for research in the *Reich*. New institutions, he wrote, should only be founded for immediate military needs that could not be covered otherwise, or to exploit the capacities of highly qualified scholars from the universities. Wegner was not the "outstanding brain" deserving an institute of his own, but a central institute was indeed a necessity of the war.[123]

Mentzel sharply rejected Krauch's proposal and, moreover, objected to new four year plan institutes generally, arguing that the RFR must be in charge of any new institute and that a *Reichsinstitut* for mathematics was already in the planning. Wegner's and Süss' proposals had appeared at a decisive point in the struggle for authority between Mentzel and Krauch. Mentzel now was forced to act, and consequently Süss was put in charge of preparations for an institute. At this point Süss struck "applied" from the title of the prospective institute. A call to a chair in Göttingen then helped Süss to acquire the *"Lorenzenhof"* at Oberwolfach in the Black Forest as the site of the new institute.[124]

The first entry in the Oberwolfach *Vortragsbuch* (book of abstracts of lectures) is dated September 24, 1944, and reports on the plans for a lecture course on the calculus. Neither this nor the immediately following entries create the impression of an eager search for military work. The institute turned out to be a place for mathematicians to survive the last months of the war and the difficult times thereafter. With the argument that printing of mathematical tables was as important as the printing of military maps, the institute was included in the list of groups to be fully maintained in face of the armament emergency program of January and February 1945. (Süss' good cooperation with the head of the physics division of the RFR, by now Walther Gerlach, helped again.)[125] By April 1945 fourteen mathematicians, some with spouses and children, and a number of assistants were at the *Lorenzenhof*.[126]

As far as mathematics was concerned, the institute was the outcome of a combined effort of Süss and his collaborators in- and outside mathematics. Süss' success depended, however, not merely on collaborators, but also upon his having competitors within the discipline. His *Reichsinstitut* would not have come into existence except by folding disciplinary conflicts into the wider struggle for political authority. In this particular game were university mathematicians only; applied mathematicians from the THs stood aside, while defending their interests. To the extent that they did take sides, they apparently preferred collaboration with Süss. Süss used military needs as argument, but was in fact concerned with autonomous, applicable mathematics. This self-definition fitted with the definition of authority of the RFR and the REM, charged with research and the universities generally, and clearly having limited influence in military R&D. Thus Süss was eventually able to leave the qualification "applied" out of the name of the new institute. On the side of the state, i.e. REM and RFR, the creation of the institute was a matter of chance in the struggle for authority and for survival, certainly not of any rational R&D policies. Besides mathematics, the list of the *Reichsinstitute* founded in the summer of 1944 includes research on mineral oil, cancer, fat, psychology, psychiatry, and pneumoconiosis.[127]

7. MATHEMATICAL SLAVE LABOR

Finally, it is necessary to describe another sort of mobilization of mathematical competence, related but opposite to "self-mobilization": the exploitation of scientific slave labor. There are two cases. One, not so well documented, arose out of the policies and practices in the *Generalgouvernement*, the western part of Poland annexed to the *Reich*; the other arose out of those of the SS covering the entire *Reich*.

In the *Generalgouvernement*, in pursuit of the goal of destroying the Polish intelligentsia, Cracow University was dissolved, its members largely liquidated. Instead the *Institut für Deutsche Ostarbeit* (IDO, literally: Institute for German Work on the East) was established for the "spiritual penetration" of this "space", and as the basis for a future great university.[128] The IDO employed prisoners of war and inmates of concentration camps in its departments. The section for astronomy and mathematics was headed by the astronomer Kurt Walter. Here Russian prisoners of war were occupied as well as, "on the average, 6 further [workers] from the KZ-camp Krakau-Plassow." The work included assignments from the mathematics group of the LFA Braunschweig, translations of Russian mathematical works for Walther's IPM, and calculations for Hasse of FEP III. In the summer of 1944 the institute had to move west. Walter's section was re-established near the Ravensbrück concentration camp.[129]

As the SS expanded progressively into military production and into the military itself, it also reached out to exert itself in military R&D.[130] The SS-foundation *Ahnenerbe* (literally: ancestral heritage), initially concerned with Germanic early and pre-history, had turned to "exact-scientific methods" by 1939, and established in 1942 its *Institut für wehrwissenschaftliche Zweckforschung* (Institute for Applied Research in Military Science).[131] This was the institutional frame for the horrifying experiments on humans in concentration camps, partly in cooperation with military and civil research establishments.[132] On May 25, 1944, Heinrich Himmler, chief of the SS, declared:

Among the Jews we have coming in now from Hungary, as well as otherwise among our prisoners in concentration camps, there are doubtlessly quite a lot of physicists, chemists, and other scientists. I order *SS-Obergruppenführer* Pohl to establish in a concentration camp a place for scientific research, where the expert knowledge of these people can be set to work for the man-power and time-consuming tasks of calculation of formulae, finishing of specific constructions, but also for basic research.[133]

The head of the *Ahnenerbe*, Wolfram Sievers, in cooperation with the graduated mathematician Helmut Fischer, from the *Reichssicherheitshauptamt*, started planning the concentration camp (KZ) institute.

Fischer, according to his own report, proposed a mathematical institute and found a young mathematics graduate to lead the institute. Karl-Heinz Boseck was an informer of the RSHA at Berlin University who was eager to get into a SS-uniform.[134] It was proposed to "commission the steering of assignments to Professor Walter [sic!], Darmstadt, who disposes of the largest calculation institute in Germany and is in contact with all bodies requesting work."[135] Walther assented to cooperate and to train Boseck for a fortnight at the IPM. Safeguarding his interests, he proposed long-term projects for the KZ-institute, while immediate war work would be kept at the IPM. Due to the bombing of the IPM, not much cooperation appears to have been realized. Via Gerlach, Süss was also informed of the project. He welcomed the plan and proposed they engage at the institute the mathematician Ernst Mohr, whom Süss presumed to be in a concentration camp. The SS followed up this proposal, but did not find Mohr. He was in fact not in a KZ: he had been sentenced to death for "defeatism" and was imprisoned in the Brandenburg penitentiary. There he survived with execution-postponements due to assignments for *kriegswichtig* work arranged by Hans Rohrbach.[136]

While Fischer toured concentration camps to find qualified inmates, Boseck was active securing materials, including desk calculators seized from all over Europe. The KZ-institute was set up at Sachsenhausen. In December eighteen prisoners worked on various assignments of the usual type, all requested by the Army Ordnance Office, which was charged one *Reichsmark* per man-hour.[137] The files, showing ordinary bureaucratic conduct on all sides, appear not so much different from those concerning Braunschweig or Oberwolfach, except for a memo of Aug. 4, 1944:

Re: Employment of prisoners for scientific work.
Reliability of the scientific work is limited. To avoid labor, results are frequently cooked up. It has been established that if certain reliefs are granted, e.g., permission to work in civilian clothes, the work was immediately more reliable. Upon proving satisfactory, one should go so far as granting leave of absence during employment for scientific work, to create the impression that those concerned are really working as scientists.[138]

The prisoners in Sachsenhausen were not "really" employed as scientists. One out of every two in Sachsenhausen was killed, and tens of thousands died during the evacuation marches in early 1945.[139] The fate of the workers at the mathematical department is unknown. Only a few names are left: Emil Peuker, Ignacio Ante-Gomez, Adolf Baasch, Helmut Berns, Guillaume Crols, William Hauben, Edouard Jozan.[140]

8. CONCLUSION

These few names, standing outside the historical narrative, remind us of the individual fates of victims. Here the historian has to pause and admit limits to rational analysis and narrative. The existential experience of those involved on either side is, in the final analysis, not accessible. There will be no simple and no complete answer to the question of how all this could happen. We will not find an explanation bringing full relief, and we should certainly not seek it by ascribing responsibility and guilt to selected individuals. Most important are the questions themselves and the careful discussion of partial explanations. Such questions and explanations concern not only history and not only Germany, but the present and the future, everywhere.

During the war it always was adverse to my pacifism to see that some of my work in the end came to serve military measures. I have tried, especially in cooperation with Magnus, to do as much work as possible of general, peaceful value, and further to save as many young scientists as possible.[141]

Thus Alwin Walther wrote, early in 1947, to Richard Courant, and in response to Courant's proposal that he come to work in the USA, Walther stressed his responsibility for his institute and his collaborators. Such an attitude and conduct deserve respect. Similarly we have to acknowledge Wilhelm Süss' attempt to put in a word for Ernst Mohr; his organization of mathematical work so as to avoid immediate military applications; and his help for many scientists.

Simply to work in any responsible position within the Nazi state meant being involved with the system, meant collaboration. Oppositional activities, even the simple attempts to help indivudals rather then to serve the state and the war, needed a certain amount of camouflage. The files do not tell everything, and sometimes they lie. Any interpretation is in danger of being unjust. But we have to raise the questions and follow up the doubts, not so much to judge individual conduct in the past, as to prepare ourselves for individual conduct – our own, and that of others – in the future.

Walther, at about the time of the quoted letter, was also writing the affirmation, quoted at the beginning of this paper, of the "autonomous life and power of mathematical ideas". And Süss, in his introduction to the *FIAT Review* volumes on pure mathematics, wrote of the "silent fostering" of the "garden of true science" during the war.[142] Such appeals to higher values appear as hollow claims to innocence and purity. Mathematicians and their discipline were entangled in the Nazi system. The "silent garden", if research was a place of retreat at all, meant fleeing reality. Certainly there were those who did, who could hardly do anything else. But this does not apply to Süss, nor to the majority of his colleagues.

Following the theme of mathematics and war into the history of Nazi Germany makes it necessary to study the involvement of mathematicians with the specific militance and destructive character of this society. The brutality and terror of Nazism were present and visible even before 1933. Adaption to the regime, even the mere maintenance of the simple normality of everyday life, was possible only at the cost of silence and suppression. Personal and political communication deteriorated. Blindness and deafness to inhumanity was the gradually aggravating syndrome. In compensation, self-respect had to be constructed otherwise. "Unpolitical" professional functioning, clinging to the restricted values of scholarly performance, was one possibility.

To proceed with the conclusion, I now turn to the epistemological question, looking at the varieties of mathematical prescriptive symbolisms and their uses. Except for aeronautics, there was little preparation for the war in terms of scientific research. Entangled in the ideological web, a few mathematicians drew on probability and statistics as a tool for the fostering or the eradication of innate qualities, German or un-German, respectively. Others invented exercises for use in schools that illustrated the usefulness and the objectivity of mathematical operations in pursuing the aims set by the regime. And some singled out *Wehrmathematik* to demonstrate and to study the symbolisms attached to weapons and warfare. A part of the mathematical enterprise demonstrated the specific "power of mathematical ideas" in the frame set by the regime, while the larger part of the mathematical enterprise, silently accepting the protecting shadow of this demonstration, did the normal work of the academic mathematician.

Only when the war had lasted far more than a year did the demand for technoscientific means rise. Multiplication, acceleration, and sophistication of weapons and operations called for mathematical knowledge and competence. Steps of sophistication can be found in R&D concerning homing devices, the V–2, lead angle calculators, or in cryptoanalysis. Sophisticated information processing and control was built into military artefacts. Sophisticated prescriptive symbolisms had to be attached and incorporated, largely mechanically as in the "geometry of a curved body" for an aiming device studied at the LFA, or in orbit integrators developed for the V–2. The multiplication of weapons in numbers and variants entailed the multiplication of theoretical and calculative tasks, calling for automation of symbolic procedures by machinery and for sophistication of methods of processing, most obviously in numerical calculation. As a result, mathematicians at the IPM, for example, became very well aware of the problem of propagation of errors.

The central phenomenon in the technoscience of warfare was acceleration. Most mathematical tasks arose from rising speed and range – the speed of projectiles, of planes, of armament production, of information, and of calculation. To speed up machinery and operations raises the problem of tighter control of effective performance. The wing flutter of aircraft or the stability of flight of supersonic projectiles are typical examples. Theoretical anticipation of performance is necessary and the conditions have to be precisely controlled by symbolically processed prescriptions for shape, material, and physical or chemical processes. The software of control, mathematics, becomes an indispensable tool operating at a high level of complexity.

With the increased military orientation brought by the war, mathematicians, in general, shifted their work from the autonomous to the applicable, or from the applied to the attached. The basic epistemological character of mathematics, however, was not given up. The study of prescriptive symbolisms was carried on in the immediate vicinity of the objects and activities that were to follow the prescriptions. With the shift back into academic autonomy after the war much of the experience and of the results was integrated into practices, and revised and synthesized in autonomous theoretical work. Conversion proved to be effective; the war had accelerated developments that had started before, especially in automatic computation and the mathematical methods for it.

The control of technical artefacts has another side: social control. This topic has not been sufficiently treated in the present paper. Social control was one of the central elements of the "spirit of the *Reich*", in race and health politics as well as in militarization and in political domination. The mathematical awareness of such a task is clearly seen in the demonstrations of the possible power of mathematics in "national-political education," and in fostering the sound characteristics of the Germans.[143] The move towards control by formal symbols can be seen in the introduction of the identity card and in the much too ambitious attempt to establish a central registration and information office for military purposes, an office that was to number and register every nut and bolt, every person and every building in the *Reich*.[144] A question to be raised for historians of mathematics is whether it would not in many cases be better to turn away from the compartmentalized history of the scholarly discipline, and to re-define the object of study as the software in technologies of control. As to the social integration of mathematics and its finalized neighbourhoods, the conditions set by Nazism were difficult. What had been established earlier in the century was functional, but now the inner contradictions of the regime and of its ideology led to various and often contradictory adaptive moves, partly forced, partly voluntary. The enterprise of mathematics suffered from purges and pressures. The network of the technoscientific elites deteriorated into groups and individuals cooperating and competing at the level of the smallest common denominator. The network was revived during the second year of the war, and in a specific way re-established in 1942 to 1943, when politicians were willing to turn to the experts. Compared to the first eight years of the *Reich*, the war presented favorable conditions that were exploited to re-integrate mathematics into the social system and to extend its ties and tasks.

The lesson taught by Felix Klein, of the necessity of a professional politics of integration in the interest of the discipline, was reinforced and specified. Not to intervene in political matters was a necessary safeguard in face of internal terror and Nazism, but it is also the rule of the modern system of science and politics. Almost every one of those few scientists who tried to be simultaneously Nazi politician and scholar was eventually rejected by both the political leadership and their academic colleagues. The succesful politicians of the disciplines, like Wilhelm Süss, kept rather to their fields of expertise. As the academic expert, Süss was allowed to maneuver between the political lines, as long as he was not identified as a clear partisan of any one of the competing authorities, and not suspected of striving for political authority himself. Süss was successful because he carefully respected this borderline with the political. The same observation applies, perhaps even more strongly, to science and the military. Mathematicians, however, hardly had any opportunity to probe that borderline.

The war showed the ability of the mathematical enterprise to engage in closer technoscientific cooperation, to move into finalized fields, and to establish, in a time of need, specialties of finalized type: institutionalized groups and collaborative networks, oriented toward problems set by warfare. Self-mobilization meant self-conversion of the enterprise to war production. Functionality of research mathematics for the aims set elsewhere in society had always been present, but the discipline guarded its autonomy. Before and during the war, gates were opened, not reluctantly, but actively. The war appears as the opportunity to show the real utility of mathematics and mathematicians.

The ideology and rhetorics of *Anschauung*, and the inborn racial qualities pertaining to mathematical work, were present and prominent in Nazism, but remained marginal in the discipline. Objectivity and functionality were what the mathematicians had to offer most of all, even partly at odds with the prevailing ideological orientations of the Nazi movement. However, in the state bureaucracy, in industry, and in the military, there were ready collaborators of functionality. True, the worship of technology was also part of the movement's ideology. But mathematicians were too removed from the practical to profit much from that side. The exercises for schools, prepared by the *Reichsverband*, presented mathematical objectivity in the technicalities of solutions to unquestioned problems, thus giving expression to the form of rational, destructive functionality that was gradually taking over. Mathematics and the natural sciences rejected the ideology of blood, soul, and race as counterproductive, but could integrate themselves elsewhere into the ideological web, subscribing to "service" with its aims not defined or debated.

I have been writing this concluding section on the day of the fiftieth anniversary of the *Wannsee Konferenz*, the one documented meeting on the coordination of the "final solution," that took place on the 20th of January, 1942. My newspaper of that day published part of the report of the conference on its front page, thus making me return to the connection with my topic.[145] The words of the document are "final solution", "natural diminution" by labor, and for the "remainder", representing the "natural selection" of the "most resistant," a "corresponding treatment." The rhetoric is that of bio-technical experts speaking the objectifying language of evolutionary biology and breeding. This type of objectivity, combined with economic rationality of exploitation of a labor force, is a means of psychological suppression. The genocide had no master-plan, and there was no clear sequence of orders. In the fragmented and socially disrupted system the functionaries fulfilled their functions and solved the problems. Hans Mommsen has analyzed the genocide as the "realization of the utopian." Eradication of Jewry had been a utopian metaphor rather than a realistic plan. In the structure and process of that society and its political leadership it gradually turned into a problem to be solved.[146]

The war was not greeted with enthusiasm. But in a similar way it turned into the problem of realizing the utopian. The metaphors were the master race, the chosen people, unlimited dominance. In the dynamic of that state and society the necessity of negotiating for peace could not be recognized. "Total war" was the consequence. Mobilization and self-mobilization of technoscience were part of this dynamic. The experts sought their function in a mixture of the higher values of Nazi Germany and of their disciplines, of professional pride and narrow-mindedness, of simple struggle for survival and simple everyday brutality – a mixture that made them unable consciously to recognize the realities and the inevitable. Mathematicians had no hand in the destruction of the Jews. But they were able to cooperate with the functionaries of the state and of the SS on the basis of blind objectivity, effectiveness, and mutually accepted self-interest. They did their best on the technologies of symbolic control, not necessarily to advance the military front, but in an effort to be on the scientific and technological front. World War II added massively to the development of this type of technology. Its history, subsuming that of mathematics, needs further study as a political historiography keeping the destructive potentials in mind.

Technische Universität, Braunschweig

NOTES

[1] The present paper extends my "Angewandte Mathematik und Anwendungen der Mathematik im nationalsozialistischen Deutschland," *Geschichte und Gesellschaft*, 1986, 12:317–347. It draws on research supported by Stiftung Volkswagenwerk, Deutsche Forschungsgemeinschaft, and Thyssen Stiftung. Translations from the German are mine if not otherwise indicated.

[2] Unpaginated introduction to vol. 3 of *FIAT Review of German Science 1939–1945*, 84 vols., publ. by the Office of Military Government, Field Information Agencies Technical (British, French, U.S.) (Wiesbaden: Dieterichsche Verlagsbuchhandlung, 1948); text in German; (identical edition for Germany under the title *Naturforschung und Medizin in Deutschland 1939–1946*); vols. 3–7: *Applied Mathematics (Angewandte Mathematik)*, ed. Alwin Walther.

[3] Klein, F., 1918. "Festrede zum 20. Stiftungstage der Göttinger Vereinigung für angewandte Physik und Mathematik," *Jahresbericht der Deutschen Mathematiker-Vereinigung*, 27: 217–228; p. 219; on the use of "pre-established harmony" see Pyenson, L., 1985. *The Young Einstein: The Advent of Relativity* (Bristol: Adam Hilger), chapt. 6: "Relativity in Late Wilhelmian Germany: The Appeal to a Pre-established Harmony between Mathematics and Physics"; Sigurdsson, S., 1991. *Hermann Weyl, Mathematics and Physics, 1900–1927*, Diss. Harvard University (Cambridge), chapt. 1.

[4] Forman, P., 1991. "Independence, not Transcendence for the Historian of Science," *Isis, 82*: 71–86.

[5] See David Edgerton's contribution to this volume.

[6] For a general, if incomplete, survey cf. Beranek Jr., W. and Ranis, G. (eds), 1978. *Science, Technology, and Economic Development: A Historical and Comparative Study* (New York: Praeger).

[7] After the war German nuclear physicists denied that they had aimed directly at the bomb. See, however, Walker, M., 1989. *German National Socialism and the Quest for Nuclear Power, 1939–1949* (Cambridge: Cambridge Univ. Pr.); Walker, M., 1990. "Legenden um die deutsche Atombombe," *Vierteljahreshefte für Zeitgeschichte, 38*: 45–74, and more recently, Walker, M., 1995. *Nazi Science: Myth, Truth, and the German Atomic Bomb* (New York/London: Plenum).

[8] The quote is from Haber's open farewell letter of 1933 to the members of his institute, when he, himself of Jewish descent, left in protest against the dismissal of Jewish collaborators; see "*...im Frieden der Menschheit, im Kriege dem Vaterlande...*": *75 Jahre Fritz-Haber-Institut der Max-Planck-Gesellschaft, Bemerkungen zur Geschichte und Gegenwart* (Berlin: Publ. by the collective of authors, 1986); see also Goran, M., 1967. *The Story of Fritz Haber* (Norman: Univ. of Oklahoma Pr.).

[9] Below, Section 6; the pacifism in a letter to Courant, Jan. 20, 1947, who tried to engage Walther for military research in the USA, folder "Walther, Alwin", Courant papers, Courant Institute Archives, New York University.

[10] For a closer analysis of the separation with respect to ethical questions see Mehrtens, H. "Verantwortungslose Reinheit: Thesen zur politischen und moralischen Struktur mathematischer Wissenschaften am Beispiel des NS-Staates," in: Fülgraff, G. and Falter, A. (eds), 1990. *Wissenschaft in der Verantwortung: Möglichkeiten der institutionellen Steuerung* (Frankfurt: Campus), pp. 37–54; Engl. version: "Irresponsible Purity: On the Political and Moral Structure of the Mathematical Sciences in National Socialist Germany," in: Renneberg, M. and Walker, M. (eds), 1994. *Science, Technology and National Socialism* (Cambridge: Cambridge University Press), pp. 324–338.

[11] Mehrtens, H., 1990. *Moderne – Sprache – Mathematik: Eine Geschichte des Streits um die Grundlagen der Disziplin und des Subjekts formaler Systeme* (Frankfurt: Suhrkamp), chapt. 6; Mehrtens H. "Symbolische Imperative: Zu Natur und Beherrschungsprogramm der wissenschaftlichen Moderne," in: Zapf, W. (ed), 1991. *Die Modernisierung moderner Gesellschaften: Verhandlungen des 25. Deutschen Soziologentages in Frankfurt am Main 1990* (Frankfurt: Campus), pp. 604–616; Rotman, B., 1988. "Toward a Semiotics of Mathematics," *Semiotica, 72*: 1–35.; Krämer, S., 1988. *Symbolische Maschinen: Die Idee der Formalisierung in geschichtlichem Abriß* (Darmstadt: Wissenschaftliche Buchgesellschaft).

[12] The "paradise" is Hilbert's, the "universe" Hardy's, cf. Mehrtens[11], pp. 25f., 514f.

[13] The difference between "autonomous" and "heteronomous" mathematics is made in Scholz, E., 1989. *Symmetrie – Gruppe – Dualität. Zur Beziehung zwischen theoretischer Mathematik und Anwendungen in Kristallographie und Baustatik des 19. Jahrhunderts* (Basel: Birkhäuser).

[14] Levinson, N., 1970. "Coding Theory: A Counterexample to G. H. Hardy's Conception of Applied Mathematics," *American Mathematical Monthly, 77*:249–258.

[15] For a comprehensive survey of the socio-political situation of mathematics in Germany at the turn of the century see chapt. 5 of Mehrtens[11] ; a small biography of Klein is Tobies, R., 1981. *Felix Klein* (Leipzig: Teubner); for Klein's political activities related to technology and especially on the *Göttinger Vereinigung* see Manegold, K.-H., 1970. *Universität, Technische Hochschule und Industrie: Ein Beitrag zur Emanzipation der Technik im 19. Jahrhundert unter besonderer Berücksichtigung der Bestrebungen von Felix Klein* (Berlin: Duncker & Humblot).

[16] Ibid. (Manegold) p. 237f., see also Ebert, H. and Rupieper, H.J., 1979. "Technische Wissenschaft und nationalsozialistische Rüstungspolitik: Die Wehrtechnische Fakultät der TH Berlin 1933–1945," in *Wissenschaft und Gesellschaft: Beiträge zur Geschichte der Technischen Universität Berlin 1879–1979*, 2 vols., ed. Reinhard Rürup (Berlin: Springer), vol. 1, pp. 469–491.

[17] Klein, F., 1908. "Über die Neueinrichtungen für Elektrotechnik und allgemeine technische Physik an der Universität Göttingen," in *Die Physikalischen Institute der Universität*

Göttingen: Festschrift (Leipzig: Teubner), pp. 189–193, p. 190.

[18] Rotta, J.C., 1990. *Die Aerodynamische Versuchsanstalt in Göttingen, ein Werk Ludwig Prandtls: Ihre Geschichte von den Anfängen bis 1925* (Göttingen: Vandenhoek & Ruprecht); Trischler, H., 1992. *Luft- und Raumfahrtforschung in Deutschland 1900–1970: Politische Geschichte einer Wissenschaft* (Frankfurt: Campus), pp. 34–70; on the KWG see Vierhaus, R. and Brocke, B. von (eds), 1990. *Forschung im Spannungsfeld von Politik und Gesellschaft: Geschichte und Struktur der Kaiser-Wilhelm/Max-Planck Gesellschaft* (Stuttgart: DVA).

[19] The "mandarins" allude to Ringer, F., 1969. *The Decline of the German Mandarins* (Cambridge: Harvard Univ. Pr.); Ringer is concerned with social sciences, for an analysis on related lines of natural scientists see chapters 1 and 8 of Harwood, J., 1993. *Styles of Scientific Thought: The German Genetics Community, 1900–1933* (Chicago: Univ. of Chicago Pr.).

[20] There is still no comprehensive study of the Göttingen system; see above, notes 3, 15, 18; further Peckhaus, V., 1990. *Hilbertprogramm und Kritische Philosophie: Das Göttinger Modell interdisziplinärer Zusammenarbeit zwischen Mathematik und Philosophie* (Göttingen: Vandenhoek & Ruprecht); Richenhagen, G., 1985. *Carl Runge (1856–1927): Von der reinen Mathematik zur Numerik* (Göttingen: Vandenhoek & Ruprecht).

[21] Manegold[15], pp. 239f.

[22] Böhme, G., Daele, W. van den and Krohn, W., 1973. "Die Finalisierung der Wissenschaft," *Zeitschrift für Soziologie*, 2:128–144; Böhme, G., Daele, W. van den and Hohlfeld, R. (eds), 1978. "Finalisierung revisited," in *Starnberger Studien I: Die gesellschaftliche Orientierung des wissenschaftlichen Fortschritts*, Max-Planck-Institut zur Erforschung der Lebensbedingungen der wissenschaftlich-technischen Welt, (Frankfurt: Suhrkamp), pp. 195–250; enlarged English translation of the latter book, including a bibliography of the debate: Schäfer, W. (ed), 1983. *Finalization in Science: The Social Orientation of Scientific Progress* (Boston Studies in the Philosophy of Science 77, Dordrecht: Reidel).

[23] Timerding, H.E. "Die Schulmathematik und der Krieg," in: Schmid, B. (ed), 1919. *Deutsche Naturwissenschaft, Technik und Erfindung im Weltkriege*, (München: Otto Nemnich), pp. 961–974.

[24] Timerding, H.E., 1915. *Analyse des Zufalls* (Braunschweig: Vieweg).

[25] See Kremer, A., 1985. *Naturwissenschaftlicher Unterricht und Standesinteresse: Zur Professionalisierungsgeschichte der Naturwissenschaftslehrer an höheren Schulen* (Marburg: Redaktionsgemeinschaft Soznat), pp. 99–107.

[26] Schubring, G. "Pure and Applied Mathematics in Divergent Institutional Settings: The Role and Impact of Felix Klein," in: Rowe, D.E. and MacLeary, J. (eds), 1989. *The History of Modern Mathematics. Vol.II: Institutions and Applications* (Boston: Academic Press), pp. 171–223; Pyenson[3], chapt. 7; Mehrtens[11], ch. 5.2.

[27] Tobies, R., 1980. "Zur Einflußnahme des Reichsverbandes deutscher mathematischer Gesellschaften und Vereine auf die Verstärkung der angewandten Mathematik in der Ausbildung," in: *Philosophie und Naturwissenschaften in Vergangenheit und Gegenwart*, No. 19 (Berlin: Humboldt Universität), pp. 66–71; Mehrtens, H., 1985. "Die 'Gleichschaltung' der mathematischen Gesellschaften im nationalsozialistischen Deutschland," *Jahrbuch Überblicke Mathematik*, pp. 83–103; Engl. transl. "The 'Gleichschaltung' of Mathematical Societies in Nazi Germany," *The Mathematical Intelligencer*, 1989, *11*, No.3:48–60.

[28] Trischler[18], pp. 110–116; see also Rotta[18], pp. 115–198.

[29] Hanle, P.A., 1982. *Bringing Aerodynamics to America* (Cambridge, Mass.: M.I.T. Pr.), pp. 53–60; Kármán, T. von, 1954. *Aerodynamics: Selected Topics in the Light of Their Historical Development* (Ithaca, N.Y.: Cornell Univ. Pr.), pp. 68–72.

[30] Kármán, T. von and Edson, L., 1967. *The Wind and Beyond* [autobiography of v.K.] (Boston: Little, Brown and Company), pp. 79–81.

[31] Mises, R. von, 1918. *Fluglehre: Vorträge über Theorie und Berechnung der Flugzeuge in elementarer Darstellung* (Berlin); for biographical information see Stadler, F. "Richard von Mises (1883–1953) – Wissenschaft im Exil," in: Stadler, F. (ed.), 1990. Richard von Mises, *Kleines Lehrbuch des Positivismus: Einführung in die empirische Wissenschaftsauffassung*

(Frankfurt: Suhrkamp), pp. 7–51.

[32] Quoted in: Bernhard, H., 1984. *Richard von Mises und sein Beitrag zur Grundlegung der Wahrscheinlichkeitstheorie im 20. Jahrhundert* (Diss. Humboldt-Universität Berlin), p. 19; see also Biermann, K.-R., 1988. *Die Mathematik und ihre Dozenten an der Berliner Universität 1810–1933: Stationen auf dem Wege eines mathematischen Zentrums von Weltgeltung* (Berlin: Akademie-Verlag), pp. 190f., 200ff.

[33] Forman, P., 1971. "Weimar Culture, Causality, and Quantum Theory, 1918–1927: Adaptation by German Physicists and Mathematicians to a Hostile Intellectual Environment," *Historical Studies in the Physical Sciences, 3*:1–115; on von Mises see pp. 48–51, 62, 80–82.

[34] Discussed as such in Mises, R. von, 1928. *Wahrscheinlichkeit, Statistik, Kausalität und Wahrheit* (Wien: Julius Springer), pp. 170–179.

[35] Mises, R. von, 1921. "Über die Aufaben und Ziele der angewandten Mathematik," *Zeitschrift für angewandte Mathematik und Mechanik, 1*:1–15; on turbulence see Battimelli, G., 1984. "The Mathematician and the Engineer: Statistical Theories of Turbulence in the 20's," *Rivista di storia della scienza, 1*:73–94, p. 87.

[36] In this case the serious political problems between the adversaries of the war were overcome with relative ease; see Battimelli, G. "The Early International Congresses of Applied Mechanics," in: Juhasz, S., 1988. *IUTAM: A Short History* (Berlin: Springer-Verlag), pp. 9–13; on international relations in science after World War I see Schröder-Gudehus, B., 1978. *Les scientifiques et la paix: La communauté scientifique internationale au cours des années vingt* (Montréal).

[37] See p.8. of Gericke, H., 1972. "50 Jahre GAMM," *Beiheft zum Ingenieurarchiv, 4*, (Berlin: Springer); see also Tobies, R., 1982. "Die 'Gesellschaft für angewandte Mathematik und Mechanik' im Gefüge imperialistischer Wissenschaftsorganisation," *NTM Schriftenreihe für die Geschichte der Naturwissenschaften, Technik und Medizin, 19*, No.1:16–26.

[38] Ibid. (Gericke), p. 7.

[39] Cf. Tobies[37].

[40] Homze, E.L., 1976. *Arming the Luftwaffe: The Reich Air Ministry and the German Aircraft Industry 1919–1939* (Lincoln: Univ. of Nebraska Pr.); Trischler[18], pp. 134–141.

[41] See Mehrtens[11], 4.1.,4.2.; Davis, M., 1987. "Mathematical Logic and the Origin of Modern Computers," in *Studies in the History of Mathematics*, ed. Esther R. Phillips (MAA Studies in Mathematics, vol. 26, Washington D.C.: The Mathematical Association of America), pp. 137–165; Hodges, A., 1983. *Alan Turing: The Enigma* (London: Burnett Books); on Zuse see Petzold, H., 1985. *Rechnende Maschinen: Eine historische Untersuchung ihrer Herstellung und Anwendung vom Kaiserreich bis zur Bundesrepublik* (Düsseldorf: VDI Verlag), p. 323.

[42] On Göttingen mathematics see Schappacher, N. "Das Mathematische Institut der Universität Göttingen 1929–1950," in: Becker, H., Dahms, H.-J. and Wegeler, C. (eds), 1987. *Die Universität Göttingen unter dem Nationalsozialismus: Das verdrängte Kapitel ihrer 250jährigen Geschichte* (München: K.G. Saur), pp. 345–373; Tollmien, C. "Das Kaiser-Wilhelm-Institut für Strömungsforschung verbunden mit der Aerodynamischen Versuchsanstalt," ibid., pp. 464–488.

[43] Siegmund-Schultze, R., 1989. "Zur Sozialgeschichte der Mathematik an der Berliner Universität im Faschismus," *NTM Schriftenreihe für die Geschichte der Naturwissenschaften, Technik und Medizin, 26*, No.1:49–68.

[44] See Ludwig, K.H., 1974. *Technik und Ingenieure im Dritten Reich* (Düsseldorf: Droste), pp. 271–288; Trischler[18], pp. 198–240.

[45] Mommsen, H. "Nationalsozialismus als vorgetäuschte Modernisierung," in: Pehle, W.H., 1990. *Der historische Ort des Nationalsozialismus: Annäherungen* (Frankfurt: Fischer Taschenbuch Verlag), pp. 31–46; Hirschfeld, G. and Kettenacker, L. (eds), 1981. *Der "Führerstaat": Mythos und Realität* (Stuttgart; DVA); see also Mehrtens, H. "Kollaborationsverhältnisse: Natur- und Technikwissenschaften im NS-Staat und ihre Historie," in: Meinel, C. and Voswinkel, P. (eds), 1994. *Medizin, Naturwissenschaft, Technik und Nationalsozialismus: Kontinuitäten und Diskontinuitäten* (Stuttgart: GNT Verlag).

[46] This is also reflected in the few surveys; Ludwig[44], Chapt. 6; Mehrtens, H. "Das 'Dritte Reich' in der Naturwissenschaftsgeschichte: Literaturbericht und Problemskizze," in: Mehrtens, H. and Richter, S. (eds), 1980. *Naturwissenschaft, Technik und NS-Ideologie – Beiträge zur Wissenschaftsgeschichte des Dritten Reiches* (Frankfurt: Suhrkamp), pp. 15–87; Renneberg/Walker[10].

[47] For a survey and sociological interpretation see Mehrtens, H., 1987. "The Social System of Mathematics and National Socialism, A Survey," *Sociological Inquiry*, 57:159–182," repr. in: Fischer, R., Restivo, S. and Bengedem, J.P. van (eds), 1993. *Math Worlds: New Directions in Social Studies and Philosophy of Mathematics* (Albany: State Univ. Pr. of New York), pp. 219–246; Mehrtens, H., 1988. "Das soziale System der Mathematik und seine politische Umwelt," *Zentralblatt für Didaktik der Mathematik*, 20:No.1, pp. 28–37.

[48] Weiß, E.A., 1933. *Wozu Mathematik?* (Bonn: Published by the author), p. 12.

[49] Hamel, G., 1933. "Die Mathematik im Dritten Reich," *Unterrichtsblätter für Mathematik und Naturwissenschaften, 39*:309.

[50] An example for adaptability of rhetorics is presented by a guide to the study of mathematics, published in 1946, where the "need of the day" is the return to a "spiritual center" by "Anschauung", the "new beginning" is to be "inspired with *Wehrhaftigkeit* (military spirit)", and the following chapters prove the text to have been written earlier, when a "watchword" like "Fight for the inner *Reich* of *Geist*!" fitted smoothly with the *Geist of the Reich*; Steck, M., 1946. *Grundgebiete der Mathematik* (Heidelberg: Winter), pp. xif., 23. The tradition, by no means restricted to mathematics, of the scholar's "manly self-denial" in his "fight" for truth reaches back to the 19th century; see Mehrtens[11], pp. 101ff., 233ff.

[51] Hasse, H. "Mathematik," in: Reichsministerium für Erziehung, Wissenschaft und Volksbildung, 1939. *Deutsche Wissenschaft: Arbeit und Aufgabe* (Leipzig: Hirzel), pp.149–151, p. 150.

[52] Steck, M., 1942. *Mathematik als Begriff und Gestalt*, (Die Gestalt 12) (Halle: Max Niemeyer), p. vii.

[53] Mehrtens, H. "Ludwig Bieberbach and 'Deutsche Mathematik'," in Phillips[41], pp. 195–241; on the counter-modern tradition of the rhetorics of *Anschauung* see Mehrtens[11], chapt. 3., 4.3.

[54] "Shape" could be translated as *Gestalt*, a concept serving frequently in ideologically adaptive philosophies, e.g., in Steck[52], where the formal "skeleton" of mathematics is said to need "flesh and blood" of *Geist* and *Anschauung* to become the "bearer of a Gestalt" (p. 14); on the collaborator of the series "Die Gestalt" in chemistry see Bechstedt, M. "'Gestalthafte Atomlehre' – Zur 'Deutschen Chemie' im NS-Staat," in: Mehrtens and Richter[26], pp. 142–165.

[55] For a more detailed analysis of ideological definition of the individual see Haug, W.F., 1986. *Die Faschisierung des bürgerlichen Subjekts: Die Ideologie der gesunden Normalität und die Ausrottungspolitiken im deutschen Faschismus* (Argument Sonderband AS 80), (Berlin: Argument Verlag), on will and character sections 5.5., 5.6.

[56] Hasse[51] ; Schmeidler, W. "Angewandte Mathematik," *ibid.* pp. 152–154.

[57] Haug[55], p. 95.

[58] One example is the statistician Siegfried Koller who was engaged in defining hereditary *Gemeinschaftsunfähigkeit* (incapacity for social behavior) and proposed legislative consequences. In 1990 the German section of the International Biometrical Society withdrew honorary membership from Koller. One delicate question of the preceding investigation was whether Koller had declined the right to live to "anti-social" individuals. The society determined that he had not proposed extermination of individuals. Nonetheless, Koller, by proposing eugenic legislative measures concerning hereditary diseases and anti-social behavior, had, in the judgement of the German Section of the International Biometrical Society, gone beyond the work of the mathematician. The whole case, together with four papers of a conference on "science and resposibility around biometrics", has been published in the journal *Biometrie und Informatik*, 1990, 21, No. 4; see especially Lorenz, R.J. "Die Arbeiten Siegfried Kollers zur Rassenhygiene in der Zeit von 1933 bis 1945," *ibid.*, pp. 196–230.

[59] On *Wehrwissenschaft* see the exemplary study of Ebert and Rupieper[16].

[60] On the division of labor between DMV and Reichsverband see Mehrtens[27]; for more details

on the DMV see Schappacher, N. and Kneser, M. "Fachverband – Institut – Staat," in: Fischer, G. et al. (eds), 1990. *Ein Jahrhundert Mathematik 1890–1990. Festschrift zum Jubiläum der DMV* (Braunschweig: Vieweg), pp. 1–82.

[61] Dorner, A. (ed), 1936. *Mathematik im Dienste der nationalpolitischen Erziehung mit Anwendungsbeispielen aus Volkswissenschaft, Geländekunde und Naturwissenschaft: Ein Handbuch für Lehrer*, herausgegeben im Auftrage des "Reichsverbandes Deutscher mathematischer Gesellschaften und Vereine" (Frankfurt, M.: Diesterweg); see Mehrtens, H. "Nationalsozialistisch eingekleidetes Rechnen: Mathematik als Wissenschaft und Schulfach im NS-Staat," in: Dithmar, R., 1989. *Schule und Unterrichtsfächer im Dritten Reich* (Neuwied: Luchterhand), pp. 205–216.

[62] A first conference on applied mathematics and the regulations for exams in mathematics was held in 1934 and is documented in *Jahresbericht der Deutschen Mathematiker-Vereinigung*, 1935, *45*; there was one lecture by a representative of the army, Lechner, G. "Soldaten und Mathematik," *ibid.*, pp. 15–17; another series of articles appeared in the same journal, 1937, *47*; in a meeting of 1938 the term *Diplommathematiker* was introduced, and commented on in soldier-rhetorics: "Research and industry demand *ganze Kerle* (real men).", "Bericht über die Jahresversammlung des M.R." *Jahresbericht der Deutschen Mathematiker-Vereinigung*, 1935, *45*, 2.Abt., p. 85.

[63] First discussed by Ulfried Geuter, *Die Professionalisierung der deutschen Psychologie im Nationalsozialismus* (Frankfurt/M.: Suhrkamp, 1984); pb. with new preface 1988.

[64] Personal communication to the author, Sept. 7, 1983.

[65] Author's interview with Karl Storck, Feb. 23, 1984.

[66] Homze[40].

[67] Trischler, H. "Self-mobilisation or Resistance? Aeronautical Research and after National Socialism," in: Renneberg and Walker[10], pp. 72–87.

[68] Trischler[18], pp. 194ff.; Tollmien[42].

[69] See below, Section 6; a list of scientific personnel in: *25 Jahre Deutsche Forschungsanstalt für Luftfahrt e.V.: DFL Braunschweig 1936–1961* (Braunschweig: DFL, 1961), pp. 7–15; on aeronautical R&D 1935–39 see Trischler[18], pp. 208–240.

[70] Schreyer, H.T. "Die Entwicklung des Versuchsmodells einer elektronischen Rechenmaschine," and "Grundschaltungen eines während des letzten Weltkrieges entwickelten elektronischen Rechners," in: Gebhardt, F. (ed), 1983. *Skizzen aus den Anfängen der Datenverarbeitung* (Berichte der Gesellschaft für Mathematik und Datenverarbeitung Nr. 143) (München: Oldenbourg), pp. 23–52; on Schreyer see Petzold[41], pp. 310–317; on Zuse, pp. 317–337; see also Zuse, K., 1970. *Der Computer, mein Lebenswerk* (München); Zuse, K. "Some Remarks on the History of Computing in Germany," in: Metropolis, N., Howlett J. and Rota, G.-C. (eds), 1980. *A History of Computing in the Twentieth Century: A Collection of Essays* (New York: Academic Press), pp. 611–672.

[71] On Walther and the IPM see Collatz, L., 1967. "Alwin Walther 6. Mai 1898–4. Januar 1967," *Zeitschrift für angewandte Mathematik und Mechanik*, 47: 213–215; Walther, A., 1936. "Zur Behandlung der Mathematik auf der technischen Hochschule," *Technische Erziehung*, *11*:1–7; Walther, A. "Wechselwirkungen zwischen Mathematik und Technik," in: Technische Hochschule Darmstadt (ed), 1946. *Die Welt des Ingenieurs: Vortragsreihe im Wintersemester 1945/1946* (Heidelberg: Carl Winter), pp. 9–32; Beauclair, W. de. "Prof. A. Walther, das IPM der TH Darmstadt und die Entwicklung der Rechentechnik in Deutschland 1930–1945," in: Gebhardt[70], pp. 53–90; Barth, W. "Alwin Walther – Praktische Mathematik und Computer an der THD," *Jahrbuch 1978/79, Technische Hochschule Darmstadt*, pp. 29–34; on the Peenemünde slide-rule see entry No. 32 in the long list of publications appended to Walther, A. and Kron, A.-W. "Nomographie und Rechenschieber" in: *FIAT Review* [2], vol. 3, pp. 119–127.

[72] On analogue calculators see part 2 (resp. 2.4. on military uses) of Petzold[41].

[73] *Ibid.*, 2.5.

[74] *Ibid.*, 68f.; on the differential analyzer see De Beauclair[72], p.59; I have not found any archival sources on the early assigments from the rocket project.

[75] Letter Walther to R. Courant, March 28, 1945, cf. n. 9.

[76] Osenberg files, Bundesarchiv Koblenz (henceforth: BA) R 26 III, 43, p. 13; R 26 III, 44; on Osenberg and the *Rückrufaktion* see Ludwig[44], pp. 243ff., 251–271; on the lack of engineers, pp. 288–300.

[77] BA R 26 III, 51, p. 28.

[78] Hayes, P., 1987. *Industry and Ideology: IG Farben in the Nazi Era* (Cambridge: Univ. Pr.), p.322; Trischler[18], p. 302.

[79] For a detailed survey see Ludwig[44], chapt. 6; this is still the best study of R&D politics in the Third Reich; on the *Forschungsführung* see Trischler[18], pp. 245–261; on Mentzel and the RFR, Zierold, K., 1968. *Forschungsförderung in drei Epochen: Deutsche Forschungsgemeinschaft: Geschichte, Arbeitsweise, Kommentar* (Wiesbaden: Steiner), pp. 237–257; on Krauch's *Reichsamt*, Ludwig[44], pp. 231ff., and Schlicker, W., 1979. "Karl Krauch, die IG Farben und die Forschungslenkung im faschistischen Deutschland: Ein Beitrag zur imperialistischen deutschen Wissenschaftspolitik," *Bulletin des Arbeitskreises Zweiter Weltkrieg* (Akademie der Wissenschaften der DDR, Zentralinstitut für Geschichte), No. 2, pp. 5–47; much quoted, but problematic in factual content and interpretation is the anonymously published article: "Größe und Verfall der deutschen Wissenschaft im Zweiten Weltkrieg," in *Bilanz des Zweiten Weltkriegs*, ed. anonymously (Oldenburg: Stalling, 1953), pp. 251–264; an interesting, if even more unreliable, source are the recollections of one of the authors of that article, a mathematician working at the *Reichssicherheitshauptamt*, Fischer, H.J., 1984 / 1985. *Erinnerungen, Teil I: Von der Wissenschaft zum Sicherheitsdienst, Teil II: Feuerwehr für die Forschung*, (Quellenstudien der Zeitgeschichtlichen Forschungstelle Ingolstadt 3, 4; Ingolstadt: Zeitgeschichtliche Forschungsstelle).

[80] Cf. note 2.

[81] Magnus, W. "Spezielle Funktionen der mathematischen Physik," *FIAT Review* [2], vol. 1, chapt. X.

[82] Wielandt, H. "Eigenwerttheorie," *FIAT Review* [2], vol. 2, Chapt. XXI.

[83] Copies of personal files from H. Wielandt in possession of the author.

[84] Letter A. Walther to Ludwig Prandtl, Aug. 20, 1942, file "Betr. Herausgabe mathematischer Werke", Nachlaß Prandtl, Archive of the Max-Planck-Institut für Strömungsforschung, Göttingen.

[85] Zwicky, F. "Report on certain phases of war research in Germany," vol. I, typescript, Archives of the California Institute of Technology, Pasadena, pp. 23–34.

[86] File "Walther" in BA R 73 (DFG), 15487.

[87] *Ibid.*, p. 34.; on work at the IPM see R. Zurmühl, "V–2 Ballistik," *FIAT Review* [2], vol. 7, ch. X.

[88] Petzold[41], pp. 85–88; Tomayko, J.E., 1985. "Helmut Hoelzer's Fully Electronic Analog Computer," *Annals for the History of Computing*, 2:227–240.

[89] Author's interview with H. Unger, March 9, 1984.

[90] Bundesarchiv-Militärarchiv, Freiburg (henceforth BA-MA) RH 8/ v. 3642; on the V–2 see Neufeld, M. "The Guided Missile and the Third Reich: Peenemünde and the Fostering of a Technological Revolution," in: Renneberg and Walker[10], pp. 51–71, and Neufeld, M.J., 1995. *The Rocket and the Reich: Peenemünde and the Coming of the Ballistic Missile Era* (New York: Free Press).

[91] E.g., with Zuse. See Petzold[41], p. 331.

[92] On details see De Beauclair[72], Petzold[41], pp. 384–387; reports on the development of mathematical machines at the IPM written in 1946/47 are in the archives of the Gesellschaft für Mathematik und Datenverarbeitung, St. Augustin, Zuse Collection 037/002, 019, 020; 046/013; De Beauclair collection B06/001, 002, 003.

[93] Petzold[41], p. 331.

[94] De Beauclair[71], p. 66.

[95] Petzold[41], pp. 317–337, the quote on p. 336.

[96] A survey, mainly based on interviews, is given in Kahn, D., 1973. *The Codebreakers: The*

Story of Secret Writing (New York: MacMillan), chapt. 14.

[97] Rohrbach, H. "Mathematische und maschinelle Methoden beim Chiffrieren und Dechiffrieren." *FIAT Review*[2], vol. 3, chapt. IX; revised engl. transl. "Mathematical and mechanical methods in cryptography," *Crytologia*, 1978, 2: 1–37, 101–121; see also his "Report on the decipherment of the American strip-cipher O–2 by the German Foreign Office," *Cryptologia*, 1979, *3*:6–26.

[98] Documents are extremely scarce; questionnaires of employees of the department "Chi" of the OKW including 9 mathematicians in BA-MA RW3/ v.777; further information from author's interviews with Hans Rohrbach, October 17 and 18, 1983, and with Wolfgang Franz, March 13, 1984.

[99] The listings are varying in precision and completeness, those of the Navy are best; BA R 26 III (RFR), vols. 3–5; a more complete list of the the Navy in BA-MA RM 24/M/697 E.

[100] Author's interview with Wolfgang Haack, November 14, 1983; in the interview he depicted his turn to war work as a clever move for safeguarding himself; I have found no other evidence on his motives. The documentation on occasion of Haack's honorary doctorate mentions his war-work briefly, avoiding military terminology and mentioning the theory of "slim bodies" and "twist-stabilized flying bodies", see "Festkolloquium aus Anlaß der Ehrenpromotion und des 80. Geburtstages von Prof.Dr.h.c. Wolfgang Haack 26./27. April 1982," (Berlin: Freie Universität, 1982), p. 10.

[101] The 'reading' of the "Rotterdam device," the 9-cm radar found early in 1943 in a British bomber downed near Rotterdam, is a striking example of the unintended but effective 'communication' across battle lines referred to in Section 1, above. Reuter, F., 1971/1950. *Funkmeß – Die Entwicklung und der Einsatz des RADAR – Verfahrens in Deutschland bis zum Ende des zweiten Weltkrieges* (Opladen: Westdeutscher Verlag), and p. 19 of H. Lux, "Technische Entwicklung und Forschung bei Telefunken während des Krieges," *Telefunken Zeitung*, 23, No.87/88: 11–26. See also M. Eckert's paper in this volume.

[102] According to the Navy list in BA-MA[99]; letter Wilhelm Magnus to the author, Sept. 7, 1983; on the Navy's project of self-guiding torpedoes in which Hasse was involved see also Aschoff, V., 1961. "Physikalische Probleme des Unterwasserkrieges erläutert am Beispiel eines zielsuchenden Torpedos," *Wehrtechnische Monatshefte*, 6: 253–266.

[103] Lists of assignments from the physics department (also responsible for mathematics) of the RFR in BA R26 III, pp. 182ff., 239ff.

[104] BA R 26 III, 8, 9; for Babelsberg see cards Bachmann, Kaluza, Rellich, Weise.

[105] Ludwig[44], pp. 241–245.

[106] Trischler[67]; for the detailed story of aeronautical reasearch 1942–1945 see Trischler[18], pp. 241–283.

[107] See Beyerchen, A., 1977. *Scientists under Hitler: Politics and the Physics Community in the Third Reich* (London: Yale Univ. Pr.), pp. 184–188.

[108] Ludwig[44], pp. 241f. The attitude described is not absent in present historiography of science; see, e.g., Hermann, A., 1982. *Wie die Wissenschaft ihre Unschuld verlor: Macht und Mißbrauch der Forscher* (Stuttgart: DVA); the chapter "Wissen ist Macht" has the thesis that Hitler could not win a war because he ignored Germany's "real strength", science.

[109] This is, so to speak, an hypothesis on the mean value from impressionist evidence; see, e.g., Trischler[18], Tollmien[42], Mehrtens[1,27] on Ludwig Prandtl; for a discussion of "self-mobilization" see also Trischler[67].

[110] Süss, W., 1943. "Die gegenwärtige Lage der deutschen Wissenschaft und der deutschen Hochschulen. Referat, gehalten auf der Rektoren-Konferenz in Salzburg am 26.8.43," (Freiburg: privately printed typescript), pp. 7f., 15f.

[111] See Seier, H., 1976. "Niveaukritik und partielle Opposition: Zur Lage an den deutschen Hochschulen 1939/40," *Archiv für Kulturgeschichte*, 58:227–246.

[112] Schappacher and Kneser[60], p. 69.

[113] The main bodies of sources on Süss' activities are two small files that I will not quote in detail, "Mathematische Forschungen", BA R73, 12976, and "Briefwechsel betreffend Herausgabe mathematischer Werke", Nachlass Prandtl, Archiv des Max-Planck-Insituts für Strömungsforschung, Göttingen; a further source are the recollections of Süss' wife, with reproductions of documents, Süss, I., 1967/1980/1983. "Entstehung des mathematischen Forschungsinstituts Oberwolfach im Lorenzenhof: Zur Einweihung des Neubaues 1967," (Oberwolfach: Mathematisches Forschungsinstitut); slightly revised Engl. version, "Origin of the Mathematical Research Institute Oberwolfach at the Countryseat 'Lorenzenhof'," *General Inequalities 2*, Proc. Sec. Int. Conf. on Gen. Ineq. Oberwolfach 1978, (Birkhäuser: Basel), pp. 3–13; "The Mathematical Research Institute Oberwolfach through Critical Times," *General Inequalities 3*, Third Int. Conf. on Gen. Ineq. Oberwolfach 1981, (Birkhäuser: Basel), pp. 3–17; Süss' papers are in possession of the director of Mathematisches Forschungsinstitut Oberwolfach and have thus far not been accessible to historians except for restricted use in the case of the anniversary volume of the DMV, cf. Schappacher and Kneser[60].

[114] Cf.[84].

[115] The report of this conference is documented with detailed comment in Siegmund-Schultze, R., 1986. "Faschistische Pläne zur 'Neuordnung' der europäischen Wissenschaft: Das Beispiel Mathematik," *NTM Schriftenreihe für die Geschichte der Naturwissenschaften, Technik und Medizin*, 23, No.2:1–17.

[116] Cf.[84].

[117] On the Heidelberg aeronautical institute see Trischler[18], pp. 273–276; Wegner's report on the institute of 1944 in the file "Wegner, Udo, research", Berlin Document Center; a file on the institute indicating the cooperation of Doetsch, Wegner and others, "Institut für Weltluftverkehr (Luftfahrtforschung Heidelberg)" UAH 8–6950, Universitätsarchiv Heidelberg; an investigation of irregularities in Wegner's function as dean gives a rather unfavourable picture of his conduct and his scientific competence: "Untersuchungen der Unkorrektheiten, die in der Naturwissenschaftlich-Mathematischen Fakultät während des Dekanats von Prof. Udo Wegner vorfielen" UAH B–1590, 2nd vol.

[118] Schappacher and Kneser[60], p. 73.

[119] Luftfahrtforschungsanstalt Hermann Göring, Braunschweig, Arbeitsgruppe für Industriemathematik, Tätigkeitsberichte 1 (31.7.1943), 2 (30.11.1942), 3 (31.3.1944), Historisches Archiv der DFVLR, Köln-Porz.

[120] List of research assignments, August 1943, Forschungsführung, Ref. F III, BA R26 III 5, p. 34.

[121] Information from Mrs. E. Gröbner, and file 12976[113].

[122] Cf.[79]; Krauch's proposal of Sept. 17, 1943, and the following correspondence in BA R21 (Rep 76), 932.

[123] Süss to Fischer (RFR) Oct. 7, 1943, BA R21 (Rep 76), 932.

[124] The story of the acquisition and of the situation at the end of the war is told by I. Süss[113]

[125] BA R 26 III, 47, pp. 69–75.

[126] See Irmgard Süss[113] on the life at the institute at that time; for the later development: Gesellschaft für mathematische Forschung und Stiftung Volkswagenwerk (ed), 1984. Mathematisches Forschungsinstitut Oberwolfach, *Anniversarium 1984* (Freiburg: Mathematisches Forschungsinstitut); also Schappacher, N., 1985. "Max-Planck-Institut für Mathematik: His-

torical Notes on the New Research Institute," *The Mathematical Intelligencer*, 7:41–52.

[127] Captured German Materials, Manuscript Division, Library of Congress, Washington, D.C., cont.109, reel 102, folder 12764 ("Reichsinstitute").

[128] Quoted in Rössler, M., 1990. *'Wissenschaft und Lebensraum': Geographische Ostforschung im Nationalsozialismus: Ein Beitrag zur Disziplingeschichte der Geographie* (Berlin: Reimer), p. 90; Rössler has been the first to evaluate the files of the IDO (BA R 52 IV); see pp. 84–91 for background and foundation.

[129] Quote and following information in the collection "Nürnberger Prozeß", Institut für Zeitgeschichte, München, NO–640, memo of Nov. 24, 1944; information on technical and chemical research also in BA NS 21, 845, memo of Jan. 1, 1945.

[130] Ludwig[44], chapt. 11.

[131] On the Ahnenerbe in general (but very brief and largely erroneous about the mathematics institute) see Kater, M.H., 1974. *Das Ahnenerbe der SS 1935–1945: Ein Beitrag zur Kulturpolitik des Dritten Reiches* (Stuttgart: DVA); on the experiments on humans see Schleiermacher, S. "Die SS-Stiftung 'Ahnenerbe': Menschen als Material für 'exakte' Wissenschaft," in: Osnowski, R. (ed), 1988. *Menschenversuche: Wahn und Wirklichkeit* (Köln: Volksblatt Verlag), pp. 70–87.

[132] On cooperation with medical research in aeronautics see Schleiermacher[131], pp. 54ff., Trischler[18], p. 267; on the cooperation of the KWI for anthropology with Mengele, see Müller-Hill, B., 1984/1988. *Tödliche Wissenschaft: Die Aussonderung von Juden, Zigeunern und Geisteskranken 1933–1945* (Reinbek: Rowohlt Taschenbuch Verlag), pp. 71–75; Engl. transl.: *Murderous Science: Elimination by Scientific Selection of Jews, Gypsies, and Others, Germany 1933–1945* (Oxford: Oxford Univ. Pr.).

[133] The account is mainly based on the small file NO–640[129]; information from that source will not be annotated in detail; further NO–1056, *ibid.*, and a file "Judeneinsatz" BA NS 21, 96. I am indebted to the late Hans Ebert, who left an unpublished manuscript "Mathematiker im KZ 1944/45" based on largely identical files in the Berlin Document Center.

[134] Fischer[79], part II, p. 101f.

[135] NO–1056[133].

[136] Personal communication from Hans Rohrbach[98]; letter Mohr to Sommerfeld Dec. 6, 1946, Sommerfeld Papers, 1977–28, A 236, Deutsches Museum, München; partly erroneous information in Ebert[133].

[137] Assignments and invoices in BA NS 21, 96; notes on the institute also in Sievers' diary BA NS 21, 11, entries of 1944, June 15, 27, July 13, 24–28, Aug. 27, Sept. 12, 13, Oct. 5, Nov. 18, 21, 22.

[138] NO–640[133].

[139] *Sachsenhausen, 1981. Dokumente, Aussagen, Forschungsergebnisse und Erlebnisberichte über das ehemalige Konzentrationslager Sachsenhausen*, 3rd ed. (Berlin: Deutscher Verlag der Wissenschaften).

[140] Boseck's report of Nov. 29, 1944, NO–640[130].

[141] Cf.[9].

[142] Unpaginated preface to *FIAT Reviews*[2], vol. I.

[143] Cf.[61].

[144] Aly, G. and Roth, K.H., 1984. *Die restlose Erfassung: Volkszählen, Identifizieren, Aussondern im Nationalsozialismus* (Berlin: Rotbuch).

[145] *Die Tageszeitung*, Berlin, 20. Jan. 1992, pp. 1, 18–19.

[146] See his careful discussion: Mommsen, H., 1983. "Die Realisierung des Utopischen: Die 'Endlösung der Judenfrage' im 'Dritten Reich'," *Geschichte und Gesellschaft*, 9:381–420.

LEWIS PYENSON

ON THE MILITARY AND THE EXACT SCIENCES IN FRANCE*

In the military-scientific complex, whose domination of physics in modern America has been probed with special acuity by Paul Forman and José Manuel Sánchez Ron, we see the twin incarnations of Athena – goddess of war and avatar of wisdom.[1] Among the cultures of the West today, the advancement of knowledge has become inseparable from destruction of the ways and means of knowing. Whence the affinity between research scientists and professional killers?

A dispassionate reader may ask whether soldiering is not an honorable profession, whether generals and admirals, resplendent in their costumes under television lights, are not simply one form of specialized manager. There are many precedents for military conquerers acting as regents or functionaries, from the times of Alexander the Macedonian and Wang Mang to Francisco Franco and Dwight David Eisenhower. Indeed, the training for military command has often consisted of a large dose of polite culture. The *epheboi* of Hellenic antiquity and the great schools of Han China both produced gentlemen who became in due course generals. For such commanders, ordering an army to execute a flanking maneuver was of a piece with undertaking an irrigation project or hunting down wild animals. Military leaders have been schooled as gentlemen down to the middle of the twentieth century. Today's soldier-administrator is certainly heir to the ruling-class customs in diverse societies, but his authority derives from more than familiarization with the classics or drawing-room etiquette. Military authority derives from a decisive ability to kill.

The West has invented a vocabulary to insulate soldiers from the havoc of war. Soldiers are enjoined against becoming vandals, thugs, assassins, or hooligans – denying to Germanic, South-Asian, Arabic, and Gaelic peoples the dignity of honorable murder. Soldiers are warned not to go berserk or run amok. They are encouraged to imagine that they serve a nation, not merely a warlord commander. In attaining a goal by the invocation of death, however, soldiers inevitably wreak indiscriminate horror. Despite injunctions against torture or rape when enslaving or "liberating" a subject population, soldiers generally effect violence of this kind. From their ability to extinguish life – immediately and without compromise – military personnel have become sanitized in our time as "specialists on violence."[2]

The association between killers and creators apparently begins with practicality. Unlike university professors or church deacons who need do nothing

P. Forman and J. M. Sánchez-Ron (eds.), National Military Establishments and the Advancement of Science and Technology. Studies in the 20th Century History, 135–152.
© 1996 Kluwer Academic Publishers. Printed in the Netherlands.

more than mumble platitudes to maintain their sinecure, military officers must lead men-at-arms to control society at large. The respect leading to victory in battle required, according to Carl Philipp Gottlieb von Clausewitz, both power and knowledge. In his view, the knowledge was of a special kind. It depended more on the *doing* that "cannot properly stand in any book" than on the *knowing* that can be displayed in print. For this reason, war was unlike physics. "Where the object is creation and production, there is the province of Art; where the object is investigation and knowledge Science holds sway. After all this it results of itself that it is more fitting to say Art of War than Science of War." Clausewitz went further. War could be seen as an extension of business activity:

We say therefore War belongs not to the province of Arts and Sciences, but to the province of social life. It is a conflict of great interests which is settled by bloodshed, and only in that is it different from others. It would be better, instead of comparing it with any Art, to liken it to business competition, which is also a conflict of human interests and activities; and it is still more like State policy, which again, on its part, may be looked upon as a kind of business competition on a great scale.[3]

The modern army in fact emerged from the premier business community of the seventeenth century, the Netherlands. Forged by Mauritz of Nassau, the Dutch army emphasized continual drill to routinize the mechanics and choreography of armed assault, constant exertion in the way of constructing fortifications to lessen the possibility of mutiny, and modular articulation in the vertical file of battalion, company, and platoon to insure coordinated action. "The creation of such a New Leviathan," William H. McNeill has emphasized, "was certainly one of the major achievements of the seventeenth century, as remarkable in its way as the birth of modern science." The "military-commercial complex" that underlay the modern army, in McNeill's view, was a signal characteristic of post-Renaissance Europe.[4]

Permanent military institutions dispensed knowledge and radiated power. Engineering (the Dutch uncle of physics) was, until scarcely six generations ago, the province of military and naval garrisons; we retain this memory in our designation of "civil engineer" for a non-military bridge-and-road builder. Surgery, the Western skill that astonished Tokugawa Japan, was a military specialty requiring technical dexterity and speed as well as detailed anatomical knowledge; professional standing armies depended on such expertise, as it was easier to patch up a seasoned soldier than to train a new one. The military ran engineering and surgical schools where the standards of instruction equalled or exceeded those in the universities. Through the nineteenth century the finest of scientists – Poincaré, Helmholtz, Michelson – received years of systematic instruction at institutions financed by professional killers.[5]

North American academic writers, schooled in the Wilhelmian virtues of secular *Wissenschaft*, have expressed surprise at the military regency over scientific research during the past half century.[6] This domination seems unusual only because it followed the cometary passage of the research university. The ascendancy of research universities, it may be emphasized, is a short-term historical phenomenon, existing in Germany for about one hundred years (roughly, 1830–1930) and in the United States for half that span (1890–1940).[7] The research university never quite succeeded in Russia, France, or Spain, not to mention much of Latin America, Africa, and Asia, and by the second third of the twentieth century, major research installations everywhere had quit the groves of academia for the *campus martius*.[8]

In the following pages, I direct attention to the persistent French military presence in the exact sciences. I argue for a revision of Robert Gilpin's evaluation that "a great cleavage has long existed between the academic world and the military establishment in France."[9] I submit that the French model of the military-scientific complex is present in much of physics and astronomy today.[10] The commonly perceived attenuation of French performance in the exact sciences over the past six generations suggests that while the military mind may excel at commanding grand scientific projects, it slowly stifles certain kinds of creative work.[11]

The perspective offered here is not easy to reconcile with the image of France as an open and tolerant nation, a tribe of free-thinking literary and artistic innovators. For this reason, it is well to emphasize that various artistic disciplines, no less than scientific ones, follow the letter of special social contracts. The notion of an artistic avant-garde resembles the picture of researchers pushing back the frontiers of learning, and there have certainly been golden ages characterized by literary, artistic, and scientific achievement in a national setting. It is nevertheless presumptuous to imagine that all the arts and all the sciences necessarily flourish or fall together.

At issue may not be the creative residues of men and women, but rather the very concept of a nation and the equivocal invention of national "character." Novelist Robert Musil, writing shortly after the First World War, cautioned that the nation was a fantasy. It "was not to be found either as race or in the form of the state, but these are in fact the two places where it has been sought." What people called a nation was a vague and contradictory spirit:

In the course of an extraordinarily confusing historical period, millions of individuals have stuck their heads into a world that they understand to very different degrees and in very different ways, from which they want completely different things, of which they see little more than the threat of their own livelihood and hear some great general noise, in which here and there something echoes that makes them prick up their ears. This enormous, heterogeneous mass, on which nothing can quite make an impression, which cannot quite express itself, whose composition changes every day as much as the stimuli that act upon it, this mass, non-mass, that oscillates between solid and fluid, this nothing without firm feelings, ideas, or resolution, is, if not itself the nation, at least the really sustaining substance of the nation's life.[12]

National needs are nevertheless articulated, and often invented, by the state, which possesses the means of punishing dissidence. Nations have indeed been given life by the state through force of arms.[13] In such circumstances, the military (and its manifestation in state-security services) will naturally seek to discipline and direct the nation's cultural productions.

Among national monuments celebrating military conquest are memorials to learning, and monstrous destroyers have included savants in their retinue. Napoleonic France, and especially its Egyptian Institute, is instructive in this regard, even though Napoleon's patronage of science continued a tradition of the *ancien régime*.[14] There is a cognate bonapartist cast to General Gabriel García Moreno, the theocratic architect of the Quito Institute of Technology (one of nineteenth-century America's most distinguished scientific schools), just as there is to General Louis-Hubert-Gonzalve Lyautey, the French conqueror of Morocco and the patron of literature and science.[15]

How shall these developments be judged? George Sarton, who saw his native land twice conquered, wrote about Tamerlane, the thirteenth-century general who erected

pyramids of skulls to remind the people of his merciless "Schrecklichkeit." When the destruction was sufficiently complete to assure the foundation of a new order, Timur built himself an imperial city, Samarqand, and posed as a patron of science and arts. The scientific or artistic creations which may be credited to his patronage are not few, yet negligible as compared with the lives and goods which he had caused to be destroyed.[16]

Tamerlane's grandson Ulugh Beg, astronomer and astronomical Maecenas, worked with al-Kashi to produce original observations of the fixed stars and planets (not mere slavish transmission of Ptolemy's longitudes), which circulated in Europe during the sixteenth century. The Samarqand Observatory, rich in learning, derived from a military regime of monstrous evil, but Ulugh Beg did not chose his grandfather. The originality of al-Kashi's *zijes* is undiminished by Samarqand's repellent origin. Pure learning seems not to be related necessarily to patronal perversions – much less to purity of bloodline.[17]

With these considerations in mind we turn to consider the French military interest in exact sciences over the past two centuries.

SAILOR, STARGAZER

Between the end of the *ancien régime* and the Fourth Republic, astronomy in France was tied to practical surveys of the earth's surface. The connection provided an excuse for maintaining a central bureaucracy to supervise the exact sciences and a justification for placing talented apprentices in permanent research posts. Measuring the land, both at home and abroad, required military precision, and the military controlled France's finest scientific school, the Ecole polytechnique. The armed forces, as a result, became an alembic for refining expertise in astronomy and physics.

The meter, called into being by the Constituent Assembly in 1790–91, required for its instantiation a detailed survey of the terrestrial meridian passing from Dunkirk to Barcelona. The surveying work proceeded intermittently, finally propelling Jean-Baptiste Biot and Dominique François Jean Arago to the Island of Formentera in 1808. The arc they surveyed gave greater precision to the standard meter, and the surveying provided lustre to the Bureau of Longitudes, the independent astronomical office set up in 1795 to collect and publish observations and generally to oversee the French observatories.[18] The Bureau came to reside at the Paris Observatory, its highly paid staff – an assembly of astronomers, mathematicians, hydrographers, and geographers – charged with organizing astronomical missions and publishing an ephemeris. By the value of its emoluments and an accompanying freedom to undertake research of a general nature, a senior appointment to the Bureau was highly prized; and the nature of French learning guaranteed that the Bureau's senior staff would be members of the Academy of Sciences.[19]

The Bureau functioned as the astronomical research institute of the Paris Academy from its inception, and its hand figured in every major French astronomical expedition over the succeeding two centuries. Yet it operated no major observatories of its own – except for the portion of the Paris Observatory that it came to control by sufferance.[20] And although it projected a strong presence in distant parts during eclipse or planetary-transit expeditions, it spent very little, if anything at all, on permanent overseas installations. Furthermore, the Bureau was by its nature an interdisciplinary fellowship that transcended the mandate of its administrative superior, the Ministry of Public Instruction. Soldiers and sailors freely entered its ranks. Military and naval assistance was indeed crucial in mounting astronomical expeditions, but multiple allegiances discouraged the emergence of a scientific school that could claim to be attached to the Bureau. The Bureau's *modus operandi* facilitated the appointment of Admiral Amadeo Ernesto Bartolomé Mouchez as director of the Paris Observatory in 1878. Mouchez then set the course for French astronomy over the next two generations with an ambitious project to compile a photographic chart of the sky's two million brightest stars.[21]

A military regimen was precisely what the Paris Observatory required. Its director at mid-century had been Urbain-Jean-Joseph Le Verrier, perhaps the greatest of French astronomers and the worst of French administrators. He sought to cut off the observatory from its quasi-feudal past by transforming it into a state bureaucracy (the Cassini family and eventually Arago had run the observatory as a private fiefdom, and when Le Verrier inherited it, the staff were bound together by blood and marriage). The annals of early nineteenth-century French science, however, were written in individual accomplishment and discovery. Le Verrier's subordinates would not suffer anonymous submersion in his bureaucracy, and his tenure as director marked the most tempestuous years of French astronomy.[22]

Mouchez's claim to Le Verrier's mantle has nothing to do with the Ministry of Public Instruction (the observatory's source of funding) and everything to do with the Navy's scientific ambitions. Naval engineers were assimilated into a separate and well-paid corps in 1814. By mid-century, under the inspired leadership of Charles-François Beautemps-Beaupré, a member of both the Bureau of Longitudes and the Academy of Sciences, the elite corps was recruiting from graduates of the Ecole polytechnique. Among Beautemps-Beaupré's catches were Magloire-Thomas Daussy, who in 1832 became secretary-librarian of the Bureau of Longitudes' observational facilities, and who rose in 1855 to take Beautemps-Beaupré's position in the Academy of Sciences; and Louis Urbain Dortet de Tessan, who in 1861 took Daussy's chair at the Academy. Other naval hydrographers who supervised major overseas expeditions and enjoyed distinguished scientific careers include Jean Jacques Anatole Bouquet de la Grye, Philippe Eugène Hatt, Maurice Rollet de l'Isle, Henri Roussilhe, and Lazare-Eugène Fichot. The elite hydrographers numbered 73 over the century between 1814 and 1914.[23]

Scientific work – whether on astronomical expedition, with the Bureau of Longitudes' *pépinière* at Montsouris (organized by Mouchez), or in the Navy's own observatory at Toulon – was also confided to career naval officers outside the corps. Mouchez was in the latter category, which also included Octave Marie Gabriel Joachim de Bernardières, Edouard Jean Pierre Marie Sylvain Perrin, Georges Ernest Fleuriais, Jules-Jean Feillet, Victor François César Beuf, Jean Alexandre Justin Paul Eugène Fitte, Georges Roux, Charles Poisson, and Edmund de la Villemarqué. Complementing the corps of naval hydrographers, these men projected French scientific presence around the world.[24]

Naval engineers and regular officers constituted the single largest reservoir of astronomical talent in France at mid-century, a circumstance freely exploited by the Ministry of Public Instruction. In 1868, for example, the naval officer and mathematician Jean-Philippe-Ernest de Fauque de Jonquières urged the Ministry of Public Instruction to lend a spectroscope to Philippe Hatt, then charting the coast of Indochina, so that Hatt could observe the upcoming solar eclipse at the nearby island of Poulo Condore.[25] When France constituted expeditions to observe the transit of Venus in 1874, Mouchez and Bouquet de la Grye were seconded from the Navy to the Ministry of Public Instruction; and the Navy sent out four teams of its own to observe the 1882 transit.[26] This practice was followed in all French overseas astronomical expeditions during the nineteenth century.

Mouchez was the man of the astronomical hour. After graduating from the Ecole navale, he quickly distinguished himself as an astronomical calculator. He rose through the ranks to become *capitaine de vaisseau* in 1868, member of the Bureau of Longitudes in 1873, member of the Academy of Sciences in 1875, and vice-admiral in 1878. The veteran of numerous foreign tours of duty (Mouchez spoke Spanish fluently), he was the Navy's premier hydrographer and expeditionary. Installed as director, he spent nearly a decade renovating practical astronomy in Paris. Then he committed resources to an ambitious and far-reaching project.

The project turned on advances in photographical astronomy. By the early 1880s, David Gill at Cape Town had demonstrated the feasibility of using celestial photography to inventory the stars; in 1885 he began a map of the southern skies; he took the photographical plates, and Jacobus Cornelis Kapteyn, professor of astronomy at Utrecht, measured them. Mouchez had inherited a star-mapping project of his own, an inventory of the ecliptic started in the 1860s by the brothers Paul and Prosper Henry at the Paris Observatory. The Henrys abandoned their catalogue when they reached the profusion of the Milky Way. They began experimenting with photography. By the middle 1880s, with a photographical refractor of their own design, they discovered celestial objects that had escaped visual astronomers. Their achievements led Mouchez to correspond with overseas colleagues about the possibility of a photographical map of the heavens. Gill urged an international congress to discuss the proposal. Mouchez was only too happy to offer Paris as its site.[27] Here was a way to supersede all the star catalogues that had been produced or were under production by foreign rivals like Friedrich Wilhelm August Argelander at Bonn, Benjamin Apthorp Gould at Córdoba, and of course Gill himself.

The organizing conference of 1887 assembled more than 50 astronomers from nearly a score of countries. They divided the sky into portions assigned to various observatories; they planned for photographical illustrations (in 22000 sheets showing stars down to fourteenth magnitude) and tables of some two million brighter stars. Uniformity would be insured by using identical telescopes and ruled photographical plates. As the project was underwritten by the Paris Observatory, the Academy of Sciences covered the cost of publishing its memoirs and bulletins (although not, of course, all parts of the catalogue). A permanent commission came into being to supervise the project, assign work schedules, direct special studies, and generally handle problems as they arose.[28]

Within several years, the commission established standards and conventions – even for the thorny issue of stellar magnitude.[29] The authorized telescope would conform to the 34 cm refractor designed by the Henrys and built by Gautier in Paris. Photographical plates would be delivered by Lumière Frères in Lyons. Stellar zones went to eighteen participating observatories. The commission had no money of its own to distribute, however. Each participating observatory had to find its own funding. United States astronomers abstained from committing any of their impressive resources to the project (and the Germans committed theirs reluctantly), but Mouchez nevertheless believed that he would be able to set the enterprise on a firm path. French projection overseas – in which the navy played a large part – confirmed his faith.

Among the eighteen observatories taking slices of the sky, three were on French soil: Paris, Bordeaux, and Toulouse. Another, Algiers, was in France's major overseas appendage. France effectively controlled two observatories in the Southern Hemisphere, Santiago de Chile and La Plata in Argentina, as French astronomers had recently arrived to take charge of each operation. A third South American observatory, that at Rio de Janeiro, had been managed earlier by a refugee from Le Verrier's tyranny and then had passed into the hands of a native of Belgium. And the Mexican national observatory at Tacubaya was also Francophile. Each of the five overseas observatories would be harnessed to a long-term project, using French astronomical technology, for the greater glory of the imperial seat. To the constellation of eight observatories owing greater or lesser allegiance to Paris, Mouchez added in five from the English-speaking world – Greenwich, Oxford, Cape Town, Sydney, and Melbourne – which he thought could be disciplined from the British metropolis. The remaining partners in the project would be pulled along by inertia or national pride.

The tale unwound otherwise. Among the French overseas astronomical satrapies, only Algiers completed its task within a generation of starting it. Santiago de Chile, La Plata, and Rio de Janeiro withdrew well before the First World War. Tacubaya persisted, completing its assignment on the eve of the Second World War. Paris collaborated with the British empire to pick up most of the defaulted areas.

It has been argued that the burden of the project undermined the health of French astronomy by focusing attention on routine operations related to stellar position instead of on exciting new areas in astrophysics.[30] By 1920, with the French portion of the *carte du ciel* still incomplete, the French astronomical community felt fatigued and discouraged. The Academy of Sciences commissioned astrophysicist Henri Deslandres, director of the Meudon Observatory, to undertake a systematic and secret enquiry into the malaise. He concluded that France would do well set off on a new path by allocating half of its astronomical resources to astrophysics, following the pattern of astronomers in the United States.[31]

We must not imagine that the sky-chart work in itself impeded creative research. Stellar inventory underlay analyses of proper motion and stellar parallax – studies directly related to cosmological issues – and this is precisely how American and British observers used their catalogues. Furthermore, as Deslandres pointed out in his secret report, the French *hexagone* had ten major research observatories; nine belonged to the state, and one (Abbadia) to the Academy of Sciences. Of the six observatories not implicated in the star-chart project, Lyons and Nice had been created at the end of the nineteenth century and endowed with reasonable facilities; by the early twentieth century, Lyons concentrated on stellar photometry. Meudon, in greater Paris, comprised a large complex devoted exclusively to astrophysics. None of these observatories had enormous telescopes under ideal observing conditions, but the means were certainly available for significant, new undertakings.

The routine nature of positional astronomy, which absorbed the lion's share of funding at French observatories up to the middle of the twentieth century, was well suited to the social structure of astronomy in France. Observatories affiliated themselves only loosely with local universities, when, indeed, they had any university affiliation at all. Astronomers advanced in their career through apprenticeship, generally working their way up in the observatory hierarchy; to obtain a *doctorat ès sciences* in observational astronomy, one had to be accomplished in the art of observing, and this meant spending years in a junior and poorly paid assistantship. *Anciens élèves* of the Ecole normale supérieure such as Félix Tisserand, Benjamin Baillaud, Ernest Esclangon, and André Danjon did rise to astronomical heights in France, as did graduates of the Ecole polytechnique like Deslandres, but the recruitment of talented science students remained something of an accident.[32] Many astronomers found their calling in the Army or Navy, and others rose to positions of authority by dogged perseverance. The guild system in astronomy, however, is not special to France. It flourished in the British empire and in the Americas, giving way early in the twentieth century to the German and Dutch model of academically trained astronomers. France, ironically, preceded the English-speaking world in requiring the doctorate of senior observatory staff.[33]

Nor can the attenuation of the astronomical discipline be the cause of the insecurity endemic to French astronomers. It is true that no clear path led to excellence in the field. One could rise to the top by starting in the Army, the Navy, the provincial universities and *lycées*, or the observatories, but in each kind of institution a prospective astronomer had to navigate around numerous obstacles and spend a great deal of time on things unrelated to the stars. And if one did discover new stars or mathematical equations, there was no respected, disciplinary journal to receive the discovery – the *Bulletin* of the French Astronomical Society being a house-organ appealing in large measure to amateurs. But with one exception – the strong military presence – these circumstances also applied to Dutch astronomers during the nineteenth century, a community that rose to world prominence in the 1920s.

The Dutch comparison is instructive, for in the Netherlands university professors freely collaborated with the Navy on a principled basis.[34] Tedium and sterility in France derived from rigid comportment and inward focus. With the pervasive presence of military discipline and with all eyes turned toward Paris, few astronomers took the trouble to keep abreast of publications in German and English periodicals. Only with the political convulsions of the twentieth century did foreign-educated astronomers find their way to French observatories – Maurice Loewy and Milan Ratislav Stefánik being exceptional, earlier immigrants. Here the contrast with the Netherlands could not have been greater. Most Dutch astronomers would have been literate, if not fluent, in German, English, and French; Dutch astronomers prized individual initiative and outward vision. By the late nineteenth century, French astronomers withdrew into collective work that contributed to the glory of institutions in Paris – not the least of which were the Ministries of War and the Navy.

MAPMAKER, SIGNALMAN, WEATHERMAN

Just as the Navy provided careers for nineteenth-century French scientists, so the Army provided special employment for scientifically inclined officers. The Navy needed charts, the army needed maps. It assigned surveying to its Dépôt de guerre. Maps issued in a continuing stream across the first three quarters of the nineteenth century, both for overseas territory and for the *hexagone*, but the army surveyors enjoyed nothing approaching the cachet of their naval counterparts. François Perrier, a senior military surveyor, set out to rectify the situation. In 1881 he succeeded in assigning geographical work to a section under the Army's Etat-major, which he captained. Six years later the Dépôt was abolished in favor of a geographical service directed by General Perrier. Under Perrier's guidance, the Army began to undertake gravimetric surveys; these continued with Perrier's successors, although the service focused increasingly on its North African possessions, linking them by topographical triangulation to French meridians.[35]

Wherever the French Army camped, it sent a team of topographers to chart the surrounding terrain. Sometimes the military geographical services received orders from the council of a local regent (Tunisia), a governor general (Indochina and Madagascar), or an occupying army (Syria), but even in cases where civilians came to carry out the surveying (Morocco, Algeria, and Tunisia), military norms were omnipresent. Military geographers attempted to engage projects relating to pure science. The most ambitious of their projects – an early twentieth-century remeasuring of the eighteenth-century arc of longitude in Ecuador – did little more than advance the military career of those, such as Robert Bourgeois and Georges Perrier (son of François Perrier), who slogged over South American hill and dale.[36] The Ecuadorian experience soured the geographical service on geophysics and astronomy. Bourgeois, who had commanded the expedition to Ecuador, rose to direct the service in 1911. He began the First World War with 30 geodesists in his service; by war's end, 12 had been killed, 3 had died through accident or disease, and a large number had retired or been transferred; of the pre-war staff, only Georges Perrier remained.[37] During the interwar years, Perrier labored in vain to reconstitute a geophysical program for the military surveyors. Scientific progress eroded his claim to head the Army's scientific elite.

Whereas in the nineteenth century the military surveyors had pioneered construction of portable telescopes and analysis of astronomical data, advances in technology lay at the root of the diminished scientific status of the geographical service. Two related developments reduced the geographers to little more than mapmakers: aviation and wireless. Aerial photography, which came to subtend mapping, depended on collaboration with the independent-minded air corps; furthermore, as we shall see, the corps's National Meteorological Office, which provided weather reports for airports, freely implicated itself in geophysical projects. Closely tied to aviation was communications technology. Wireless, which provided the precise time signals that were essential for mapping, was the domain of the Army's communications branch, whose resident expert was Gustave-Auguste Ferrié. He became the Army's senior scientist.

After having graduated from the Ecole polytechnique and the Ecole d'application at Fontainebleau, Ferrié served as a lieutenant at Grenoble and Besançon. Soon after he made captain in 1897, he was transferred to the central office of military telegraphy. In 1898 he began experimenting with Hertzian waves, and the following year he assisted Guglielmo Marconi in sending wireless signals between England and France. The Ministry of War lost no time in having Ferrié organize a military radiotelegraphic service; over the next 30 years Ferrié directed the service's expanding fortunes. He wrote technical manuals, installed radios on airplanes and balloons, established radio communication with French colonies abroad, and lectured tirelessly to popularize his medium. At Ferrié's insistence, radio played a part in France's invasion of Morocco in 1908, and at his command the Eiffel Tower became a giant radio mast for broadcasting time signals. In close collaboration with astronomers at the Paris Observatory, he immediately made use of the time service to determine longitude differences in France and between Paris and Washington. Appointed lieutenant colonel in 1914, he set up a scientific research center that during the war employed a number of talented physicists, among whom were Henry Abraham, Eugène and Léon Bloch, Léon Brillouin, Marius Latour, Lucien Lévy, Edmond Rothé, and Ernest Labrouste. Colonel in 1915, brigadier in 1919, General Ferrié received the Osiris prize of the Academy of Sciences (FF 100000) and then took Alfred Grandidier's chair there in 1922. By this time he belonged to every major French committee exercising authority in the exact sciences.[38]

As Daniel Berthelot's successor in presiding over the section of terrestrial magnetism and atmospheric electricity in the Comité français de géodésie et de géophysique, Ferrié exercised discretionary power over funding. He corresponded with Charles Maurain – France's premier geophysicist and dean of the Sorbonne's Faculty of Sciences – about most geophysical mapping projects in the 1920s.[39] When the *Annales de l'Institut de physique du globe et du Bureau central de magnétisme terrestre* strained Maurain's means (in 1927 the budget for his institute and its three dependent observatories at St-Maur, Val-Joyeux, and Nantes was FF 52000), he enlisted Ferrié's support to

ask the Caisse des recherches scientifiques for a special grant of FF 20000
to cover the cost of publication; the grounds for involving Ferrié concerned
Maurain's project to compile and publish a new magnetical map of France.[40]
Maurain and Ferrié's schemes also implicated French physicists, such as
Daniel Berthelot, Louis Dunoyer, and Aimé Cotton. The Comité français de
physique actually carried a budgetary attribution, late in the 1920s, for magnetical surveying – the proper ownership of which occasioned complicated
discussions between the geophysicists and the physicists.[41]

Maurain and Ferrié worked together for nearly 20 years on questions
relating to aviation and radio, and from their mutual interests they formed
common cause against outsiders who might threaten their authority. Notable
rival outsiders were military geophysicists in the French Air Corps. Among
kinds of warrior there is no solidarity.

The Air Corps retained the weather as a prize of the First World War.
Before then, meteorology (and related global phenomena such as magnetical and seismological disturbances) had passed from the astronomers to a
Central Meteorological Bureau in Paris. The Central Bureau, under Eleuthère
Elie Nicolas Mascart and his successor Alfred Angot, recorded data world-
wide by leaning heavily on consular agents and military physicians, the latter
of whose meteorological interest outlived the early nineteenth-century mias-
matic theory of disease.[42]

With the appearance of trench warfare, poison-gas attacks, and military
aviation during the First World War, meteorology came to interest the high
command. After the disastrous Champagne offensive in 1915, the Army had
the director of its Geographical Service, General Robert Bourgeois, take
effective control of the Central Meteorological Bureau and its network of
observing stations. The Army also established a Meteorological Service of
its own in the Air Corps under *lieutenant de vaisseau* Jules Rouch, a graduate
of the Ecole navale who had been charged with oceanographical, meteorological, and magnetical measurements on the second Antarctic expedition
of Jean-Baptiste-Etienne-Auguste Charcot. In 1917 Jules Rouch took charge
of the Navy's meteorological service, the army's being reorganized under
Colonel Emile Delcambre.[43] In 1918 Bourgeois assimilated the Army's meteorological service into his civilian network.[44]

The weather was an enemy that would sign no armistice. Jules Rouch wrote
during the summer of 1918:

To use a comparison from current affairs, the synoptic map, showing us the disposition of
atmospheric forces, is analogous to a map giving the exact position of enemy forces. The latter
document would not perhaps reveal the true intentions of the enemy in a precise way; it would
not always allow predicting with certainty the succession of events. But where is the general
who, having it in his possession, would neglect to consult it.[45]

To defeat this inhuman foe – to predict its next move – required a continuing,
military effort in synoptic meteorology. Alfred Angot's retirement in 1920

allowed the military to keep in peace what it had gained in war, albeit with a few administrative modifications. To the new Office national météorologique, directed by Delcambre and placed with the Undersecretariat of the Air in the Ministry of the Interior, went not only Angot's bureau but also the Meteorological Service of Air Navigation and the Central Services of Military Aviation.[46]

Colonel (subsequently, General) Delcambre watched over French weather everywhere. His rise, and that of his office, derived directly from the remarkable growth of military and civilian aviation, especially overseas. In the late 1920s Delcambre tried to dominate meteorology in North Africa and the Middle East; he succeeded in placing agents at the head of Algerian and Lebanese weather forecasting.[47] Delcambre's successor Philippe Wehrlé sent an expedition deep into the Algerian Hoggar for the International Geophysical Year of 1932/33.[48] And by the 1930s weather forecasting for air transport had come to absorb much of the attention of geophysicists in Indochina.[49] The French National Meteorological Office was not the only force behind these developments, but it radiated broad influence because its department, the Air Corps, evolved into an independent ministry in 1928.

A MEASURE OF MILITARIZATION

The French Army and Navy underwrote a large part of astronomy, geophysics, and meteorology through the fall of the the Third Republic. Professional killers did more than simply authorize expenditures. They had a determining voice in matters of scientific discourse by their presence in the Academy of Sciences. The Academy's structure reserved seats of honor for them.

During the nineteenth and twentieth centuries, the Academy had two divisions. The first division, *sciences mathématiques*, featured sections devoted to geometry, mechanics, astronomy, geography and navigation, and general physics; the second division, *sciences physiques*, comprised sections of chemistry, mineralogy, botany, rural economy, anatomy and zoology, and medicine and surgery. Each of these eleven sections had six members. In addition, there was a division of ten *académiciens libres* and, in the twentieth century, six-member sections for non-residents and for industrialist scientists.

In the first division (the exact sciences), the section of geography and navigation was the domain of military and naval engineers. The last non-military member of the section, Alfred Grandidier, was replaced by General Gustave-Auguste Ferrié after the First World War. To the warriors' *chasse gardée* that constituted one-sixth of the Academy's exact-sciences division must be added men admitted to the other sections or divisions as tribute to their military activity.[50] As nearly all the latter academicians were observers of the physical world rather than speculators about ideal mathematical realms (the mathematician Vice-admiral Jean-Philippe-Ernest de Fauque de Jonquières is a notable exception),[51] warriors exercised a strong voice – and sometimes a stentorian one – in the Academy's deliberations about physics and astronomy.[52]

Let us now rephrase the question raised at the outset. If military penetration into research universities during the middle of the twentieth century reflects the desire of generals and admirals to reassert their authority as custodians of nature and interpretors of natural law, whence the original inspiration? Why should warriors have a professional interest in abstract knowledge?

Clausewitz provides part of the answer: Science may provide technology for military use. To remain connected to the source of such potential gain, generals and admirals keep scientists on retainer.[53] The observation is insufficient, however, to explain an acquiescence to supervise extraordinarily expensive projects of little practical import – such as measuring an arc of longitude in Ecuador, sending dozens of tons of scientific equipment into the southern Sahara, or placing a man on the moon. The complete answer must go beyond matériel.

The evidence impels us to seek a part of our answer in a notion no more abstract than that of "potential utility." We must invoke the prestige of pure learning. From this perspective, the patronage of science is indulgence money – not to say wergild – for professional killers, who would like to persuade credulous onlookers that an army and a navy exist to preserve and strengthen the very institutions they destroy during the greatest horror humanity can experience. Chaperoning an eclipse expedition or measuring atmospheric ozone – morally unexceptionable projects undertaken with extensive military support – offer an inverted and entirely distorted image of military rule. These kinds of undertaking serve to mask the terror that reigns in times of war. It is just this "prestige racing" (to borrow Walter A. McDougall's expression) through the support of disinterested research which served to enhance American military fortunes during the disastrous Indochinese war of the 1960s.[54] Viewing the Apollo moon landings on television, or hearing a military concert-band, we are lulled to feel against all reason (as some astronomers at the observatory of Ulugh Beg might have felt) that war is the most effective cultivator of science, art, and "the best fruits of human nature."[55]

University of Louisiana, Lafayette

NOTES

* This article is based on my book, *Civilizing Mission: Exact Sciences and French Overseas Expansion, 1830–1940* (Baltimore/London, 1993); permission for use is acknowledged from The Johns Hopkins University Press. The following abbreviations are used in the notes: *Astron.*: Philippe Véron, *Astronomes français 1850–1950* (MS.; St-Michel l'Observatoire, 1989); *CRAS*: Paris, Académie des sciences, *Comptes-Rendus*; *DSB*: *Dictionary of Scientific Biography*, Gillispie, C.C. (ed), (New York, 1970–80); *DBF*: *Dictionnaire de biographie française*,

Baiteau, J., Barreux, M., Prevost, M., *et al.* (eds), (Paris, 1933+); PAN: Paris, Archives nationales; PIPG: Paris, Institut de physique du globe, Archives; VM: Vincennes, Archives de la marine, "Etat-Major."

[1] Forman, 1987. "Behind Quantum Electronics: National Security as Basis for Physical Research in the United States, 1940–1960," *Historical Studies in the Physical and Biological Sciences, 18,* 149–229; Sánchez Ron, J.M., 1992. *El Poder de la ciencia* (Madrid), pp. 197–374.

[2] Lasswell, H.D., 1941. "The Garrison State," *American Journal of Sociology, 46,* 455–68, for "specialists on violence." The term seems better as a description of private mafias. Paul Forman kindly drew this article to my attention.

[3] Clausewitz. Transl. Graham, J.J., 1940. *On War* (London), *1,* 117–21.

[4] McNeill, 1982. *The Pursuit of Power: Technology, Armed Force, and Society since A.D. 1000* (Chicago), pp. 133, 117.

[5] On the early nineteenth-century Japanese reaction to European surgery: Vos, K., 1989. *Assignment Japan: Von Siebold, Pioneer and Collector* (The Hague).

[6] The recent literature on the American military-industrial complex, and its connection with the academic world, is largely silent about nineteenth-century antecedents and French precedent, to be discussed shortly. No word is given to these issues in a recent symposium: Mendelsohn, E., Smith, M.R. and Weingart, P., 1988. "Science and the Military: Setting the Problem," in their edited volume, *Science, Technology and the Military* (Dordrecht), pp. xi–xxix. For the army as a technological innovator: Smith, M.R. (ed), 1985. *Military Enterprise and Technological Change: Perspectives on the American Experience* (Cambridge, Mass.), especially the historiographical remarks by the editor (pp. 1–37).

[7] A perceptive account of the rise of the American research university is Nathan Reingold, "Graduate School and Doctoral Degree: European Models and American Realities," in: Reingold and Rothenberg, M. (eds), 1987. *Scientific Colonialism: A Cross-Cultural Comparison* (Washington), pp. 129–49. The research university rose dramatically between 1880 and 1900, just at the time that the military and naval slice of federal science funding fell from 27% to 7%. Reingold, N. and Bodansky, J.N., 1985. "The Sciences, 1850–1900: A North Atlantic Perspective," *Biological Bulletin, 168* supplement: *The Naples Zoological Station and the Marine Biological Laboratory: One Hundred Years of Biology,* pp. 44–61, on pp. 53–4.

[8] Japan, where the Samurai ethic pervaded Western science late in the nineteenth century, is a notable exception to this (and other) twentieth-century generalizations. Nakayama, S. Transl. Dusenbury, J., 1984. *Academic and Scientific Traditions in China, Japan, and the West* (Tokyo), pp. 205–6.

[9] Gilpin, 1968. *France in the Age of the Scientific State* (Princeton), p. 258.

[10] To identify such commerce between learned professors and professional killers is not to assert that scientific *discourse* – what appeared in research publications – derived from military needs. I do not dispute the contention of Bernhelm Booss and Jens Høyrup, who see little connection between the discourse of mathematical science and military practice in nineteenth-century France. Booss and Høyrup, 1984. *Von Mathematik und Krieg* (Marburg), appearing in preliminary form as *Tekst* N° 64 (1983) of the Institut for Studiet af Matematik og Fysik samt deres Funktioner i Undervisning, Forskning og Anvendelser, Roskilde; Høyrup, 1994. *In Measure, Number and Weight* (Albany) has an English translation.

[11] Science and the military in Revolutionary France have recently been illuminated in an exhaustive study about ordnance by Gillispie, C.C., 1992. "Science and Secret Weapons Development in Revolutionary France, 1794–1804: A Documentary History," *Historical Studies in the Physical and Biological Sciences, 23,* 35–152.

[12] Musil, R., 1990. *Precision and Soul: Essays and Addresses,* ed. and trans. by Pike, B. and Luft, D.S. (Chicago), pp. 110–111.

[13] Crawford, E., 1992. *Nationalism and Internationalism in Science, 1880–1930: Four Stud-*

ies of the Nobel Population (Cambridge), p. 31, for Józef Pilsudski's contention that the state invents the nation, not the reverse.

[14] Gillispie, C.C., 1989. "Scientific Aspects of the French Egyptian Expedition, 1798–1801," *Proceedings of the American Philosophical Society, 133*, 447–74.

[15] García Moreno and Lyautey are treated in my *Civilizing Mission: Exact Sciences and French Overseas Expansion, 1830–1940* (Baltimore, 1993), pp. 141–44, 310–13.

[16] Sarton, 1948. *Introduction to the History of Science, 3: Science and Learning in the Fourteenth Century* (Washington), p. 1101.

[17] Even in post-medieval Spain, which "made a fetish of purity of blood – *limpieza de sangre,*" knowledge seems not to have been keyed rigorously to the family origin of an author. Lanning, J.T., in TePaske J.J. (ed), 1995. *The Royal Protomedicato: The Regulation of the Medical Professions in the Spanish Empire*, (Durham, N.C.), pp. 175–200.

[18] Bassot, L., 1891. *La Géodésie française* (Paris), *separatum*, p. 7. Zupko, R.E., 1990. *Revolution in Measurement: Western European Weights and Measures Since the Age of Science* (Philadelphia), pp. 137–75, on the meter and its adoption in France; Heilbron, J.L. "The Measure of Enlightenment," in: Frängsmyr, T., Heilbron, J.L. and Rider, R.E. (eds), 1990. *The Quantifying Spirit in the Eighteenth Century* (Berkeley), pp. 207–42

[19] Bigourdan, G., 1928. "Le Bureau des longitudes: Son histoire et ses travaux, de l'origine (1795) à ce jour," Bureau des longitudes, *Annuaire pour l'an 1928* (Paris), pp. A.1 – A.72, continued in each of the next five years of the *Annuaire: 1929*, pp. C.1–C.92; *1930*, pp. A.1–A.110; *1931*, pp. A.1–A.151; *1932*, pp. A.1–A.117; *1933*, pp. A.1–A.91. Crosland, M., 1967. *The Society of Arcueil: A View of French Science at the Time of Napoleon I* (Cambridge, Mass.), pp. 209–13.

[20] According to Bigourdan, "Bureau des longitudes"[19], over the first half of the nineteenth century, the distinction was blurred between Bureau initiatives and Observatory functions. The two institutions assumed an independent character with the formation of a civil service at the observatory.

[21] *DSB* and VM, "Mouchez," *s. v.* Mouchez was born in Madrid, and according to the VM his birth and marriage registrations bear the Spanish Christian names.

[22] *DSB*, "Le Verrier," *s. v.*

[23] Rollet de l'Isle, M., 1950. *Etude historique sur les ingénieurs hydrographes et le Service hydrographique de la Marine (1814–1914)* [*Annales hydrographes, 1 bis*] pp. 3–41; *CRAS, 50*, (1860), v, gives Daussy's Christian name as Pierre.

[24] Feillet went to direct the naval school (and its associated observatory) at Santiago de Chile; Beuf received direction of the La Plata Observatory in Argentina; Fitte was called to direct the Phu-liên Observatory at Haiphong; Roux directed the Moroccan meteorological and geophysical service. After they retired from the Navy, Poisson received direction of the Ambohidempona Observatory on Madagascar and de la Villemarqué the direction of the Zô-Sè Observatory near Shanghai.

[25] PAN, F17/3008, E. de Jonquières to Minister of Public Instruction, 14 Mar 1868; *DSB*, "de Jonquières," *s. v.*, where, however, his Christian names are in an order different from that given in *CRAS, 130* (1900), 8.

[26] PAN, F17/2928, 1, Direction politique, Ministry of Foreign Affairs to Minister of Public Instruction (Jules Simon), 7 May 1873, for the seconding; VM, BB4/1342, Institut de France to Minister of the Navy, 9 May 1883, for the naval teams under Fleuriais, Hatt, Bouquet de la Grye, and de Bernardières.

[27] *DSB* and *Astron.*, "Mouchez," *s. v.*; *Astron.*, "Henry," *s. v.*; Rayet, G., 1887. "Notes sur l'histoire de la photographie astronomique," *Bulletin astronomique, 4*, 165–75, 262–72, 307–20, 344–60, 449–56, on pp. 249–53 for the Henrys and the triumphs of their celestial photography. The origins of the *carte du ciel* project in Lankford, J. "The Impact of Photography on Astronomy," in: Gingerich, O. (ed), 1984. *General History of Astronomy, 4: Astrophysics*

and *Twentieth-Century Astronomy to 1950, Part A* (Cambridge), pp. 16–39, on pp. 25–9.
[28] Lankford, "Impact of Photography"[27].
[29] Weaver, H.F., 1946. "The Development of Astronomical Photometry, Period III," *Popular Astronomy*, 54, 287–299, 339–51, 389–404, on pp. 291–4.
[30] Lankford, "Impact of Photography"[27], p. 32; *Astron.*, "Introduction," citing Jules Baillaud in 1936.
[31] Deslandres, "Rapport sur l'inspection des observatoires français décidée par l'ordre ministériel du 29 avril 1922." A copy of this document was kindly lent to me by Philippe Véron, St-Michel l'Observatoire.
[32] The pivotal role in science of an education at the Ecole normale supérieure and the Ecole polytechnique has been elaborated in many studies, for example: Zwerling, C. "The Emergence of the Ecole normale supérieure as a Center of Scientific Education in the Nineteenth Century," in: Fox, R. and Weisz, G. (eds), 1980. *The Organization of Science and Technology in France, 1808–1914* (Cambridge), pp. 31–60; Shinn, T., 1980. *Savoir scientifique et pouvoir social: L'Ecole polytechnique 1794–1914* (Paris).
[33] Véron, P. "Introduction," in *Astron.*, where 1910 is given as the date for requiring a doctorate of all senior observatory staff. Véron argues that astronomy in France sustained great damage from capricious recruitment practices.
[34] Pyenson, L., 1989. *Empire of Reason: Exact Sciences in Indonesia, 1840–1940* (Leiden), pp. 1–3, 181–3. On the interaction between the most important Dutch university observatory and the Dutch navy: Dekker, E., 1990. "Frederik Kaiser en zijn pogingen tot hervorming van 'het sterrekundig deel van onze zeevaart,'" *Tijdschrift voor de geschiedenis der geneeskunde, natuurwetenschappen, wiskunde en techniek, 13*, 23–41.
[35] France, Ministry of National Defense and War, *Le Service géographique de l'armée: Son histoire, son organisation, ses travaux* (Paris, 1938), under the supervision of Géraud Maurice Marie Joseph de Fontanges, pp. 35–62. For a survey of an earlier period: Konvitz, J., 1985. *Cartography in France, 1660–1848: Science, Engineering, and Statecraft* (Chicago).
[36] *Service géographique* [35], pp. 63–120.
[37] Perrier, G., 1929. "Les Travaux actuels de la section de géodésie du Service géographique de l'armée française," in Prague, Masaryk Akademy, Section of Natural History and Medicine, *Publication scientifique*, N° 47 (Prague), 35–46, on p. 37.
[38] "Travaux et publications du Général Ferrié," *La Météorologie, 76* (1933), 355–62; Brenot, P., "Le Général Ferrié," *ibid.*, 363–7; Bourgeois, R., "Le Général Ferrié," *ibid.*, 373–6; Belhague, "Paroles prononcées aux obsèques du Général Ferrié," *ibid.*, 376–80; Fabry, C., "Gustave Ferrié," *ibid.*, 408–11; Ocagne, M. d' "Le Général Ferrié," *ibid.*, 411–13; Jouast, R., "Le Général Ferrié," *ibid.*, 114–21.
[39] PIPG. Maurian to Ferrié, 18 Feb 1929, for example, where Maurain discusses the distribution of certain funds.
[40] PIPG. Maurain to Ferrié, 2 Jul 1927 and appended proposal of 1 Jul 1927.
[41] PIPG. Maurain to Ferrié, 8 Dec 1927.
[42] A recent, general account in Fierro, A., 1991. *Histoire de la météorologie* (Paris).
[43] Maurain, C., 1950. *La Météorologie et ses applications* (Paris), p. 18; *DBF*, "Delcambre," *s. v.*; "La Carrière du Commandant Rouch," Les Amis du Musée océanographique de Monaco, *Bulletin trimestriel*, N° 41 (1957), 5–20. Rouch paints a vivid picture of his wartime meteorological exploits – where science was not much in evidence – in "L'Offensive du 16 avril 1917," L'Amicale des Ecoles de perfectionnement de la marine, *Bulletin*, 30 Jun 1936, 8 pp.
[44] Rouch, J., 1921. "Historique du Service météorologique aux armées (mai 1915 – février 1918)," *Revue maritime*, pp. 167–80; Schereschewsky, P., 1920. "La Météorologie militaire pendant la guerre: Historique sommaire," *L'Aéronautique, 1*, 531–5; Vincennes, Service historique de l'Armée de la terre, 5N 263, "Service météorologique," Bourgeois to Minister of War, 8 Jan 1918.

[45] Rouch, J., 1918. "La Prévision du temps," *Revue scientifique*, *56*, pp. 391–400, on p. 400. In 1918–19 the Norwegians came up with a new cyclonic model featuring the "front"–inspired by trench warfare. Friedman, R.M., 1989. *Appropriating the Weather: Vilhelm Bjerknes and the Construction of a Modern Meteorology* (Ithaca), pp. 122–37. The earliest use of the word "front" in meteorology, however, came from the geophysicist Ernst August Ansel in his doctoral dissertation of 1913 at the University of Göttingen. Krgian, A.K. Transl. Hardin, R., 1970. *Meteorology: A Historical Survey* (Jerusalem), p. 212; Pyenson, L., 1985. *Cultural Imperialism and Exact Sciences: German Expansion Overseas, 1900–1930* (New York and Berne), pp. 86–7.

[46] Dettwiller, J., 1982. *Chronologie de quelques évènements météorologiques, en France et ailleurs* (Paris) [Direction de la météorologie, *Monographies*, N° 1], pp. 41–2; Maurain, C., 1921. "Répertoire de laboratoires français, suite (1)," Direction des recherches scientifiques et industrielles et des inventions, *Bulletin officiel*, 2, 569–75, on pp. 574–5; Fierro, *Météorologie* [42], pp. 241–6.

[47] Lasserre, A. "Historique de l'Institut de météorologie et de physique du globe de l'Algérie," in: Seltzer, P. (ed), 1946. *Le Climat de l'Algérie* (Algiers), vii–xxiii, on pp. xv–xvi; Combier, C., 1933. "La Climatologie de la Syrie et du Liban," *Revue de géographie physique et de géologie dynamique*, 6, 319–46; Combier, 1931. *Les Jésuites en Syrie 1831–1931, 8: L'Observatoire de Ksara et services de météorologie* (Paris), p. 25.

[48] Wehrlé, 1934. "La Participation de l'Office national météorologique à l'année polaire internationale 1932–33," *La Météorologie*, N° 107, 1–29.

[49] Bruzon, D.M.E., 1940. "Note sur le Service météorologique colonial français dans le Pacifique," Sixth Pacific Science Congress, *Proceedings* (Berkeley), *3*, 751–4, on p. 753.

[50] Warriors outside the section of geography and navigation: Admiral Mouchez in the first division's astronomy section, Vice-admiral Abel Aubert du Petit-Thouars, Vice-admiral Jean-Philippe-Ernest de Fauque de Jonquières, Marshall Ferdinand Foch, and Marshall Jean-Baptiste Philibert, *comte* de Vaillant, all as *académiciens libres*. Membership in the Academy of Sciences is given annually in the first number of the *CRAS*. My discussion is based on the years 1860, 1880, 1900, 1920, and 1939; it excludes foreign associates and *correspondants*.

[51] *DSB*, "de Jonquières," *s. v.*

[52] The point is absent in a general survey by Crosland, M., 1978. "The French Academy of Sciences in the Nineteenth Century," *Minerva*, *16*, 73–104.

[53] Forman, "Behind Quantum Electronics"[1] p. 220 for the techniques and the retainer in another context.

[54] McDougall, 1985. *The Heavens and the Earth: A Political History of the Space Age* (New York), p. 390 for prestige racing, p. 305 for the political significance of prestige.

[55] Maistre, J. de. In: Sorokin, P.A., 1928/1964. *Contemporary Sociological Theories* (New York), p. 350.

EDUARDO L. ORTIZ

ARMY AND SCIENCE IN ARGENTINA: 1850–1950

INTRODUCTION

For a considerable period of time the armed forces have been a most important element in the composition of power in Argentina. From a privileged position, the Army not only has influenced the development of politics and economics in Argentina but, directly or indirectly, it has also had a considerable impact on its cultural and scientific life. This paper is a brief historical survey of connections between the national objectives enunciated explicitly or implicitly by theoreticians of the Argentine armed forces and the path followed in that country by some branches of science and technology.[1]

In the first two decades of the nineteenth century the Army was already a user of some forms of that era's technology, but towards the 1870s science was also added to its recognized needs. In the last quarter of the century and particularly towards the turn of the century, the process of modernization and technification of the Argentine armed forces was supported on advances made previously by renovated scientific and technical departments of the national university. By the turn of the century the armed forces had a group of officers trained in engineering, mainly at the national universities, and a few abroad. Interesting changes began to take place in the Army and Navy then in areas related to technology and science. A large international scientific congress was organized to commemorate the centenary of Argentina's independent life; army and navy experts were sufficiently numerous and active for the congress to have specific sections on military and naval sciences. The outbreak of the First World War made the armed forces aware of their critical dependence on foreign supplies: a policy of import substitution began to attract the attention of technically trained Argentine officers. After the First World War Argentina's armed forces were flooded with armaments from abroad at sale price; they discouraged the development of local armament manufacture.

The period immediately following the First World War was one characterized by renovation and reform, when many new ideas on social, literary, artistic and scientific matters entered Argentina. There were also important changes in the political front. Towards the end of the 1920s the first signs of the changing world economic situation had reached Argentina. Its force brought the first coup d'état by the armed forces on September 6, 1930. In these years the Army began to break its dependence on the national universities for advanced technical training. This was attempted through the development of an advanced technological center within the Army itself. Aware of the

153

P. Forman and J. M. Sánchez-Ron (eds.), National Military Establishments and the Advancement of Science and Technology. Studies in the 20th Century History, 153–184.
© 1996 Kluwer Academic Publishers. Printed in the Netherlands.

strategic implications of industrial development, the Army began to express concern with the direction industrial development was taking in Argentina, particularly after the beginning of the Second World War. A second coup d'état followed in 1943; after the end of the war the armed forces began to consider the possibility of Army-sponsored scientific research in specific areas, such as nuclear physics, rocket propulsion, and telecommunications, which they regarded as essential to their activities. Adoption of policies along these lines had important consequences for the pattern of development of science in Argentina and for the interactions between its military and scientific communities.

EARLY USES OF SCIENCE BY THE ARGENTINE ARMED FORCES: THE NINETEENTH CENTURY

The Early Years

An interaction between the incipient scientific and military communities of Argentina can be detected from Colonial times, but it was with the Wars of Independence, in the second decade of the nineteenth century, that such coupling became more visible, more articulated and more interesting. Several armies organized to fight these wars used technology to an unprecedented degree. The most interesting example is that of José de San Martín's Ejército de los Andes, which took war against Spain across the Andes. Setting it up was the most elaborate logistic and technological project ever developed in that area up to 1814. Through it a serious attempt was made to transfer to Argentina strategies and technologies currently used by European armies in order to compete successfully with those of Spain.[2] On account of its achievements, this early attempt of army technification and of organization based on modern European standards left a deep mark on the conception of the armed forces in Argentina. This fact did not escape the attention of Bartolomé Mitre,[3] an historian of these campaigns and also a general of the Army and President of Argentina from 1862 to 1868. From that high office Mitre began a process of re-modernization of the Argentine armed forces.

The end of the Wars of Independence is marked by a definite effort to enhance the position of science in Argentina by relating it more closely to Europe through importation of foreign scientists. European scientists were invited to Buenos Aires to teach at a new university created there. However, from the early 1830s to the early 1850s Argentina lived through a repressive period, characterized by dictatorial rule and emigration of scholars,[4] in which cultural life in Argentina did not fulfill promises of development the earlier decades suggested. The adoption of European models was rejected as inappropriate to the country's needs.

Natural Sciences in Argentina: Darwinian Controversies and Military Campaigns

Both science and the Army began to be developed in Argentina in a structured fashion after the 1860s, imitating once again European models. This process accelerated in the following decade. In it a group of young natural scientists emerged in Buenos Aires who supported enthusiastically Darwin's scientific ideas.[5] In 1879 General Julio A. Roca conducted a military campaign, called Expedición al Desierto,[6] against Argentine Indians. It was Argentina's parallel to the North American conquest of the West. Its objective was to push the aboriginal inhabitants of Central-Southern Argentina further south and west, towards the Andes. It also was the prelude to a large immigration effort which resulted, up to the beginning of the First World War, in the addition to Argentina of some three million European settlers.

Roca was accompanied in his expedition by a group of natural scientists of German origin who were then teaching at the University of Cordoba. On their return they wrote an extensive and detailed report on the observations made and the material collected during the campaign; their report covered zoology, botany and geology.[7] In the introduction to that report, geologist Adolfo Doering, who helped the Army identify the terrain which should support the Indians' life, stated that: "General Roca's inspiration comes from the modern advances made by the science of war, which has ceased to be an art subjected to whims or martial instincts [of generals, having now] become subjected to fixed rules and the severe methodology of experimental science".[8] The campaign itself was developed "with the serene regularity of a physics experiment, organized in a laboratory to manifest known laws, rather than with the contingencies of an enterprise of war".[9] Even if the decisive role played by geology in the campaign may have been exaggerated in Doering's pronouncements, it cannot be denied that it played an important part and, above all, that it showed in a dramatic way to those in power the potential uses of scientific knowledge. No doubt that was a primary objective in Doering's mind.

By opening that large territory the military campaign had a profound effect, not only on the economy and future demography of Argentina, but also on the development of its sciences. In particular, the advance of natural sciences in Argentina was directly affected by the necessity to provide a detailed description of the new territories. The subsequent subdivision of land had an impact on the development of cartography, geodesy and observational astronomy, and therefore on applied mathematics. It also helped consolidate a small community[10] capable of handling and maintaining delicate instruments, from which experimental physics and astronomy benefited. The results of the campaign also had a large influence on the development of new scientific institutions in Argentina, particularly those concerned with descriptive sciences.

The Beginnings of Physics and of Modern Engineering in Argentina

Electrical telegraphy[11] was introduced in Argentina in the mid 1850s and was well developed by the time of Roca's expedition. He made extensive use of it in his campaign.[12] Telegraphy contributed to the early establishment in Argentina of one of the chapters of the then modern physical sciences: electricity. Another form of communication, the railways, was responsible for the reception of the theory of heat and thermodynamics. Engineer Manuel Bahía ran a special school for state telegraphers, for which he wrote the first comprehensive treatise on applied electricity published in that country.[13] Both Army and Navy became interested in this new tool and soon created a telegraphers' corps.

To set up an engineering school in Buenos Aires European teachers, Italian on this occasion, were contracted for a period of twenty years. Graduates first came out in 1870; among them we should mention Valentín Balbín,[14] who became the leading Argentine mathematician of the nineteenth century and played an important role in the development of science and engineering in his country. By the end of the 1870s most of the chairs at the engineering school of the university of Buenos Aires had already passed into the hands of young Argentines; some of them had received postgraduate training in Europe.

Modernization and Early Technification of the Armed Forces

It was also in the mid 1860's that the Army started a process of modernization stimulated by president Bartolomé Mitre.[15] At the time of the Desert campaign the Argentine Army was dominated by the Remington rifle and the carbine for cavalry. A series of internal and external conflicts which followed that campaign pressed the Army for the acquisition of new and more mechanically sophisticated arms. American Gatling machine guns had been used in the war with Paraguay[16] while Krupp retrocharge guns were incorporated in 1871.

In 1868 the author and educator Domingo Faustino Sarmiento began a six-year term as President. He brought an end to the war with Paraguay, and, among his many contributions to Argentina's cultural and technical infrastructure, in 1872 Sarmiento created a Military College[17] and a Naval School.[18]

A more centralized structure was gradually adopted in the organization of the Army which allowed for the creation of new specialized training schools within it. In December 1879, following the Desert Campaign, Roca's secretary to the expedition, Lieutenant Colonel Manuel José Olascoaga, became head of the new Military Cartographic Board.[19] In the same decade the Argentine

Navy consolidated its fleet with acquisitions made in Great Britain. In the Desert campaign the Navy played a support and logistic role. With its new ships the Navy started to move from the rivers and coasts into the Atlantic Ocean. This move provided stimulation for still more advanced technical support. Officers oriented to science and technology were in close contact with the small Argentine scientific community and shared with it some of its communication networks. Lieutenant Colonel José I. Garmendia published papers in *Anales Científicos Argentinos*[20] discussing the uses of artillery and automatic weapons in a modern war, with particular reference to their use in the Franco-Prussian war of 1870.

Consolidation of Physics and Engineering in Argentina

In 1872 an Argentine Scientific Society[21] was founded; in 1876 it began publication of *Anales*, which took over from the journal mentioned before. Concurrently with the presence of foreign teachers, mainly French, Spanish, German and Italian, the transfer of culture to Argentina was accelerated, from Colonial times, sending individuals for periods of study abroad. After 1870 a group of young university graduates began to extend this process to areas related to engineering.

Balbín was sent out for postgraduate training in engineering in England. There he also acquired a solid knowledge of advanced mathematics. After his return he helped to reform and update engineering and mathematical studies at his university. He introduced in Buenos Aires advanced mathematical theories as well as the new methods of grapho-geometrical computation. Jorge Duclout, an engineer of French origin trained at the Polytechnikum, Zürich, started teaching elasticity theory and the construction of metallic bridges with large spans. In the last decade of the century mathematical analysis began to be used fairly regularly in papers in *Anales* or in technical journals.

Impact of the Consolidation of Physics and Engineering on the Armed Forces

From the 1870's a number of Army officers were officially designated as engineers. A decade later young Army officers started to follow courses of lectures at the University of Buenos Aires, which became the main source of advanced engineering and science training for the armed forces. A new large gunpowder factory opened in Argentina in 1883 and in the following year an engineering section was added to the High Command[22]; it became the Military Engineering School[23] in 1886. Its teaching syllabus was coordinated with that of the Colegio Militar and the university engineering school[24] to take full advantage of the courses given there. There were definite benefits from this instruction; the syllabus of the military engineering school was updated and improved in 1893.

In the decade of the 1880s several civil engineers from the local university started to do technical work for the armed forces on a regular basis. Balbín worked for the Navy between 1888 and 1891[25]; he supervised dredging of a channel leading to the Navy's works on the banks of the Luján river.

From the mid 1890's border problems with Chile became more severe and there was a real possibility of war between the two countries. In those days Balbín published a treatise on military fortifications.[26] By the end of the decade Teobaldo J. Ricaldoni, a physicist and inventor, experimented on wireless telegraphy with the assistance of the workshops of the national telegraph company and the Navy. By 1900 he managed to make radio transmissions to a distance of 9 km. Ricaldoni advised the Navy on electrical equipment and on different systems of radio telegraphy offered by foreign firms. In the early 1890s he designed a submarine; a parliamentary committee discussed financing its construction in 1893 and, subsequently, the lower chamber voted a fairly large sum of money for it. However, the submarine was never built.

By the late 1890s the Army had in its ranks an interesting group of officers who had graduated in engineering at the university. A particularly large group of officers graduated at Buenos Aires University in the year 1903; they accounted for one in four graduates. This group included Agustín P. Justo, who later became War Minister and President of Argentina, and Enrique P. Mosconi, a passionate defender of his country's natural resources. The path of study abroad, opened by Balbín in the field of engineering, began to be adopted by the armed forces in the 1880s for specific fields of military or naval training. Pablo Ricchieri was sent to the Belgian military academy. However, the national university continued to be used by the armed forces for scientific and technical training.

In the 1880s the Navy started to widen its international relations sending cadets for further training abroad. In these years it moved from sail to steam propulsion and from wood to steel-plated ships. Considerable advances were made on the artillery of warships in these years, particularly following introduction of the torpedo. A torpedo school[27] was created in Argentina in 1883. Due to the requirements of this section of the Navy, technology developed fast, especially electricity and the chemistry of explosives.

In 1881 the State of Buenos Aires's government imported a French naval officer, Francisco Beuf, a former director of the Toulon naval observatory, as director of the Nautical School established near the new city of La Plata[28] in proximity and associated to a new Central Hydrographic Office,[29] which was beginning to update the charts drawn by Robert Fitzroy during the voyage of the *Beagle*. Beuf became director of both institutions, and soon after also of a new astronomical observatory.[30]

(All were attached to the Comandancia General de Marina, which became an independent Naval Ministry[31] in 1898.) In 1882 Argentina agreed to take part in an international astronomical network for the observation of the transit of Venus. In that connection a decision was taken to establish this second large astronomical observatory, in addition to the one in Córdoba, close to the new city of La Plata and hence to Buenos Aires, in the River Plate area, where the Navy then concentrated its efforts. Sarmiento, and other Argentine statesmen who followed him as presidents of Argentina, regarded the humid Pampas as the key to the future of Argentina; the defense of the River Plate, its natural access, was a major responsibility of the navy. Beuf was director of this Naval Observatory; a Navy officer was the second in command.[32]

In 1895 Guillermo Villanueva, a fellow student of Balbín, with twenty-five years of technical experience in design of railways, roads and harbours, was made Secretary of State for the Army and Navy.[33] This appointment shows the acute concern of government in these years for the technical advancement of the armed forces. During Villanueva's time in office his department advanced in several directions. The presence of Villanueva and of other scientists and engineers in top decision making positions of successive Argentine governments of the late nineteenth and early twentieth century suggests that the Argentine scientific community was acquiring a new social dimension and a more sophisticated structure. Societies and journals on science, engineering and industrial manufacturing created in that period had a national resonance for a considerable number of years. A number of national and Latin-American congresses started to be held in Argentina from precisely that period.

In the same years a number of Argentines went abroad to study new branches of engineering, in particular, electrical engineering. Villanueva attracted some of the best of them to the electricity departments of the Navy. Electricity was initially used to power spotlights to identify targets and also for lighting in warships, but soon was also applied to torpedo propulsion and to power machinery. The hydrographic office conducted studies related to sea and river navigation using delicate instruments; geomagnetic observations were started by the Navy in the far south of Argentina in the same period. Border conflicts with Chile, no doubt, were a factor in the Navy's move into the Atlantic and behind scientific explorations in Patagonia.

Manuals and elementary training books began to be regularly produced by officers of the armed forces in the 1880s. This practice had already been established for secondary and university textbooks on science. They were usually compiled using mainly French sources (as in Spain), but in some cases British and German references were also used. In the early 1890s Captain Antonio García Reynoso designed a new revolver at the army's arsenal. A more advanced design was built at the Belgian arms factory of

Mr. Pieper, in Liège; the firm acquired full rights over it. A description of García Reynoso's revolver was published by the *Revue de l'Armeé Belge* and translated by *Revista Técnica* in 1896. The editor of the Argentine journal emphasized the importance of technical work like García Reynoso's for the advancement of the Army and the need to stimulate and widen its scope.[34]

Anales of the Sociedad Científica continued publishing articles on scientific matters relevant to the armed forces, but specific journals began to appear in Argentina from the 1880s. The *Boletín* of the Centro Naval, the navy officers' club, started publication in 1882 and it initially represented the interests of a group of young technically oriented officers. *Revista Militar*, edited by the Círculo Militar, the army officers' club, followed only in 1901. Military journals in other neighbouring countries date from exactly the same period: 1899 in Brazil and 1903 in Uruguay.

ENGINEERING AND THE NEW PHYSICAL SCIENCES IN ARGENTINA: TECHNIFICATION OF THE ARMED FORCES IN THE EARLY TWENTIETH CENTURY

The New Century and the Beginnings of Modern Research in Experimental Physics in Argentina: The Influence of Germany

By the turn of the century problems related to energy and communications began to be seriously considered in Argentina. Oil was accidentally discovered in the South in 1907.[35] Installed electrical power in industry doubled every few years while telegraphic lines already covered most of the territory of Argentina and connected it with most countries abroad. The expansion of studies in the experimental sciences was favoured in these years, particularly those in chemistry and electricity, for theoretical[36] as well as for practical reasons. With this expansion, physics and mathematics, already recognized in the last decades of the century as important tools for engineering, acquired a much higher scientific status.

A new national university at La Plata, a modern city some 50 km from Buenos Aires, was reorganized in the early years of this century following the example of British and American universities. Experimental science was given a very high priority at it: the university motto was *Pro Scientia et Patria*. A considerable effort was made at La Plata to promote the development of electricity, the new form of energy and the key to communications, from a modern scientific point of view. Initially Ricaldoni was made director of its Physics Institute. He began to equip it and continued with his experiments, now on X-rays and radioactivity. A doctorate in physics was later created there with a syllabus clearly oriented towards theoretical studies on electricity.[37]

This was an important juncture in the history of science in Argentina: the exact sciences slowly began to take a more central position on the scientific stage, until then occupied exclusively by the sciences of observation. This marks the beginning of a departure from an exclusive interest in scientific activities related to the traditional needs of agriculture and cattle farming and a reorientation which focused interest more intensely on areas which could be of direct interest to the development of local industry.

As in the 1870s, when German scientists were brought to the seventeenth-century university of Córdoba to revitalize its old institutions, German-speaking teachers of experimental science were brought the new university of La Plata. Emil Bose, an able German physicist who specialized in electro-chemistry, came to take charge of that science in 1909 and was given further facilities to complement the equipment already ordered by Ricaldoni. The construction of the new Physics Institute was described by Bose's Danish wife, also a physicist, in *Physikalische Zeitschrift* in 1911.[38] Bose was clearly a more formidable scientific personality than his predecessor. He was a protégé of Walther Nernst, whose contacts with Argentina became closer through Bose. Bose died only two years after his arrival and was replaced in 1914 by the German physicist Richard Gans, a world authority on the theory of magnetism, who remained in his post until 1925. Meanwhile, lectures on theoretical physics were given in Buenos Aires by Camilo Meyer, a French physicist who emigrated to Argentina in the mid 1890s. In 1912 two students from La Plata, trained at Bose's institute, Teófilo Isnardi and José B. Collo, were sent to Germany to do research work, under Nernst.

Modernization of the Armed Forces in the New Century

Pablo Ricchieri was made director of ordnance on his return from Europe in 1898; he played a significant role in the process of renovation of armament in the 1890s, which allowed for further steps in the road towards professionalization and technification of the Argentine Army. In 1891, he had suggested the adoption of the German Mauser rifle by the Army; it gradually displaced the American Remington while the Gatling machine gun was being replaced by Colt and Maxim models.[39]

The possibility of a war with Chile was a factor promoting the modernization of the armed forces. Chilean armed forces had been trained by German officers and had given proof of their proficiency in a recent encounter with Perú. However, the quest for modernization has much deeper roots, which go back to the early efforts of San Martín to introduce European models and Mitre's efforts to technify the armed forces.

This process of technification of the armed forces, which took place in the last decades of the nineteenth century, would have been a much more formidable task, if at all possible, without the technological infrastructure created by the development of engineering and related sciences in Argentina. Advances in education and in communications, territorial expansion through the Desert campaigns, university advancement, the promotion of science and technology and army technification were all, as was the policy of mass immigration, different sides of a global policy of modernization of Argentina.

Foreign Influences in the Armed Forces

German cartographers, specialists in geodesy and engravers of maps did work closely related to the needs of the Army for a considerable time after the Desert campaign and even before it. French (and also Italian, Spanish and Austro-Hungarian) former officers were also incorporated into the Argentine Army or Navy as advisors or to head or organize specific, usually technical, services. Besides those foreigners imported or hired locally on a contract for specific technical tasks, massive immigration had brought to Argentina not only workers, but also many men with a good technical training, including a few graduates from elite European technical schools.

A procedure used by the Army at the turn of the century to accelerate the transfer of skills was, as earlier in the case of the University of Buenos Aires engineering school, the importation of foreign instructors, in addition to sending graduates for further study abroad. In 1900 a War Academy[40] was inaugurated by General Roca, then president of Argentina for a second term. He engaged officers from Germany for a further stage in the process of making the armed forces a body with a more homogeneous professional training and, consequently, for its modernization. This training contributed to creating within the Argentine Army an elite officer corps which greatly affected its organization. As had happened earlier in the engineering school of the University of Buenos Aires, foreign instructors at the War Academy were replaced by Argentines in less that two decades. Perhaps the outbreak of the First World War and the recall of foreign instructors to take up duties at home prevented another sort of conflict from arising due to the eagerness of their Argentine students to replace them.[41]

The Military Engineering School[23] was reorganized in 1904 as a school of applied science and technology,[42] annexed to the Escuela de Guerra. Although it never aspired to reach the technical standards of the national universities, delicate optical and acoustical measurements were made at laboratories of this school. An electro-mechanical laboratory was also created for the communications corps.

After intense controversies, German arms manufacturers became the main suppliers of the Argentine Army.[43] From 1910 the Mauser model 1909, a modern and reliable weapon, displaced older models and the products of other manufacturers. Also German ship works received orders for modern turbine-propelled destroyers for the Argentine Navy.

In his second presidency, Roca encouraged the further development of telegraphy and in 1902 a national law extended regulations for telegraphy to radio telegraphy. In these years problems of telegraphy and radio telegraphy attracted the attention of some of the best technically qualified armed forces officers. Shortly after the return of Isnardi and Collo from Germany, the Naval School[19] invited them to join its staff. In this indirect way, the German teachers of La Plata contributed also to the progress of the Argentine Navy. Although the Navy's contacts with Great Britain were closer, Germany, Italy, the USA and Spain tried to became arms suppliers. Contacts with the USA became closer, as the American Navy gave Argentina training facilities in its naval units. Ultimately, the USA rivalled Britain as a supplier of naval equipment.

German Assistance in the Development of Science and the Armed Forces in Argentina

No doubt the prestige of the German Army after 1870 had a lot to do with Argentina's choice of it as a model army.[44] But that was not the only reason. Germans were keen to offer direct and efficient help,[45] sending officers to Argentina to do work which was appreciated and which later would be remembered by local officers. They also accepted a regular flow of young Argentine officers into their own academies and units. Such arrangements would have been far more difficult, if not impossible, with Belgium, France or Britain. German actions in these matters were part of a broad policy of cultural imperialism,[46] which included admission of foreigners to work for a doctorate in its universities. Clearly Germany aspired to a more dominant position in world affairs and used all possible channels, for all were interconnected. Thus, Germany, like other leading European powers, did its best to make armament sales a companion of military training.

The agents of European powers who engineered agreements for the transfer of science and technology to Argentina,[47] presented them to the metropolitan administrators as, ultimately, substantial contributions to political influence. Local foreign communities in Argentina were in a position to play a most important role in this process as they knew what Argentina needed and what the metropolitan powers could supply.

However, it is a debatable matter whether tangible political interests were objectively advanced through the export of science, always a slow business in a rapidly changing world. Political alignments, no doubt, answer to far more formidable forces than admiration for technical, scientific or artistic ability, even when it does operate. Answers to these questions are even more paradoxical if we take into account the views expressed by those who actually participated in the transfer of science from Germany to Argentina from 1870 to the 1920s.[48] They do not seem all to have been obedient carriers of the Imperial flag. Some among them held radical views, very far from those officially sanctioned at home.[49]

Involvement of Argentina with foreign nations took a variety of forms. Particularly interesting are the patterns of association of Argentina with Spain, a country with around one million nationals residing in Argentina, and with Great Britain, a dominant economic power, with large political influence. In the first case contacts in the fields of science and technology are characterized by long-lasting joint cooperation agreements. In the second no official efforts can be detected to communicate through science with Argentina, in proportion to the large investment there. However, Great Britain effectively exerted its influence through private firms. Large transfers of technology were made through these channels. In particular, the Navy used its placement of large orders to gain access to technical training in work shops and ship yards of large British companies in the 1890s.[50]

The Army and Navy at the International Scientific Congress in Argentina, in 1910

In 1910, at a time of unprecedented prosperity while also the number of immigrants entering from Europe reached new heights each consecutive year, Argentina celebrated the centenary of its independence from Spain. Among the events included in these celebrations were several large meetings, one of which was a Congreso Científico Internacional Americano.

The Congreso lasted for two weeks in July, in Buenos Aires. It was sponsored by over a hundred foreign institutions, including the main academies and universities of Europe and the USA, with delegates from a number of countries. The meetings attracted close to 1500 participants. Some five hundred papers were read and discussed and over one hundred resolutions and recommendations were passed on the basis of the discussions.

In structure and announced aims this scientific meeting had strong similarities with those organized in Europe by various national associations for the advancement of science.[51] A wide and modern concept of science was adopted for this centenary meeting, which allowed economical and statistical studies as well as geography, psychology (meaning also psychiatry and mental health) and the social sciences to be included in its agenda alongside the classical sciences and engineering. An even more interesting fact in the context of our discussion was the inclusion, in full parity with natural sciences, of sections on military sciences and naval sciences. By 1910 the Army already had a sufficiently large group of qualified men to be represented in a meeting of this nature. Among them were several engineers graduated at Buenos Aires University: Dellepiane, Justo, Mosconi and also Mesa, Perlasca, Uriburu and Velasco.

Although military and naval sciences had their specific sections in the Congreso, many topics relevant to the Army or Navy were discussed outside them; mainly in subsections on engineering and chemistry. These discussions identify numerous areas of common interest for scientists, technologists and army and navy officers in the early years of this century.

The Beginnings of Aviation in Argentina and the Response of the Army

Just before the First World War civil aeronautics began to be developed rapidly in Argentina, where experiments on aerostatic navigation had already been conducted for some years. The Aero Club Argentino, a private organization, was created in Buenos Aires in 1908. Jorge Newbery, who had received university training in engineering in the USA and had worked in the electricity department of the Navy on his return to Argentina, and the engineer and physicist Jorge Duclout played a leading role in the Aero Club and in the development of aviation in Argentina. Physicist Emil Bose had made extensive trips in balloons in Germany; in Argentina he advised Newbery on scientific matters. In 1912 military[52] and civil[53] aviation schools were created on the basis of aircraft lent by the Aero Club Argentino. Military aeronautics was institutionalized further in 1920, when the army aeronautical service was created; engineer Colonel Mosconi was made its director.[54]

ARMED FORCES, SCIENCE AND INDUSTRY AFTER THE FIRST WORLD WAR

Impact of the First World War on the Cultural Life of Argentina

Society and culture underwent considerable changes in Argentina during the four years of the First World War. Through emigration, Europe had become less remote. A new generation of Argentineans, sons of European immigrants, participated increasingly in every aspect of Argentina's life. And this brought important changes in the political scene. In 1916 a series of governments of different shades of conservatism was brought to an end in consequence of a new law which allowed for a larger participation of the population in national elections. However, in diplomacy and foreign relations, Argentina consistently kept a neutral position during the war and afterwards at the League of Nations.

The new social movements, particularly pacifism, socialism and the Mexican and Russian revolutions, had a considerable impact on Argentina's intellectuals. Such influences quickly filtered to the universities through a movement called Reforma Universitaria. This movement regarded social problems and their consequences for Argentina's projected future as topics requiring a more detailed consideration by the universities. It was also thought that an emphasis should be put by the universities on problems directly relevant to the needs of the country.

In 1917 the Spanish mathematician Julio Rey Pastor followed the philosopher José Ortega y Gasset in the chain of visitors brought to Argentina through an interesting scientific cooperation agreement with Spain.[55] Both of them, as also a number of other visitors who came under the same scheme, remained close to Argentina. Rey Pastor settled there permanently in 1921, and he included among the topics of his lectures a course on the physical and engineering applications of functions of a complex variable which had importance in aeronautics.

Through the decade of the 1920s the role of a research scientist began to be perceived in Argentina in more dynamic terms. Physiologist Bernardo A. Houssay is the archetype of a new figure, the research scientist, which emerged in that period. He was a university professor who regarded research work as a full-time *professional* occupation. He practiced what he preached.

In the 1920s Houssay, Rey Pastor[56] and a few other scientists and engineers created new schools, imperfect perhaps, but showing already a modern and sophisticated approach in which intellectual achievement and the quality of production began to be the main consideration for its hierarchical structuring. Personal controversies did not disappear, but ceased to be predominant as drivers of scientific discussion. An effort to communicate their results to the international scientific community was another characteristic of these new clusters.

The university reform movement created conditions favourable for foreign scientists to visit Argentina and stimulated these visits. Among those invited were two leading pacifist scientists from Berlin: physiologist Georg Friedrich Nicolai, and Albert Einstein. Nicolai settled in Córdoba in 1922, and Einstein, invited in 1922, visited Argentina in 1925 and lectured on relativity theory.[57] That these visits were supported enthusiastically shows that there existed alternative ways of viewing Germany among Argentine intellectuals.

Impact of the First World War on the Argentine Army and Navy

The outbreak of the First World War confronted the Argentine armed forces with their critical dependence on industrial imports for their supplies. Among the officers concerned with this situation was prominent engineer Colonel Mosconi, who advocated a consistent national policy of import substitutions and began a long campaign to develop industrial awareness among his army colleagues. The tide turned after the end of the war, when the victorious countries of Europe and the USA found themselves with a stockpile of unwanted armament. Local production could scarcely compete with this fresh availability of modern armament at sale prices.

In October 1922 Marcelo T. de Alvear was elected president of Argentina for a period of six years; he entrusted the War Ministry[58] to Colonel Agustín P. Justo, who had graduated in 1903 in civil engineering at the University of Buenos Aires and who had long served as director of the Colegio Militar. Although there was no real evidence of danger of war, Justo was authorized to start a vast program of modernization of the Army and Navy.[59] The sums authorized were in excess of a year's national budget. Offices for the armament acquisition missions were opened in Paris and Brussels.

Aftermath of the War: Reforms and Technification of the Army Under Justo

Mosconi's industrialization campaign concentrated on the organization of the Argentine national oil company, Yacimientos Petrolíferos Fiscales or YPF, which was put under his command. He managed to achieve a very considerable success. Production increased sharply and by 1926 the first national oil refinery was in operation at La Plata. These developments not only had an impact on the economy of Argentina, but also on its perception of the possibility of self sufficiency. Furthermore, they affected significantly the development of chemistry and of chemical engineering in that country. YPF began to be regarded as a symbol of Argentina's technical independence.

The progress of the Argentine air force did not slacken with Mosconi's new posting. In 1926 a national airplane factory[60] was created, which soon began to produce aircraft under foreign license and to undertake the production of spare parts for servicing the growing fleet of military aircraft.

Improved quarters for the Naval School was another result of Justo's increased armed forces budget. It acquired modern scientific laboratories which made possible a more advanced and practical training of naval officers. To make this expansion in teaching possible, physicists Teófilo Isnardi and José B. Collo brought a second wave of science teachers to the Naval School. They were selected from former students and scientific collaborators, mainly from the university of La Plata, which was close to the seat of the Naval School, in Río Santiago. Several of them established valuable contacts with naval officers, which were of consequence some years later.

A NEW STAGE: ADVANCED TECHNOLOGICAL TRAINING WITHIN THE ARMY

Manuel N. Savio: Training within the Army

Manuel N. Savio, born in Buenos Aires in 1892, belongs to a new generation of army technical officers whose training was not directly connected with the national universities. Men of his generation would give a new sharp emphasis to the teaching of advanced technology within the army. Furthermore, the armed forces would increasingly incorporate into their strategy an effort to acquire and, later, to encourage the development of areas of technology and science they regarded as important to their activities. Such policies would have important consequences in the decades that followed.

Savio entered Argentina's Colegio Militar in 1909[61] and after graduation he attended a postgraduate course on military engineering offered by the Colegio. There he lectured on topics in engineering from 1917. Savio was thirty years old and an army Captain in 1922 when Alvear became President of Argentina. The following year he was attached to the Argentine mission for the acquisition of military equipment in Europe and remained there until 1926. It was during this period in Europe that Savio became acquainted with the new Taylorist industrial organization techniques: the concepts of scientific management and time-and-motion studies.

The Army's Advanced Technical School: Escuela Superior Técnica del Ejército

After Alvear finished his presidential period in 1928, the previous president, Hipólito Yrigoyen, was elected for a second period. However, his political charisma proved insufficient to contain a wave of unrest which swept the country. (This unrest had political as well as economic roots similar to those resulting in political reversals in various countries concomitant with the Great Depression.) In that climate, on September 6, 1930, General José Félix Uriburu staged a coup d'état and took control as *de facto* President; in an unprecedented move he dissolved Parliament.

Savio, now a Lieutenant Colonel, was given the task of upgrading the advanced postgraduate courses given at the Colegio Militar into a fully fledged, modern engineering school to be called Escuela Superior Técnica del Ejército. This Advanced Military Technical School was a more formidable development for much the same purposes as the specialized Military Engineering School started some 50 years earlier, after the end of the Desert campaign. Through this new technical school it was possible for the army to train officers up to a fairly advanced technological level at their own quarters and on the special topics they needed.

Behind this decision to break away from the framework of the national universities was presumably the wish to limit interference by the political discussions which were then taking place at the universities. In its academic staff the Escuela Técnica shared with the University of Buenos Aires several scientists and technologists. However, the Army was now in a position to exercise its discretion in the selection of those who would be invited to become members of its staff and those who would not.

The Early 1930s

In November 1931, formally ending the cycle of the military coup d'état, Agustín P. Justo, now a General, was elected President of Argentina for the period 1932–1938. Even if his promotion to high office had more to do with his being a member of the armed forces than with his engineering training Justo was, nonetheless, the first engineering graduate from Buenos Aires

University to reach the Presidency of Argentina. A technical background, no doubt, had become an extra asset for a politician of the early 1930s.

The process of modernization of the Army and Navy of the 1920s, based on the acquisition of new equipment, had exposed a number of officers to details of modern technology on an unprecedented scale. A decade later a number of interesting new initiatives began to take shape. The question of achieving self-sufficiency in some areas of armament production resurfaced and was now seen as directly dependent on the production of steel. In this way steel production and military defense began to be perceived by a new group of young officers as two sides of the same problem. The question of adapting peacetime local industrial production to the conditions of war was the main problem of a new discipline: Industrial mobilization. That was precisely the subject Savio began to teach in a course of lectures he offered at the Escuela Técnica. His students included some of the most technically gifted officers of the time. Furthermore, several other young officers had European experience similar to Savio's in the Argentine arms acquisition commission. Their exposure to details of foreign armaments production gave them a clear picture of the magnitude of the technical skills involved in their production, one which was probably more accurate than that of the more idealistic pioneer army engineers of twenty years earlier. They realized that parts of such production could easily be done at home to reduce, to some extent, the dependence on supplies from abroad at a critical time. They also perceived that other items definitely could not be manufactured locally. Furthermore they had mapped clearly which were which. The policy of arms purchases abroad had, indeed, led to unexpected consequences.

For his course on industrial mobilization at the Escuela Técnica, Savio wrote in 1933 a small book: *Movilización industrial*[62] which became his textbook and a confidential army document.[63] In it, a concept of war, new in Argentina, was developed. Savio took into account the deep technical and economic interconnections existing between army and industry, and through it, with technology and science.

The role of the Escuela Técnica and of its students was central to his industrial mobilization plan and drove them to develop close ties with industrialists. In his inventory of the manpower available in Argentina, Savio assigned an important role to the graduates of the Facultad de Ciencias Exactas, Físicas y Naturales, that is, to engineers and chemistry, physics and mathematics graduates of the university of Buenos Aires. However, the dominant place was now being taken by the graduates of the Escuela Técnica who, in his view, should form an elite available for consultation to the high command.[64] The group of experts Savio was trying to create in his field has some points in common with the new research schools we have referred to as emerging in the decade of the 1920s in specific fields of science and technology in Argentina.

In his report Savio advocated the creation of a technical board for military industrial production,[65] which would have, for the industrial effort, the same role as the High Command for the war effort. This organization would, to be sure, control a collection of military factories. But it would be also a coordinating body between civil and military industry. Over the decade of the 1930s *Revista Militar* published a number of papers in which army officers discussed industrial mobilization and, more generally, the interface between army and industry.

APPROXIMATIONS AND CONFRONTATIONS: THE DOCTRINE OF ARMY-SPONSORED SCIENTIFIC RESEARCH

Foreign Scientists in Argentina: The Anti-Fascists.

After the military coup d'état of 1930, government officials made an effort to reduce the volume of immigration into Argentina. Nevertheless, during the following decade a conspicuous group of highly trained scientists and humanists arrived in Argentina. They came from Spain, because of the Spanish Civil War[66]; from Italy,[67] because of Fascism and racial laws; and from Germany, Austria and Central European countries because of Nazism.[68] Most of these men and women entered the country through the efforts of a very small but dynamic group of radical Argentine intellectuals who shared with them views and values. Besides, they also perceived clearly, and emphasized, the contribution these scholars could make to the intellectual life of their country. In fact several of them contributed to revitalize Argentine universities, particularly outside the conglomerate Buenos Aires-La Plata.

Savio's Military Factories Board

In the early 1940s a number of small and medium sized military factories were united under the new Dirección General de Fabricaciones Militares, created to promote and coordinate industrial efforts in areas of military interest. General Savio, the theoretician of industrial mobilization, was made its first director. As with the Escuela Técnica, there are direct links between Savio's new organization and the innovations of the 1880s. Fabricaciones Militares was in a way a modern development of the early gunpowder factory of 1883. The military cartographic board of 1879 was transformed in 1942 into a military institute for geography, the Instituto Geográfico Militar, and given the responsibility for the construction of a new and more detailed map of Argentina. It developed into the most important cartographic and geographic center in that country.[69]

New Initiatives for Scientific and Industrial Research

In the early 1940's public awareness of the importance of scientific research increased substantially. There was then a surge of proposals for the promotion of scientific and industrial research, some of which reached provincial legislative chambers or even the national parliament. Influential national newspapers, such as *La Nación* and *La Prensa*, as well as professional journals, published leaders or signed articles on problems related to scientific research and higher education. Some of these suggested Argentina create a national agency to sponsor scientific research, as other more advanced countries had already done.

Also in the early 1940's, the national oil company YPF created a research department and opened a large and well equipped chemistry research laboratory in Florencio Varela, near Buenos. A year later, a committee was created by the national government for the promotion of exports and for industrial and commercial development, which suggested again that a technological research institute should be created.

As the Second World War progressed there was all over the world a greater awareness of the role played by scientific research and, particularly, by the chemical sciences, in the course followed by that war. Later, such emphasis moved to include electronics and then nuclear physics. Such perceptions struck Argentina forcefully and to some extent neutralized resistance to the admission of foreign scientists and technologists who had left Europe before and during the war. As we shall see, it also helped, after the war, to admit a second wave, with political associations different from those of the former.

The Institute of Industrial Studies

In the 1890s the Unión Industrial Argentina, an association of proprietors of industrial establishments, expressed its concern for technical education by supporting an early attempt to create a private university in Buenos Aires, the Instituto Politécnico. Some fifty years later the Unión created an institute to stimulate contacts between scientists, engineers and industrialists. It was opened in 1942 as an Institute of Industrial Studies: the Instituto de Estudios y Conferencias Industriales.

Its Honorary President was Dr. Carlos Saavedra Lamas, a patrician figure, a Nobel laureate and the Rector of Buenos Aires University. Its members included some of the most distinguished technologists and scientists Argentina then had; leading industrialists and a high ranking navy officer. Engineer José A. Gilli, an expert on Taylorism and a main promoter of secondary school technical education, was the Institute's executive director. Two engineers, Torcuato Di Tella, director and founder of a leading Argentine mechanical and electrical industrial concern and Ricardo M. Ortiz, a specialist in harbour design and operation and a university professor, were its secretaries.

Among the activities of the Institute was the organization of a program of seminar lectures on industrially related problems which attracted some of the leading scientists and design engineers of Argentina and also leading personalities from industry and the armed forces.

One of these lecturers, Casimiro Lana Sarrate, who had recently arrived in Argentina as an exile from Spain, had been trained in metallurgy in Berlin and at M.I.T.; he had direct experience in the construction of car and aircraft engines. Lana Sarrate addressed a critical question: The chances of survival of Argentine industry after the end of the war. He pointed out that because of the lack of scientific research, the more advanced industries in Argentina functioned by either paying royalties to a foreign company[70] or by an economic association with a foreign industrial group.[71]

However, the desire for discussion of problems of common interest between liberal enlightened army officers and scientists and technologists should not be exaggerated. It was, indeed, a thin layer of the armed forces which favoured and understood the value of such exchanges for the technical development of their forces. This layer was made up of officers in highly technical sections of Army, Navy and the Air Corps, and, as such, a small minority within the armed forces. The same was true for the civilians, where the large mass of *professionals* was equally heterogeneous: scientists and design engineers were also a small minority within it. The vulnerability of such minorities was soon to be tested.

Confrontation between the Army and the University after the Military Coup of 1943

In the early 1940s the Army showed, increasingly, signs of unrest. Several lodges were operating within it which favoured the idea of a new army takeover. One of these lodges was the Grupo de Oficiales Unidos or GOU, members of which were definitely sympathetic to the Axis. As young officers, several of them had played a role in the first coup d'état of 1930. Prominent among members of the GOU were the Lieutenant Colonels Enrique P. González and Juan Domingo Perón.

A second coup d'état on June 4, 1943, put the Army again in control of government. After some internal struggle Perón asserted his authority over González within the GOU and, with government support, slowly emerged as a powerful and charismatic leader. In 1944 Perón became Vice-President of the military government and from this high office he started a campaign which would result in his election to the presidency of Argentina in February 1946.

Four months after the 1943 coup d'état a group of intellectuals signed a petition asking the government to return the country to constitutional law. Although this initiative followed unofficial exchanges which suggested the government would consider it with benevolence, it was answered with unusual violence. The military government ordered the expulsion from state positions of all those who had signed it. After a period of conflict and violent protests by students, the political pendulum moved against the more regressive elements in government; the university returned to its former authorities and those expelled were reinstated. But not for long.

The military authorities had good reason to perceive the universities as a source of opposition, one difficult to control. Three months after Perón was elected President, in May 1946, the universities were taken over by officers designated by the government, called Interventores. Their task was a renewed and more thorough purification of the universities than ever before. Over one thousand teachers were dismissed or resigned in solidarity with those dismissed.

TABLE 1. Purification of Argentine universities: 1943–1946

University purified	Positions lost in University	% of positions in Sci., Tech. and Medicine lost	Positions lost in annexed schools
Buenos Aires	149	45%	–
Córdoba	302	82%	22
Cuyo	31	26%	16
La Plata	193	33%	80
Litoral	429	90%	–
Tucumán	38	60%	20
Totals	1142	Average 56%	138

Data in Table 1 have been taken from a list compiled by an association formed by the teachers dismissed by government; it was published in 1947.[72] Although not complete,[73] it gives an impressive picture of the effect of purification. Table 1 gives also a rough estimate of the percentage of lost positions related to science and technology, including medicine. Some universities had annexed secondary schools; we give separately, in the last column, the number of teachers lost by these schools. A few academics held more than one position (sometimes at different universities, mainly at Buenos Aires and La Plata) which reduces slightly the total figure of academics and teachers affected to just over one thousand. Purification also reached alarming proportions among teachers of secondary schools not associated to a university, as was the case for most of them.

There was talk of organizing a private university taking advantage of the high quality of the academics available, but there were difficulties, as the government had to validate the diplomas of graduates and was not prepared to do so. Among the alternatives discussed, the least ambitious but finally the most successful was the creation of private research laboratories where a small number of gifted young graduates could be accommodated. A private research institute created in Buenos Aires for Houssay was imitated in the cities of Córdoba and Rosario, where the universities suffered badly. It seems that government pressure on private contributors was less severe where medical research was performed than in other areas. The Rockefeller Foundation[74] contributed to the financing of some of the private research institutes, transferring grants given to scientists when they were working at national universities to their new private institutes. Aid was also given for buildings, equipment and training; it was concentrated on medicine and biology, but was also applied to a few other fields.

Purification of the Argentine universities when fascism was already defeated caused deep embarrassment to the Argentine government, which was perceived as directly inspired by fascist ideology.[75] This was, however, the only way the government saw as effective to control dissent in the universities and, by way of example, elsewhere. A number of professional and learned institutions also suffered the impact of this policy. The Institute of Industrial Studies sponsored by the Unión Industrial ended abruptly as the Union itself was taken over by the government.[76]

It seems clear that technical sectors of the Army hoped for a closer cooperation with civilians in the fields of science and technology, which were vital to them. The enthusiastic cooperation of several leading technical army officers in the works of the Institute of Industrial Studies, which connected them also with local industry, suggests that this attitude existed at the highest technical levels of the armed forces. Clearly the new political order dictated by other, reactionary sectors of the Army, was eroding such options. In response, the leading technical officers developed an assertive new doctrine on the relations between Army and science.

The Doctrine of Army-Sponsored Scientific Research

In parallel with the attempts made by those affected by purification, General Savio and some industrialists, and through them also some members of parliament, tried to find alternative solutions. Their objective was to minimize the impact of the purification on the fabric of the Argentine scientific community, particularly on that part which was relevant to technology. One of the possibilities considered was to revive the idea of creating national scientific research institutes. These could attract, perhaps more selectively,[77] some of those separated from, or unhappy with the situation of, the universities. An important official initiative was a project commissioned and supported by Savio, who was advised by Isnardi and perhaps others. The project reached Parliament by the end of 1946 but did not result in concrete developments. Other research projects were submitted and even approved by the National Parliament, with the same lack of success. The original project of Savio was submitted again two years later, in 1948, in the form of a research institute for physics and chemistry and approved in September, 1948.[78] Now, however, it was to be under military control, a condition which even in the government-controlled Senate had to be explained. Although approved, it did not materialize. Savio died in July that year.

The doctrine of military control began to be adopted more generally. It was applied also to several services; responsibility for meteorological surveying for air and sea navigation was transferred in 1949 from the National Meteorological Office to the special services annexed to the Secretaries of Aeronautics and the Navy respectively. The general thrust imparted to the actions of government by the more influential sections of the armed forces was breaking bridges carefully built over a number of years by more technically qualified sections of the Army. These moves altered sensibly the relationship between the armed forces and Argentine scientists.

They also created a serious problem. An alternative had to be found to satisfy the Army's need for new technologies in a world in which sophisticated weapons were acquiring a dominant position. Paradoxically, and unlike the previous war, the end of the Second World War looked more and more, as time passed, like an episode of an unfinished saga, highlighting the need for the development of even more advanced weapons. The Argentine government firmly believed there would be a Third World War which, in their view, would enhance further the economic position and perhaps also the international political standing of the country. The idea of military-sponsored scientific research became accepted. The case of more developed countries where such an ideology had already been adopted[79] was used as an example, but without fully realizing, perhaps, the great importance of university research in these countries as background to military-technical developments.

A New Wave of Scientists

An alternative to unreliable local scientists presented itself to General Perón's government in the form of scientists and technologists from the countries defeated in the Second World War who, for one reason or another, wished to emigrate but could not, or would not, to countries which had won that war. Late in 1947 the Argentine government began importing scientists and technologists. This policy was not, however, without its dangers, as is well shown by the Richter case: the announcement by President Perón in March

1951 that the Austrian born scientist Dr. Ronald Richter had been able to achieve controlled nuclear fusion at his secret laboratory in West Argentina.[80]

Richter promised to produce energy through a simpler and cheaper process than that pursued in the USA, which Argentina could not afford to undertake, either technically or financially. His project may have been presented to or perceived by Argentine officials as a yet unused result of German research into nuclear energy,[81] achieved towards the end of the war, perhaps only tested at laboratory scale, but already certain of success.[82] This had been the case for Kurt Tank's jet propulsion work, which he continued with success in Argentina.[83]

The Richter fiasco resulted from the authorities' faith in secrecy and in the sufficiency of their own judgment, even in highly complex technical matters,[84] coupled with a deep distrust of the university, that was transferred directly to the local scientific community. Even some distinguished scientists, who could have conducted a serious program in areas relevant to nuclear research, such as Richard Gans, Kurt Fränz, Walther Selmann-Eggeberth, Hans Joachim Schumacher, Manlio Abele and others imported from Germany or Italy after the end of the Second World War, were kept outside the development of the fusion project. Richard Gans was consulted on Richter's fusion claims only when evidence against them was unquestionable.

More understandable, perhaps, was the exclusion of the theoretical physicist Guido Beck from the atomic energy project. He was an anti-fascist who arrived to Argentina in 1943; without doubt he was the most experienced specialist on nuclear physics residing then in Argentina. Among the local experts who on political grounds were not trusted to be part of this project were such scientists as Félix Cernuschi, Ernesto Galloni, Rafael Grinfeld, Cecilia Mossin Kotin,[85] and, above all, Enrique Gaviola, the one scientist who actively tried to interest the authorities in developing an atomic energy research program.

These cases give some insight into the approach then being used by government to evaluate scientific projects. Richter's proposal, like Tank's and others accepted at the time, may have been perceived by the authorities as a very concrete project with a clear end result, hopefully achievable with a finite amount of money and within a definite period of time. This was the type of research the government seems to have been able to grasp and been prepared to fund if sufficiently attractive. The research institutes sponsored by Savio and others were, on the contrary, long-term projects without such definite promise of tangible, politically visible success. They, however, would have upgraded the scientific and technological standards of the country, what would have been necessary if the nuclear project were to be widely developed in Argentina.[86] The former perception is consistent with the objective-oriented "flowchart" approach to reality which was then prevalent in the planification circles of the Army which were close to government.

President Perón's claim of controlled fusion was impossible to sustain[87] and resulted in international discredit of the authorities and, internally, in a loss of face. In consequence control of atomic energy research passed from the hands of the Army to those of the Navy, technologically better connected. A new institution for nuclear research was created, the National Atomic Energy Commission, and Argentine scientists were incorporated into it. Although still an institution which sponsored the development of science in areas of military interest and under armed forces control, it was more open than Richter's secret atomic project in remote Bariloche. There was another important difference: the new institution was not a *project* in the sense indicated above. However, it preserved the aura of nuclear energy and a diffuse promise of, perhaps, atomic weapons if required one day.

As a result of these changes substantial progress was made in the training of scientific personnel in modern techniques. Another important source of scientific and technological advanced training in Argentina in these years was the Instituto Radiotécnico.[88] It had been organized by the University of Buenos Aires as an institute for the training of electronic engineers in 1942 with an input of navy officers among its students[89] and later financially reinforced by a closer association with the Navy. After 1951, the new Instituto attracted some leading experts, such as Gans and Fränz. Besides electronics, it promoted research on chapters of complex function theory, statistics and probability, numerical calculation, the theory of servomechanisms and the mathematical theory of circuits.

After General Perón was deposed by a third coup d'état, in June 1955, official commissions[90] were established for the judicial investigation of particularly questionable affairs. The report of the Richter commission, a most important document on this affair, was compiled with the technical assistance of physicists Isnardi, Collo and Galloni in the last months of 1955 and published in 1958.[91]

Achievements and Unresolved Questions.

However, if crudely measured in terms of Nobel laureates, the main achievement of scientific research done in Argentina in the early postwar period was not in the areas mentioned above, but in biology, biochemistry and physiology, open-ended areas which had been left out of armed forces projects, as irrelevant to them.

Besides being outside the government's interest, Houssay and other scientific leaders in this field were also out of any national research system. The experimental medicine groups were deeply affected by the mid-1940s clashes between the Army and universities and were separated from the latter. In the years of Peronism they survived in the much freer, but also very much more modest, atmosphere of privately supported research institutes.

These institutes were not in a position to offer the salaries[92] or the facilities of the large armed-forces-controlled research institutes. However, they achieved outstanding scientific results. Houssay received a Nobel Prize in Physiology and Medicine in 1947. Luis Leloir, trained in the same private institute received, some years later, the Nobel Prize for Chemistry.

Although striking a balance for this period is still a complex matter, it seems fairly clear that the military-sponsored research effort, even considering its undoubtedly positive fallout in training and modernization, may not have been the best possible choice for the development of science in Argentina in that period.

From 1956, the universities, under new management, started claiming back research which had moved or developed outside them. From armed-forces-controlled institutions, such as the Atomic Energy Commission, they drew a number of scientists and managed to keep their young science graduates within the university research system. This process, together with attempts, not always successful, to attract Argentine scientists who had emigrated abroad, made possible a rapid development of science within the University of Buenos Aires. Attempts were also made in the 1960s to develop research through private institutions, the most successful of them being the new Bariloche and Di Tella foundations.

However, the question of responsibility for research important to the armed forces was not fully resolved in the decade 1956–66. Following a fourth coup d'état in 1966, there were once again damaging clashes between the Army and the universities, with the exact sciences departments of the University of Buenos Aires among those most seriously affected.[93] Thereafter, Argentina's science policy became volitile and erratic, and, consequently, the country has suffered severely in recent decades from a continuing exodus of young science graduates to the USA and Europe. This remains an unresolved problem for Argentina.

University of London, London

REFERENCES

Bahía, M.B., 1894. Curso de electrotécnica, Buenos Aires.
Bahía, M.B., 1888. "Historia de la telegrafía eléctrica", Anales de la Sociedad Cientíifica Argentina, 26, 78–90.
Balbín, V., 1895. "Fortificaciones de campaña"; (serialized also in Revista Técnica), Buenos Aires.
Beyerchen, A.D., 1977. Scientists under Hitler, Politics and the Physics Community in the Third Reich, New Haven and London: Yale University Press.
Bose, M., 1911. "Das Physikalischen Institut der Universität La Plata", Physikalische Zeitschrift, 12, 1230–1243.

Bose, M. and Walter, B.L., 1966. "Historia de las comunicaciones" in: Historia Argentina Contemporanea, III, Buenos Aires.

Burzio, H.F., 1968. Historia del torpedo y sus buques en La Armada Argentina, 1874–1900, Serie B, Historia Naval Argentina, Nr. 12, Buenos Aires.

Cáceres Freyre, J., 1979. "Los científicos en la expedición militar de Julio A. Roca." Logos, Revista de la Facultad de Filosofía y Letras, IX, 14, Buenos Aires, 91–125.

Cannon, W.B., 1947. The way of an investigator (Argentine edition, La ruta de un investigador, Buenos Aires: Ediciones Siglo Veinte)

Cárcano, R.J., 1893. Historia de los medios de comunicación y de transporte en la República Argentina, I-II, Buenos Aires.

Comisión Científica Agregada al Estado Mayor General de la Expedición al Río Negro (Patagonia) realizada en los meses de Abril, Mayo y Junio de 1879 bajo las ordenes del General D. Julio A. Roca (en tres volumenes con 16 láminas) Informe Oficial, Buenos Aires, 1881.

Comisión Nacional de Investigaciones, 1958. "Comisión Nacional de Energía Atómica", in Casos de la Segunda Tiranía, Comité Nr. 12, Buenos Aires, pp. 42–101.

Ejército Argentino, 1951. Reseña del Instituto Geográfico Militar, Buenos Aires.

Federación de Agrupaciones para la defensa y progreso de la Universidad democrática y autónoma, Avasallamiento de la Universidad Argentina, Buenos Aires, 1947.

Forman, P., 1987. "Behind quantum electronics: National security as a basis for physical research in the United States, 1940–1960."Historical Studies in the Physical and Biological Sciences, 18, 149–229.

Fundación Standard Electric Argentina, 1979. Historia de las comunicaciones argentinas, Buenos Aires.

García Reynoso, A., 1986. "Revolver sistema A. García Reynoso", Revista Técnica II, Nr. 20, 104–107 + Plates, [Microfiche 1896–97 2:4 in The Humboldt Library edition].

Loyarte, R., 1924. "Física." in : Sociedad Científica Argentina: Evolución de las Ciencias en la República Argentina, Buenos Aires.

Lütge, W., Hoffmann, W. and Körner, K.W., 1955. Geschichte des Deutschtums in Argentinien, Buenos Aires.

Lütge, W., Hoffmann, W., Körner, K.W. and Klingefuss, K., 1981. Deutsche in Argentinien, (2nd edition of Geschichte des Deutschtums in Argentinien), Buenos Aires.

Mariscotti, M., 1985. El secreto atómico de Huemul, crónica del orígen de la energía atómica en la Argentina, Buenos Aires.

Marti Garro, P.E., 1982. Historia de la artillería argentina, Buenos Aires.

Mitre, B., 1903–1907. Historia de San Martín y de la emancipación sudamericana (según nuevos documentos), I–VI,3rd. edition, Buenos Aires). (first published in three volumes, Buenos Aires, 1887–1888.)

Newton, R., 1977. German Buenos Aires, 1900–1933, social change and cultural crisis, Austin.

Olascoaga, M.J., 1881. Estudio topográfico de la Pampa y Río Negro, 2nd edition, Buenos Aires. (first published 1880).

Ortiz, E.L., 1980. "Professor António Monteiro and contemporary mathematics in Argentina", Portugalia Mathematica, 39, XIX–XXXII.

Ortiz, E.L., 1982. "La polémica del Darwinismo y la inserción de la ciencia moderna en Argentina", Closing Lecture, in Congreso de la Sociedad Española de Historia de las Ciencias, Madrid, Actas I, 89–108.

Ortiz, E.L., 1988a. "Las relaciones científicas entre Argentina y España a principios de este siglo, la Junta para Ampliación de Estudios y la Institución Cultural Española." in: Sánchez Ron, J.M. (ed), La Junta para Ampliación de Estudios e Investigaciones Científicas, II, Madrid, II, 119–158.

Ortiz, E.L., 1988b. "El Krausismo en el marco de la historia de las ideas y de la ciencia en Argentina", in: El Krausismo y su influencia en America Latina, Madrid, 99–135.

Ortiz, E.L., 1988c. "Introduction." in The Works of Julio Rey Pastor, 8 vols., London.

Ortiz, E.L., 1991. "Las relaciones científicas entre Argentina y España, convergencias institucionales entre 1870 y 1910." Opening Lecture,in: Real Academia de Ciencias, España and Academia Nacional de Ciencias, República Argentina, II Encuentro Hispano Americano de Historia de las Ciencias, Madrid, 343–356.

Ortiz, E.L., 1993a. "Some Remarks on science and imperialism", Review paper for the Symposium 'The Structure of International Scientific Exchanges', XIX International Congress of History of Science, Review papers, Zaragoza (in press).

Ortiz, E.L., 1993b. "El contexto Europeo de la Revista matemática de Valentín Balbín: 1889–1893", in: Asúa, M. de (ed), La ciencia Argentina: Perpectivas históricas, Buenos Aires, 86–109.

Ortiz, E.L., 1994. "Mathematics in the Iberic and Ibero-American world.", in: Grattan-Guinness, I. (ed), Encyclopedia of the History and Philosophy of Mathematics, London, 1505–1511.

Ortiz, E.L., 1995. "A convergence of interests: Einstein's visit to Argentina in 1925", Ibero-Amerikanisches Archiv, I (in press).

Picciuolo, J.L., 1979. "Misión científica y técnica de la campaña de Roca. Consecuencias para el proceso de ocupación del territorio nacional." in Epopeya del desierto en el Sur Argentino, Buenos Aires: Círculo Militar, Biblioteca del Suboficial, Vol. 698, 201–218.

Potash, R.A., 1969. The army and politics in Argentina, Stanford University Press.

Pyenson, L., 1985. Cultural Imperialism and Exact Sciences: German Expansion Overseas 1900–1930, New York.

República Argentina (Senado), 1948. Diario de Sesiones de la Cámara de Senadores de la Nación.

República Argentina (YPF), 1932. Desarrollo de la industria petrolífera fiscal, Buenos Aires, 429 pp.

Rodríguez, Colonel A.G., 1964. "Reseña histórica del ejército argentino (1826–1930)", Secretaría de Guerra, Dirección de Estudios Históricos, I, No., 1, Serie II, Buenos Aires, 144 pp.

Rodríguez Carracido, 1911, "El problema de la investigación científica en España", Opening Lecture, Asociación Española para el Progreso de las Ciencias, Actas.

Rouquié, A., 1978. Pouvoir militaire et société politique, en République Argentine, Paris: Presses de la Fondation National des Sciences Politíques. Translated as: Poder militar y sociedad política en la Argentina, Buenos Aires, 2 vols., 1981 and 1982, pp. 345 and 461.

Savio, M.N., 1933. Movilización Industrial, Buenos Aires: Ejército Argentino, Escuela Superior Técnica, 1933; reprinted in Obras del General Manuel N. Savio, pp. 14–173, Buenos Aires, 1973.

Schiff, W., 1972. "The influence of German armed forces and war industry on Argentina." Hispanic American Historical Review, 1880–1914, 52, 436–455.

Siegrist de Gentile, N. and Martín, M.H., 1981. Geopolítica, ciencia y técnica a través de la Campaña del Desierto, Buenos Aires.

Walker, M., 1989. German National Socialism and the quest for nuclear power, 1939–1949, Cambridge University Press.

Westerkamp, J.F., 1975. "Física." in Sociedad Científica Argentina, Evolución de las Ciencias en la República Argentina, Vol II, Buenos Aires.

NOTES

[1] I wish to thank Paul Forman, Smithsonian Institution, Washington, Ignacio Klich, University of Westminster, London, and Horacio Reggini, National Academy of Sciences, Buenos Aires, for their valuable suggestions on an earlier draft of this paper.

[2] It should be noted that some of the rebel military leaders in these wars, including San Martin, received their training in élite Spanish units and had fought in Europe during the Napoleonic wars.

[3] Mitre, 1887–1888, vol. I, p. 129; volume II, pp. 115–119.

[4] A theme that would recur in Argentina's history.

[5] Ortiz, 1982 and the references given therein.

[6] Later Roca was elected President of Argentina in two different periods, the first of them immediately after the end of this campaign, from 1880 to 1886 and again from 1898 to 1904.

[7] Comisión Científica Agregada al Estado Mayor General de la Expedición al Río Negro (Patagonia), volumes I–III, 1881.

[8] Ibid., volume I, p. ix.

[9] Ibid., volume I, p. x. Such scientific view of the military arts was consistent with scientism and the positivistic views prevalent in Argentina at the time.

[10] With roots in the Colonial period.

[11] Bahía, 1888. Cárcano, R.J., 1893. Bose, W.B.L., 1966. Fundación Standard Electric Argentina, 1979.

[12] Olascoaga, 1880. Picciuolo, 1979. Cáceres Freyre, 1979. Siegrist de Gentile and Martín, 1981.

[13] Bahía, 1894.

[14] Ortiz, 1993b, 1994.

[15] Rodríguez, 1964. Marti Garro, 1982.

[16] The war with Paraguay lasted from 1865 to 1870.

[17] Colegio Militar.

[18] Escuela Náutica, later Escuela Naval Teórico Práctica and finally Escuela Naval Militar.

[19] Oficina Topográfica Militar.

[20] The first broad scientific journal edited by Argentines.

[21] Sociedad Científica Argentina

[22] Estado Mayor.

[23] Escuela de Ingenieros Militares.

[24] Decree of February 5, 1886; Archivo de la Universidad de Buenos Aires, Correspondence, 1886.

[25] Ministerio de Guerra y Marina, Dirección de Marina.

[26] Balbín, 1895.

[27] On the torpedo in the Argentine navy see the documented work of Burzio, 1968.

[28] La Plata became the capital of the State of Buenos Aires, as Buenos Aires became the capital of the Republic.

[29] Oficina Central de Hidrografía.

[30] Observatorio de Marina, later Observatorio Naval.

[31] Ministerio de Marina.

[32] The slow development of this naval observatory prior to its absorption by the University of La Plata may be related to its specific tasks.

[33] Ministro de Guerra y Marina.

[34] García Reynoso, 1896. From 1895 engineer Enrique Chanourdie edited *Revista Técnica*, an engineering journal directly related to applications, which served also as a forum for the technical discussion of large engineering projects. A microfilm edition of *Revista Técnica*

is currently being prepared by The Humboldt Library, an editorial project based at Imperial College, London.

[35] República Argentina (YPF), 1932.

[36] Through the ideological influence of positivism and scientism in academic circles. See Ortiz, 1988b.

[37] Initially the undergraduate physics course differed only slightly from that of electrical engineering.

[38] On Bose see also Loyarte, 1924, and Pyenson, 1985. An edition of the works and correspondence of Emil Bose is being prepared by the author for The Humboldt Library.

[39] Rodríguez, A., 1964, Chapter 5.

[40] Escuela Superior de Guerra.

[41] This is clear from pre-1914 ESG documents, particularly from its *Memorias*.

[42] Escuela de Aplicación de Artillería e Ingenieros.

[43] The diplomatic side of this process has been discussed in a documented paper of Schiff, 1972, which has taken into account information from dispatches from foreign affairs departments of the countries concerned.

[44] There were, of course, those who opposed this view within the armed forces and preferred the French model.

[45] Lütge, Hoffmann and Körner, 1955, and the new enlarged edition, Lütge, Hoffmann, Körner and Klingefuss, 1981, is a standard reference on Germans in Argentina; Newton, 1977, has studied in detail the activities of the German community in Argentina contributing interesting new perspectives.

[46] Pyenson, 1985 studied German scientific expansion overseas in the period 1900–1930 from the angle of cultural imperialism, discussing and comparing the activities of German scientists and the incidence of German institutions in an extensive geographical area which includes Argentina as well as the South Pacific and China.

[47] This may be true more generally.

[48] The same holds for other nationalities.

[49] I shall mention only three examples over the 1870–1930 period:

1) the mining engineer and mineralogist Hermann Avé-Lallemant, who arrived in Argentina in the 1870s; he was one of the first Argentine marxists and was active in Argentine left-wing politics;

2) physicist Jakob Johann Laub, who arrived in the early 1910s, became an Argentine citizen in 1915 and associated himself with the more radical group of supporters of president Yrigoyen's party; he attempted to play a role in Argentine international relations from at least 1916 to the end of the 1920s;

3) the physiologist Georg Friedrich Nicolai, arrived in 1922 and became a leading figure in the left-wing Argentine *intelligentsia* until well into the 1930s, even after he was forced to move from Argentina to Chile because of his radical views.

   The relatively high proportion of Jews among German speaking scientists in Argentina deserves also special consideration. It may point to discrimination or career delays at home which, again, is not fully compatible with the acceptance of contemporaneous patterns of social behaviour in Imperial Germany. See Ortiz, 1993a.

[50] Ortiz, 1988a, 1991, 1993a.

[51] This is not surprising as the Sociedad Científica, in charge of that meeting played in Argentina, among others, the role of an association for the advancement of science. See Ortiz, 1991.

[52] Escuela de Aviación Militar.

[53] Escuela de Aviación Civil.

[54] Servicio Aeronáutico del Ejército.

[55] Ortiz, 1988a, 1988b.

[56] Ortiz, 1988c.

[57] Ortiz, 1995.

[58] On army and politics in Argentina see the well documented studies of Potash, 1969 and Rouquié, 1978, where extensive references on this topic can be found.

[59] An argument used for this policy was that since there was no danger of war, prices of armament would not be artificially inflated.

[60] Fábrica Militar de Aviones.

[61] Archivo General del Ejército, Buenos Aires, File 12178.

[62] Savio, 1973.

[63] Savio's sources were mainly publications in the French language; Italian sources were also used.

[64] Ibid., p. 148.

[65] He called it Dirección Técnica de las Fabricaciones de Materiales de Guerra, a slightly different name, Dirección de Fabricaciones Militares, was adopted when this organization was created.

[66] Among them the mathematician Luis A. Santaló and the engineer Casimiro Lana Sarrate.

[67] For example, the mathematicians Beppo Levi and Alessandro Terracini and the physicist Andrea Levialdi.

[68] Among them was the nuclear physicist Guido Beck, who arrived through Portugal in June 1943.

[69] Ejército Argentino, 1951.

[70] The industrial conglamerate developed by engineer Torcuato Di Tella had such a relationship with Westinghouse.

[71] The glass works owned by León Fourvel Rigolleau who, like Di Tella, was a member of the Council of the Unión Industrial Argentina which created the Institute of Industrial Studies had an association with Corning Glass.

[72] Federación de Agrupaciones para la defensa y progreso de la Universidad democrática y autónoma, 1947.

[73] The document may have been compiled in Rosario, as purification of the local Universidad del Litoral is perhaps more accurately documented than at any other university; this may be seen from Table 1.

[74] Rockefeller Archive Center, New York, Documents in 1.1, 301, 3, 15.

[75] An important example is that of Nobel Laureate Walter Cannon, 1947, p. 180.

[76] In this period small seminar meetings in private homes, a remarkable form of underground culture began to spread. Many young students benefited from it, securing some minimal degree of continuity to culture in Argentina.

[77] This policy would be repeated in Argentina, after new coup d'état dislocated again its scientific community.

[78] República Argentina, (Senado), 1948; September 15, pp. 1938–1946.

[79] Forman, 1987, and references therein.

[80] This project was under the supervision of Colonel Enrique P. Gonzalez, an officer we mentioned before as a close associate and once competitor of General Péron for the leadership of the military lodge GOU. Before his assignment he was placed in the key position of head of the office of immigration, from which it was possible to provide documents to foreigners the government may wish to bring into the country.

[81] For nuclear research in Germany in that period see Beyerchen, 1977 and Walker, 1989 and references therein.

[82] Perhaps a project done in parallel with the better known German nuclear effort.

[83] The name of Tank has been suggested as Richter's contact with Argentina (the name of other persons has also been suggested); the success of his own work may have added credibility to Richter's claims. Before moving to Bariloche, Richter worked in Córdoba, where Tank had his laboratories. Later Tank apparently dissociated himself from Richter.

[84] Fusion was not the only extravagant project of the time, it seems that an anti-gravitation paint, "developed during the war", was sold in Germany to officials of the same government for a considerable sum of money.

[85] Trained in Great Britain and the USA; in Spain; in the USA and in France respectively.

[86] The government could not have expected to depend exclusively on Richter and his very small group of collaborators for the development of fusion, if actually achieved.

[87] The official investigation of Richter's affair was reported in a most important document prepared in the last months of 1955: Comisión Nacional de Investigaciones, 1958. For details on Richter's affair see Westerkamp, 1975 and, particularly, Mariscotti, 1985, which contains valuable and extensive information, mainly from Argentine sources. Mariscotti's book is a serious attempt to develop the results of the 1955 commission through further interviews with some of leading personalities of the time and new documentation. The testimonies of Colonel González and of physicist Enrique Gaviola, their views on the affair and on their respective roles some thirty years earlier, even if given with the benefit of hindsight, are interesting and revealing of their intense personalities.

[88] This listing is not exhaustive; another interesting institution, the Departamento de Investigaciones Científicas, was created in 1951 by the Universidad de Cuyo, in Mendoza; see Ortiz, 1980.

[89] Training abroad was difficult then because of the war.

[90] Comisión Nacional de Investigaciones.

[91] Comisión Nacional de Investigaciones, 1958.

[92] Either in medicine or in other fields (such as the humanities), these institutions often were not in a position to pay salaries to all its members.

[93] Its Physics Department lost 85% of its personnel according to Westerkamp, 1975; the Mathematics Department suffered similar losses.

JAVIER ORDOÑEZ AND JOSÉ M. SÁNCHEZ-RON

NUCLEAR ENERGY IN SPAIN:

FROM HIROSHIMA TO THE SIXTIES

The Spanish Civil War (1936–1939) provoked a social, cultural and economic collapse, the effects of which were to be felt long after the end of the fighting. These effects were aggravated by Spain's isolation that followed the defeat of the Axis in World War II (1939–1945). And to the political and economic isolation suffered by Spain after 1945 must be added scientific and technological isolation.[1] It was not only that the Spanish scientists found themselves discriminated against by the international community on account of the character and policies of their government, but also the Falangist ideology and the practical requirements of their country facing a world dominated by the Allied victors imposed on them a system of self-sufficiency. Thus, rather than participating in fields of basic research common to the international community, Spanish scientists favoured applied research, particularly such as was aimed at covering the basic necessities of consumption (e.g., fuel). Two further circumstances contributed significantly to creating a situation that was far from ideal for designing a scientific policy comparable to that which had been, or was then being, implanted in other countries; firstly, some of the greatest Spanish scientists found themselves separated from the officially promoted science sector, either because they were in exile abroad or because they were in an enforced interior "exile"; secondly, the victors in the civil war desired to, and, in one way or another, tried to, install their political and religious ideologies in science as practiced in Spain.[2]

For these reasons, the analysis of Spain's reaction to the greatest scientific-technical challenge left by the Second World War – nuclear energy – is a most interesting historical problem.

1. SPAIN IN THE INTERNATIONAL POLITICAL ARENA AFTER WORLD WAR II

Due to the country's battered economic and industrial condition as well as to the ideological principles of the Falangist Party, Spain under the Franco regime was unable to take advantage of the opportunities offered by the five years of European conflict to develop an industry supplying raw materials and manufactured goods to the warring countries. Further, although officially Spain maintained a neutral position and a posture of non-aggression during the Second World War, its unambiguous ideological commitment was to the

185

P. Forman and J. M. Sánchez-Ron (eds.), National Military Establishments and the Advancement of Science and Technology. Studies in the 20th Century History, 185–213.

Axis forces; indeed, Spain sent a military force to the Russian front: the "División Azul" (Blue Division) comprising some 18000 men.

Of course, the Allies were well aware of Spain's position in the international arena. Near the end of the war, on March 10, 1945, President Roosevelt wrote to the new US ambassador in Madrid in the following terms:[3]

Having been helped to power by Fascist Italy and Nazi Germany, and having patterned itself along totalitarian lines, the present regime in Spain is naturally the subject of distrust by a great many American citizens, who find it difficult to see the justification for this country to continue to maintain relations with such a regime. Most certainly we don't forget Spain's official position with and assistance to our Axis enemies at a time when the fortunes of war were less favorable to us, nor can we disregard the activities, aims, organizations and public utterance of the Falange, both past and present.

These actions cannot be wiped out by actions more favorable to us now that we are about to achieve our goal of complete victory over those enemies of ours, with whom the present Spanish regime identified itself in the past spiritually and by its public expression and acts.

The fact that our Government maintains formal diplomatic relations with the present Spanish regime should not be interpreted by anyone to imply approval of that regime and its sole party, the Falange, which has been openly hostile to the United States and which has tried to spread its Fascist party ideas in the Western Hemisphere. Our victory over Germany will carry with it the extermination of Nazi and similar ideologies.

And, indeed, the arrival of the Allied victory only increased Spain's isolation.

In June, 1945, the international conference held in San Francisco to constitute the United Nations Organization, approved unanimously the proposal of the Mexican delegation that participation in the new organisation be disallowed "to the states whose regimes have been established with the help of military forces belonging to the countries which have waged war against the United Nations, as long as those regimes are in power."[4] The resolution obviously referred to Spain, although no nation was named.

One month later, in July 1945, H.S. Truman, J.V. Stalin and C.R. Attlee, meeting in Potsdam, jointly condemned Spain and reiterated that it was impossible for Spain to join the United Nations. On this occasion the reference to Spain was explicit:[5]

The three Governments feel bound ... to make it clear that they for their part would not favour any application for membership put forward by the present Spanish Government, which, having been founded with the support of the Axis Powers, does not, in view of its origins, its records and its close association with the aggressor States, possess the qualification necessary to justify such membership.

Finally, in December, 1946, the General Assembly of the United Nations passed a resolution barring the Franco Government from membership in the UNO's specialized agencies, and recommending that all members of the organization immediately recall from Madrid their Ambassadors and Ministers Plenipotentiary accredited there. This resolution marked the lowest point in Spain's political isolation following War World II.

During these years the public posture of the United States government towards Spain was consistent with the United Nations resolutions. However, in private it was somewhat different. Thus Dean Acheson, Secretary of State, wrote on 18 January 1950 to the chairman of the Senate Foreign Relations Committee pointing out that the US delegation had serious doubts about the wisdom and efficacy of the actions recommended in the December 1946 resolution, but that it had voted for the resolution "in the interests of harmony and of obtaining the closest possible approach to unanimity in the General Assembly on the Spanish problem."[6] Acheson went on to say that "we have stated on a number of occasions that we would favor the amendment of the 1946 resolution of the General Assembly to permit specialized agencies to admit Spain to membership." There were, however, limits to the American government's support of Spanish participation in the international political arena: "It is difficult to envisage Spain as a full member of the free Western community without substantial advances in such directions as increased civil liberties and as religious freedom and the freedom to exercise the elementary rights of organized labor."

But America's coolness towards Franco's Spain was not maintained on the economic level; as Acheson pointed out in his letter, the US Government favoured an economic policy "based on purely economic, as distinct from political grounds." Therefore, there would be no objections to private business and banking arrangements, as well as trade activities. Spain would be free "to apply to and with the Export-Import Bank for credits for specific projects on the same basis as any other country."[7]

Soon, however, the political objections made in Acheson's January 1950 letter began to fade away in light of the increasingly polarized international political situation. (The Korean war, it will be recalled, began in June 1950.) Spain, whose Government was fiercely anti-communist, increasingly appeared to the United States as valuable ally in a strategically important location. In particular, the US Navy and Air Force pressed their Government to strengthen its relations with Spain.[8] No doubt all these factors contributed to the United Nations' annulment on 4 November, 1950, of its 1946 resolution. Thereupon, Spain was permitted entry into the organisation.[9]

2. WITH NEWS OF THE ATOMIC BOMBS, NUCLEAR PHYSICS ARRIVES IN SPAIN

In the plans and proposals for the reconstruction of Spanish *nacional* science under the first Franco governments – that is, until the middle forties –, nuclear studies and research do not appear anywhere, neither in the civil nor the military sectors. This circumstance is, in fact, only a continuation of that which existed prior to the Civil War. It is quite true that the first third of the century is considered to be a period of relative success in Spanish physics. With some exaggeration, it has been called a "Silver Age".[10] But nuclear physics was not among the areas of research cultivated. Before nuclear energy burst on the world in the summer of 1945 as a power that promised to modify the world political balance and change the industrial life of the nations possessing it, very little attention had been paid to it in Spain.

The absence of established groups dedicated to nuclear physics does not mean, however, that no attention at all had been directed to this branch of physics. Nearest to this field was the research carried out on atomic spectroscopy in the Laboratorio de Investigaciones Físicas (Physics Laboratory) of the Junta para Ampliación de Estudios e Investigaciones Científicas (Board for the Support of Scientific Studies and Research). The Director of the spectroscopy section, Miguel Catalán, who had discovered multiplets whilst working in London in the early 1920s, was the principal exponent of this research field from which also some information about nuclear physics could be drawn.[11]

But as an institution closely associated with the ideology and government of the Spanish Republic, the Junta and its Physics Laboratory did not survive the overthrow of the liberal regime. In its place Franco set the Consejo Superior de Investigaciones Científicas (CSIC) (Higher Council of Scientific Research), founded in November, 1939, the most important institution dedicated to research in Spain following the Civil War. It is not necessary to enter into a detailed examination of this institution to discover the deeply conservative, catholic and nationalistic, ideology which prompted its establishment.[12] Thus, we can read in the preface to the law which created the CSIC:

In the most decisive turning points in its History, the Spanish nation has centred its spiritual efforts on the creation of a universal culture. This must also be the most noble ambition of present day Spain, which, in the face of past poverty and paralyzation, now feels the stirring of the will to renew its glorious scientific tradition.

Such a task must be based, above all, on the restoration of the classic and Christian unity of the sciences, destroyed in the XVIIIth century. To do this, we must repair the divorce and discord which exists between speculative and experimental science and we must promote the harmonious development of those branches of the tree of science as a whole along with the existence of others. We must create a counterbalance to fight the exaggerated and solitary specialisation of our time, returning the sciences to a regime of sociability, a frank and safe return to the imperatives of coordination and hierarchy. We must, finally, impose upon the order of culture the essential ideas which have inspired our Glorious Movement, where the purest lessons of universal and catholic tradition lie side by side with the demands laid down by modernity.

As retrograde as this preamble may sound, we cannot deny that for decades the CSIC was practically the only institution which carried out any kind of research in Spain.[13] Nuclear physics, however, was not one of the branches to be initially encouraged.

True, as early as 1903 a Laboratory of Radioactivity had been set up (renamed, 1910, the Radioactivity Institute). But the limitations and interests of its founder and director, José Muñoz del Castillo, who was professor of Inorganic Chemistry in the Universidad Central (Madrid), led the Institute not towards atomic physics, but into the production of maps showing Spain's radioactive zones – which indeed were more extensive and substantially richer than those in any other western European country. It was not unreasonable, therefore, that in 1940 the Radioactivity Institute became a part of the Instituto Nacional de Geofísica (National Geophysics Institute).

We will return shortly to this Institute – and to radiactive ores as a distinctive Spanish national resource – but first let us see what notice was taken in Spain of the lessons learned in Hiroshima and Nagasaki, and of what was then immediately revealed of the scientific and industrial effort behind those cataclysmic explosions.

The news of the dropping of the atomic bombs was naturally reported promptly in the Spanish press. However, what interests us particularly is the reaction in "academic" circles to the news. Among these, two examples are especially revealing.

On September 1st, 1945, *Ibérica*, the most important scientific information magazine in Spain, published an article, "The Atom Bomb," written by Francisco Maldonado, the Director of the Chemistry Section of the Viladot Oliva Laboratories.[14] Dated August 12, the article had thus been written in the very week in which the two, differently constructed, atomic bombs of uranium and plutonium exploded over Japan. The novelty and surprise that the event represented for the author, and, presumably, for all the other Spanish scientists, becomes immediately clear. What it also reveals, however, is a fair knowledge of such basic pre-World War II literature on nuclear energy and disintegration preluding – but not including – the discovery and investigation of the fissioning of uranium by neutrons.

Before beginning this article, I should like to point out to my readers that it has been written scarcely a week since the world was rocked by the sensational news [of the dropping of the atomic bomb]; consequently, there are almost no sources of information available to me. Articles and newspaper reports tend to contradict one another.

The only thing which seems to be completely agreed upon is that the bomb is based on the disintegration of uranium.

Consequently, upon developing the question, I have based my article on the following scientific works: Publications from the Atomic Disintegration Laboratory of Cockroft and Walton, in Cambridge, Anderson of Chicago, Dirac of Cambridge, from Bothe and Kolhörster, from Rossi of Padua, etc.

As this quote suggests, the atomic bombs gave rise to a veritable thirst for atomic knowledge. Just what sort of knowledge would satisfy that thirst is suggested by our second example.

Towards the end of 1945, a lecture on "Atomic energy. Its characteristics and application for military ends" was delivered by José Ignacio Martín Artajo, a Jesuit professor of Electrotechnology in the Catholic Institute of Arts and Industry (ICAI),[15] a prestigious institution kept by the Jesuits and dedicated to technological education. In the published version of the paper, the author points out that "an extraordinarily large number of High Ranking [military] Officers, Professors and Engineers attended the conference ...; many other people who were interested in the matter were unable to attend, but in one way or another, it has been brought to my attention that they wish to have this paper at their disposal, considering it to contain a clearly expressed technical summary of the activities now under way following the manufacture of the atomic bomb, which, likewise, they consider not to have been included in those numerous articles ... that have so far been written on the subject".[16]

Two features of Martín Artajo's paper are worth noting: on the one hand, the emphasis which is placed on "the deepest secrecy that surrounded and still surrounds the manufacture of the atomic bomb with regard to the military use of atomic energy. Thus, we must rely on what those privy to the secret may desire to let us know." Immediately following this declaration of scientific dependency, the author had no qualms in acknowledging that the only sources to which he had had access were certain American magazines and the "official report published by Henry DeWolf Smyth for the cultured American people."[17] This last statement shows that politics and ideology notwithstanding, the sympathies of the Spanish right had begun to change. Though the United States was one of the countries then contributing to the political and economic isolation of the Franco regime, American science, and through it also the American people, were being complimented in Spain.

We have, therefore, a situation where, long before the "Atoms for Peace" program, of which we shall talk later, atomic energy, or, rather, atomic *power*, began to promote sympathies for the United States, at least in so far as Spain was concerned. More concretely, it can be seen that the influence of the famous "Smyth report", the United States Government's official report, published in the summer of 1945, which explained the US effort with regard to nuclear research for the purposes of war, was not limited to the geopolitical area of the United States, but also affected countries such as Spain, which were situated on what we might refer to as the scientific and political 'periphery.'[18]

3. TOP SECRET: EPALE

Notwithstanding the widespread interest – including that of military officers – aroused in 1945/46 by the Americans' introduction of nuclear weapons, it was not until 1948, and then only in response to a foreign initiative, that Spain set up a policy and executive body in the field of nuclear science, technology, and natural resources: EPALE/JIA. In that year, Francesco Scandone, then professor at the University of Florence, was invited to give a course of lectures on "Interference filters", "Antireflecting plates" and "Phase microscopy" in the recently created "Daza Valdés" Institute of Optics of the Consejo Superior de Investigaciones Científicas. Scandone asked the participants in the course if he might be able to speak to someone who could inform him about uranium deposits in Spain. The director of the Optics Institute, José María Otero Navascués, was a Navy Artillery engineer; Armando Durán, Head of Section for Geometrical Optics and System Calculus of the Institute, was attending Scandone's lectures. Both these men would later play a significant role in the history of nuclear energy in Spain. Durán in particular surmised that Scandone was not asking his question in a totally disinterested way. They were not wrong. In conversation with Durán, the Italian professor explained the interest that a group of Italian researchers had in nuclear studies. It was obvious that the Italians wanted to develop a nuclear research programme but had no uranium deposits of their own. Given that the allies still looked upon the Italians with a certain amount of mistrust,[19] Spain was apparently the only available source during these early postwar years.

Durán was well connected among the powerful of that period. He visited a friend, General Juan Vigón, who since 1940 had been Minister of the Air Force. Collaboration between Italians and Spaniards was begun as a result of these contacts and conversations. In order to provide legal and financial backing to the group of Spanish researchers who would dedicate their efforts to nuclear subjects, a company was founded: Nominally a private company, EPALE (*E*studios y *P*atentes de *A*leaciones *E*speciales) [Studies and Patents for Special Alloys]. EPALE was administratively covered by a secret decree which denominated the company Junta de Investigaciones Atómicas (JIA, Atomic Research Board). Franco signed the decree on September 6, 1948. Its preamble explained the opportunistic considerations underlying the formation of the Board:[20]

The possibilities of exploiting atomic energy for industrial purposes have awakened, in all parts of the world, not only an interest in keeping all research in this important field of Science within the utmost secrecy, but has also awakened a greed for acquiring the basic raw materials needed to exploit this future source of energy, materials which might exist in other countries. As there is the possibility of the existence of these materials not being duly known and controlled, there is also the possibility that they might be exported in the same way as other minerals, which have far less value.

This kind of radioactive mineral does exist in our country, which makes it necessary, through considerations of economic imperative and national security, that we know of its existence and that we prepare a team of technicians who are qualified in the surveying and use of it, with a view to exploiting atomic energy by means of a technical interchange with foreign countries and collaboration with those countries that are dedicating their efforts towards studies and gaining of experience in the matter.

The Junta de Investigaciones Atómicas depended directly on the Presidency of the Government and had the following aims:

a) To encourage the research necessary to determine the siting and extent of Spanish uranium deposits, together with other radioactive minerals which might possibly be used in the production of atomic energy.

b) To study the domestic possibilities, or, otherwise, through an interchange with other countries, of making a profit from these minerals and of transforming the mineral into pure oxide on an industrial scale.

c) To establish relations with other foreign bodies, with a view to forming a team of Spanish scientists qualified with modern knowledge regarding surveying for radioactive minerals and the industrial profit to be made from atomic energy.

d) To make use of, on an experimental scale, the material necessary for the production of atomic energy.

e) To prepare and project the building of an experimental thermonuclear pile.[21]

f) Those activities that the Board might consider necessary to progress with the experiments in the application of atomic energy.

In this list of the purposes of the Board, two points can be emphasized. Firstly, there is an evident desire to prevent the country from being drained of valuable radioactive minerals through ignorance or a lack of foresight.[22] Behind this concern there lay not only the scarcity of useable uranium minerals at that time, but also the rhetoric of self-sufficiency of the Franco regime. Secondly, there is a clear recognition of the lack of Spanish scientists qualified in atomic and nuclear physics. In those days of political isolation, the possession of some scarce raw material which was acceptable to other countries was considered to be a matter of great strategic importance.[23] Added to this were the industrial interests that were to have been the concern of the Junta de Investigaciones Atómicas, for in the second article of the aforementioned Decree it was suggested that the Board might be converted into an industrial company.[24]

The organic structure of EPALE was very simple. As chairman of the board of directors of the fictitious company, the government named the prestigious engineer and scientist Esteban Terradas, who was, at this same time, also intervening in the direction of the National Aeronautics Institute (Instituto Nacional de Técnica Aeronáutica).[25] The other members of the board were José María Otero, Manuel Lora Tamayo (professor of organic chemistry; he would become minister of Education from 1962 to 1968), Armando Durán and José Ramón Sobredo. Its offices were in the Optics Institute of the CSIC and the two strong men of the Institute, Otero and Durán, were those who most influenced the Board.

Terradas served as a bridge between EPALE and the university, and some of the courses which were given during 1949 and 1950 in furtherance of the Company's interests were developed in the Physics and Mathematics Seminar directed by Terradas at the university. Examples of these activities are courses such as: "An introduction to nuclear physics" and "The theory of nuclear reactors" by Carlos Sánchez del Río, and "Quantum mechanics" and "Neutron moderation" by María Aránzazu Vigón. The former, together with the physicist, Ramón Ortiz Fornaguera, had benefitted from EPALE grants to begin their education in atomic and nuclear physics; they had visited foreign centres, especially Italian ones (the Nuclear Physics Institute in Rome and the Centro de Informazione, Studi ed Esperienze, Milan).[26]

In the experimental area, EPALE was endeavouring to obtain, in as short a time as possible, all the techniques necessary to build and operate a slow neutron uranium reactor. Thus, in the EPALE report corresponding to 1949–50, we find that work was being carried out on "The preparation of counters", "Installation to prepare BF_3", "Ionization chambers" (two had been built),

"Tests with nuclear plates" (Alpha particle tracks in photographic plates), "Accelerator project" (Cockroft-Walton type), "Electronic circuits" (three general power supplies were set up to provide the voltage necessary for the remaining ordinary electronic devices, together with various high voltage generators, amplifiers and a scaling circuit), "A Discriminator" (useful for obtaining the curves expressing the number of particles of various energies incident upon a particle detector), and "Counters". The report also made clear the isolation of Spanish physicists: it was pointed out that one of the reasons for the slowness with which the project was being developed was that they did not even know of possible suppliers for instruments needed in the laboratory.

4. THE NUCLEAR ENERGY BOARD

The secret period of the Junta de Investigaciones Atómicas did not, however, last very long. Certainly the research that was being carried out did not justify such measures. But the opening up of information on nuclear activities in Spain coincided with modifications in the political situation; particularly the change in the attitude of the United States towards Spain. Thus, late in 1950 we begin to find newspaper articles on the subject. The break with secrecy came as declarations by José María Otero published in the newspaper *La Vanguardia*:[27]

It is time that people knew – says Professor Otero – what is being done, and what is planned to be done with regard to atomic research in Spain. Remaining silent on the subject can do no more than encourage our enemies to fantasize, especially considering that, at the present moment in time, we have nothing to hide, nor have we to cover up any negligence. Logically, nuclear and atomic research is the responsibility of the Consejo Superior de Investigaciones Científicas and, more particularly, the Physics Council [Consejo de Física]. Due to the nature of these studies, several chemists have been incorporated into the Council.[28]

This first period of Spanish atomic and nuclear research, if not wholly secret, was certainly discreet. Its end coincided, approximately, with the death of Terradas (9 May 1950). Indeed, on 21 October, 1951, through a Decree-Law, EPALE/JIA was superseded by the Junta de Energía Nuclear (Nuclear Energy Board; JEN).

Although formally a new institution, the JEN was based on the old structure, persons and experiences of EPALE. In keeping with the provision in the founding Decree of EPALE which foresaw its transformation into an industrial company, the JEN was attached to the Ministry for Industry. Its first chairman was the same General Vigón, whom we encountered at the inception of EPALE, while Otero Navascués reappears here as vice-chairman and director general. On Vigón's death, in May, 1955, the chair was assumed by another military man, General Hernández Vidal, who remained in the post until 1958. It is a characteristic of the Spanish Nuclear Energy Board that although it was not part of any military department, for many years all its general directors were drawn from the military.

According to its Charter, the JEN had the following main purposes:

- Mineral surveying to discover radioactive deposits.
- Exclusive exploitation of said deposits.
- Obtaining, preparing, conserving and treating the minerals and chemical products necessary for nuclear research and its applications.
- Obtaining, distributing and employing of radioactive isotopes.
- Training of technicians and scientists.
- Establishing relationships with foreign organisms dedicated to the same fields of research.
- To execute and carry out those studies, works, exploitation plants and installations which might be required. To create pilot plants and prototype factories.
- To advise the Government in all matters relating to atomic energy. In particular, to propose the pertinent legislation.

As can be seen, the Board enjoyed an authentic monopoly with regard to nuclear energy in Spain. Consider, for example, "radioactive mining," one of the chief directions of the JEN's efforts. The Decree-Law founding the Board envisaged the creation of mining areas, upon the suggestion of the JEN. If necessary, these areas could be considered "top secret military property." Exploiting of radioactive minerals could be done only by the Board. Individuals were obliged to inform the Board of any deposits they knew or discovered (in the latter case, they would receive an indemnity for the costs incurred in the discovery). Furthermore, the Decree established the conditions by which the Board could expropriate land where mines were discovered.[29] The first mine to be exploited was in Sierra Morena. In 1954 a new town was created, Santa Bárbara de la Sierra, which was inhabited at first by 250 families, consisting of technicians and workmen. Thus, Spain began its own production of uranium bars on a semi-industrial scale, becoming the third country in Europe, after the United Kingdom and France, with a chemical treatment pilot plant.

The history of the JEN has yet to be written, and it is not the task of this paper to do so. It should merely be pointed out that, on the whole, it is possible to differentiate various phases during this first period of the Board's existence.[30] The first phase goes from the founding of the Board to the death of Juan Vigón (1951–55). The main effort at that time, leaving aside the surveys to discover mineral deposits, was directed towards building a heavy-water moderated, zero-power reactor, using Spanish natural uranium as fuel. Furthermore, the training of personnel for later work was begun.

The second phase runs from the summer of 1955 until the end of 1958. During this phase, the main activity of the Board was the building, with US aid, of a research reactor. This phase opens with the signing of the aid agreement, and closes, in November 1958, when General Franco inaugurated the "Centro Nacional de Energía Nuclear Juan Vigón" (Juan Vigón National

Nuclear Energy Centre), in Madrid's University City. Until that time, the JEN's personnel had worked in various offices and laboratories belonging to the CSIC or the Madrid University, a clearly unsatisfactory arrangement.

5. NUCLEAR ENERGY INFORMATION TRAVELS ABROAD

We have already mentioned the changes in the attitude of the US Government towards the Franco regime, changes that helped Spain gain admission to the United Nations organisation late in 1950. It would be another three years, however, before Spain and the United States signed their first diplomatic agreement – on September 26, 1953. Considering what Spain had to offer to the Americans, it is not surprising that the agreement dealt with mutual military assistance.[31] The United States undertook "to contribute to the effective air defense of Spain and to improve the equipment of its military and naval forces," while Spain would authorize the United States "to develop, maintain and utilize for military purposes, jointly with the Government of Spain, such areas and facilities in territory under Spanish jurisdiction as may be agreed."[32]

At this same time the ground was being prepared for the agreement, two years later, providing U.S. assistance to Spain in the field of nuclear energy. On December 8, 1953, President Dwight D. Eisenhower presented, before the General Assembly of the United Nations, his programme for the peaceful use of atomic energy: "Atoms for Peace."[33] Eisenhower proposed that an international agency be set up, under the auspices of the United Nations, to which the United States would contribute both fissionable material and reactor technology. As a result of this proposal, the International Atomic Energy Agency was established, with offices in Vienna. Further, on August 30, 1954, Eisenhower signed a new law on atomic energy, permitting the United States to provide information and aid to friendly countries by means of bilateral agreements; the law previously in force, the so called McMahon bill of December 20, 1945 (known also as the "Atomic Energy Act of 1946"), put severe restrictions on the transference of nuclear know-how to other countries.[34]

Eisenhower's motives behind "Atoms for Peace" were varied. One of them was, certainly, propaganda.[35] The Spanish case shows this facet of "Atoms for Peace" to perfection.

On July 19, 1955, Lewis Strauss, Chairman of the Atomic Energy Commission, Walworth Barbour, Deputy Assistant Secretary of State for European Affairs, and the Spanish Ambassador, José María de Areilza, signed an agreement in Washington DC for cooperation "concerning civil uses of Atomic Energy."[36] The agreement was to enter into force on the day it was signed, and remain in force until July 18, 1960, when it could be subject to renewal.

The agreement obliged the United States to provide Spain (which desired "to pursue a research and development program looking toward the realization of the peaceful and humanitarian uses of atomic energy") uranium enriched in the isotope U–235, "as may be required as initial and replacement fuel in the operation" of the *research* reactors which Spain "may, in consultation with the Commission", decide to construct and operate.[37] The quantity of uranium enriched in the isotope U–235 transferred by the Atomic Energy Commission (AEC) to the Government of Spain should not "at any time be in excess of 6 kilograms of contained U–235 in uranium enriched up to a maximum of twenty percent of U–235, plus such additional quantity as, in the opinion of the Commission, is necessary to permit the efficient and continuous operation of the reactor or reactors while replaced fuel elements are radioactively cooling in Spain or while fuel elements are in transit." It was also stated that when the fuel elements containing U–235 leased by the AEC required replacement, they shall be returned to the Commission unaltered after their removal from the reactor.

The articles of the Agreement make it clear that the United States Government retained, through the AEC, effective control of nuclear matters in Spain.[38] Initially only research reactors were considered in the negotiation of the agreement. However, in the wake of the Atomic Energy Act of 1954, which charged the AEC, for the first time, with "bringing nuclear reactor technology into the marketplace,"[39] language was added expressing an expectation "that this initial Agreement for Cooperation will lead to consideration of further cooperation extending to the design, construction, and operation of power producing reactors." No doubt, the inclusion in the Agreement of a provision that "private individuals or private organizations in either the United States or Spain may deal directly with private individuals and private organizations in the other country" helped achieve the desired openness of the Spanish market to private American companies involved in the construction of power reactors.[40]

Indeed, we know that there were US firms interested in profiting from the Spanish interests in nuclear energy. An example in this direction is provided by Theodore von Kármán, who since 1948 maintained close relationships with the Instituto Nacional de Técnica Aeronáutica (National Aeronautics Institute; INTA).[41] On May 5, 1957, von Kármán wrote to an unidentified "Teddy", offering him some information related to the prospects of nuclear energy in Spain. Apparently, a group of "about 15 Spanish firms (banks and industrial organizations) ... will make up an organization for general study of the nuclear energy question; they are corresponding with almost every one in the business, G.E., Westinghouse." The Chase National Bank was also supposed to participate (the vice-president of the Madrid branch of the bank took part also in the conversations). Von Kármán learned of this through his aeronautical contacts, people associated with INTA and Julio de La Cierva, a cousin of the famous inventor of the autogyro and owner of Manufacturas Metálicas Madrileñas, S.A. Von Kármán, always keen on making some profit, told his correspondent that:[42]

we talked over the question among us and came to the result that for us it would not have much sense to hang on this large group. We thought that if you decide to form a study-company with us, based on the future application of the NDA reactor the following could be achieved:

1) During the development time (about 18 months) of the reactor our joint company could make the first serious study of the possible application of such reactor system in Spain; for production of electric power, for production of power for irrigation, eventually for making sea water usable for agricultural purposes ...

2) In the second stage the study company would determine possibilities of partial manufacturing of the reactor in Spain and the erection of the installation.

But returning to the Agreement between Spain and the United States, it must be pointed out that it was just one among the many that the US Government signed at the time. Treaties identical, line by line, to the one with Spain, were signed in Washington DC, with Turkey (June 10), Israel (July 12), China (July 18), Lebanon (July 18), Colombia (July 19), Portugal (July 21), Venezuela (July 21), Denmark (July 25), Philippines (July 27), Italy (July 28), Argentina (July 29), Brazil (August 3), Greece (August 4), Chile (August 8) and Pakistan (August 11).[43] For American diplomats connected with nuclear energy agreements, the summer of 1955 was, no doubt, a busy one.[44]

However, not all countries received the same treatment. The exceptions were: Belgium, Canada, Switzerland and the United Kingdom. The reasons for these exceptions were diverse.

Belgium had, of course, the Congo uranium ores, and had collaborated with the United States and Great Britain since 1940, providing uranium as well as ensuring effective control of its mines, so that its uranium should be not available to Germany during World War II. With the end of the war, Belgium had continued supplying "a vitally important quantity of uranium" to the United States and Great Britain,[45] which, on their part, had assisted Belgium in the establishment of a research and development program directed to the peaceful use of atomic energy. Taking into account this relationship, it is natural that Article I of the Agreement signed in July 1955 stated that "the Government of Belgium will receive from the United States Atomic Energy Commssion, in the field of the peaceful uses of atomic energy, information and materials on terms as favorable as any other major uranium supplying country except Canada." In particular, Belgium would receive technological information for the construction of specific reactors for the power program which the Belgian Government wanted to implement in Belgium, in the Belgian Congo, and in Ruanda-Urundi. While the agreements signed with countries like Spain, contained few and rather general specifications, in the case of Belgium it was explicitly mentioned that the Belgians were to receive assistance from the

AEC in the following areas: specifications for reactor materials, properties of reactor materials, reactor components, reactor physics and engineering technologies, and environment safety considerations. Moreover, the AEC was to allow selected security-cleared personnel from Belgium to participate in the construction and operation of the first civilian power reactor, the PWR reactor at Shippingport, Pennsylvania, and of such other reactors as might be agreed. Futher, the Commission was obliged to transmit to the Belgians all essential information connected with the design, construction and operation of a thermal, heterogeneous, pressurized light or heavy water (boiling or non-boiling) reactor, should the Belgian Government decide to construct such reactor, and should the Commission have found that such reactor was sufficiently developed to be of practical value for industrial or commercial purposes. This latter condition shows that the AEC sought to retain control over the development of nuclear energy for industrial purposes.

The exchange of information of military significance related to the design or fabrication of atomic weapons was explicitly excluded. There was, however, the possibility that the AEC should communicate classified technical information to Belgium if it were required for the construction of any specific reactor, but with the significant proviso that the Belgians were seriously considering the construction of such a reactor in Belgium, the Belgian Congo, or Ruanda-Urundi, *and* that private industry in the United States be permitted to undertake the construction and operation of the same type of reactor.

Of course, Belgium also had obligations to the United States. The AEC received an option to purchase any special nuclear material produced in Belgian territories. Moreover, Belgium agreed "not to transfer to any country other than the United States or the United Kingdom any special nuclear materials produced in Belgium, the Belgian Congo, or Ruanda-Urundi unless the Government of Belgium is given assurance that the material will not be used for military purposes," and further agreed "to consult with the United States on the international significance of any proposed transfer of any uranium and thorium ores or special nuclear materials to any country other than the United Kingdom."

Clearly Great Britain was a case apart in matters of nuclear energy. The British had cooperated closely with the Americans – although the Americans had not always cooperated equally closely with the British – from the very first days of the Manhattan Project.[46] Among the many agreements American diplomats signed in the summer of 1955, there were two (on June 15) between the United States and the United Kingdom on nuclear energy matters. The first dealt only with the *civil* uses of atomic energy, and did not differ essentially from the accord with Belgium (with the obvious exception of the provisions referring to the uranium ores that the Belgians had and the British lacked).[47] However, the second agreement had no counterpart with the Belgians: it dealt with "cooperation regarding atomic information for mutual defense purposes."[48] The basic point of the Agreement was that both countries "will from time to time make available to the other Government atomic information which the Government making such information available deems necessary to

(a) the development of defense plans;
(b) the training of personnel in the employment of and defense against atomic weapons;
(c) the evaluation of the capabilities of potential enemies in the employment of atomic weapons."

However, all transfers of atomic information would be limited by the provisions of the US Atomic Energy Act of 1954 and the UK Atomic Energy Act of 1946. This meant that there will be no transfers of atomic weapons or special nuclear material.[49]

The Agreement with Canada did not differ essentially from that on civil matters with the British (indeed, both treaties were signed the same day).[50] Of course, the "unique tradition of cooperation between Canada and the United States" – as stated in the Agreement – helped. Still more important, however, was the fact that Canada possessed what the M.A.U.D. Report denominated "the largest supplies of uranium" (together, of course, with the ones in the Belgian Congo).[51] By 1955, the US had been engaged for several years in atomic energy programs with the Canadian corporations (wholly owned by the Canadian Government) Eldorado Mining and Refining Limited and Atomic Energy of Canada Limited.[52]

Finally, a word about the case of Switzerland. The specificity of this case arose from the special position of Switzerland, whose neutrality and geographical situation made of it an ideal place to host international meetings.[53] As the Soviet Union charged that Eisenhower's "Atoms for Peace" of December 1953 amounted to little more than isotopes and fertilizer, the US Government made several moves. One of them was the decision to allocate to an international agency 100 kilograms of fissionable material to serve as fuel in an experimental nuclear reactor for peaceful projects. Almost at the same time it was decided to call an international conference on the peaceful uses of atomic energy. Said conference was to take place in Geneva, during September 1955, under the auspices of the United Nations.[54]

The Agreement signed with Switzerland took into account these facts. Thus, it was stated that at the conclusion of the Conference the Americans would sell, for 180 000 dollars, to the Swiss a research type nuclear reactor "which the Government of the United States of America will construct, locate, and operate in Geneva ... in connection with the aforesaid international conference, together with the reactor building, associated machinery, and exhibits."[55] The rest of the treaty did not differ from the accords with the majority of countries.

6. THE SPANISH EXPERIMENTAL REACTOR

Returning to the Spanish case and the matter of American control: every significant element of choice in one of the two major components of JEN's program in the 1955–58 phase – the construction of a research reactor – was preempted by the Americans' requirement that the fissionable material loaned could be used only as fuel in a "swimming pool" type reactor. Given, however, Spain's backwardness in nuclear science and technology, it could not – and showed no inclination to – do otherwise. The 3 megawatt reactor – JEN-1, as it came to be known, whose core, containing less than 4 kg of uranium-235, was cooled by a stream of water of 16 000 litres per second – came into operation on October 9, 1958, with several dignitaries, including the Secretary of Industry, present. As required by the July 1955 agreement, its basic parts were supplied for free by the United States, as its contribution of half the total cost of the reactor, some 350 000 dollars. JEN-1 was used especially for tests of the effects of neutron and gamma ray bombardment on materials and to obtain short and medium lived radio isotopes for use in medical, agricultural and industrial applications.

During JEN's first two decades, at least, there is no doubt that this institution – in other words, the agency which controlled the entire Spanish nuclear development program – was completely dependent on the supplies, materials and know-how of the United States. In the long term, all dependency is negative, and Spain never became more than a mere consumer in this field. However, it is only fair to recognize that, without the American aid, the albeit modest complex of installations just on the frontier of Madrid University would never have been set up.

For their part, the United States obtained a return which, whilst it is not quantifiable, was by no means insignificant. "Atoms for Peace", for example, was presented to the public without criticism or objections. The US Embassy in Madrid published a brochure which was widely distributed, in which Eisenhower figured prominently with his supposedly pacifist message: "The United States promises before you and, thus, before the whole world, to help to resolve the terrible atomic dilemma, to dedicate to the full, both its heart and its intellect to finding the means by which the miraculous inventiveness of Man be put to the consecration of human life, and not its destruction." At the same time that Eisenhower was pronouncing this message to the Spanish people, the Spanish territory itself was dotted by American Air Force bases, into which, and out of which, atomic bombs were being carried – four of which landed in the sea off the coast of Palomares.[56]

7. NUCLEAR ENERGY AND INDUSTRY

We have mentioned that the July 1955 agreement between the United States and Spain included some references to the application of nuclear energy to industry. It is appropriate, therefore, to pay some attention to the question of how Spanish society viewed the idea of introducing nuclear energy to industry.

Early proof of the interest nuclear energy held in Spanish political circles can be found in a series of eighteen lectures which were held in Madrid between May and June, 1958, coinciding with the exhibition "The atom and its peaceful applications".[57] What is interesting is that both the exhibition and the conferences were organised by the "Sindicato de Agua, Gas y Electricidad" (Water, Gas and Electricity Union).[58] In the opening speech, Daniel Suarez Candeira, national head of the Union, clearly showed his faith in the possibilities of this new source of energy:

So that our program for energy requirements might be efficiently developed and fulfilled, it is absolutely necessary that we have our own nuclear industry. And this is precisely the principal purpose of the plan and program designed by the General Directorship and the Nuclear Energy Board.

This desire to be able to count on a Spanish nuclear industry had, as its first objective, the production of electricity by nuclear power stations. It was estimated that the consumption of electricity for 1965 would be double what it was in 1957. Some 28 or 29 million kw-hrs would need to be produced. "And if, in those circumstances" – Joaquín Ortega Costa pointed out[59] – "we have a very dry year, we should have to produce some 8500 million kw-hrs of heat. This is always supposing that we have doubled our hydro-electric production, a hardly probable eventuality ... To provide, with nationally produced fuel, such a considerable production of heat from coal mining is an enterprise which we can hardly hope to succeed in, given the short amount of time at our disposal." Evidently, these words pointed clearly towards nuclear power as a solution, more especially when it was pointed out that this source of power was also economically viable:[60]

At present, the unit cost for installing a nuclear power station is between 2 and 2.5 times as much as a traditional power station. It is, of course, a considerable difference, but, on the other hand, the repercussions of fuel consumption of a nuclear plant are only a third that of more traditional power stations. When we consider high annual usage, one fact counterbalances the other and, thus, in our country, where coal is expensive, or rather, where useful calories are expensive ... the costs would be practically identical.

Following this line of thought, it was expected that the first nuclear power stations producing electricity would be in operation by 1965.

It was not only in the production of electricity that possible profit was seen. In a lecture ("Applications of radioactive isotopes to industry") given during the aforementioned cycle, José Miguel Gamboa Loyarte, head of the Section of Radiochemistry in the CSIC and of the Isotopes Department of the JEN, mentioned that "in 1953, statistics were drawn up to figure out the savings to the industry of the United States through using radioactive isotopes. The figure arrived at was 100 million dollars a year. In 1956, further information increased this figure to between 255 and 485 million dollars ... it is envisaged that, in coming years, the use of radioactive isotopes will become more and more extensive and that their use by the North American industry will represent savings of some 5000 million dollars."[61] Benefits from the use of radioactive isotopes could be seen everywhere: from the manufacture and packaging of cigarettes to the study of wear and tear on machinery, not to mention the discovery and exploitation of oil deposits.[62]

8. "ATOMS FOR PEACE" IN SPAIN: THE IMPACT OF ATOMIC ENERGY ON SPANISH SOCIETY DURING THE SIXTIES

The beginning of the sixties saw a change in the Spanish economic and social situation. In an attempt to give Spanish politics a coating of legitimacy, the Franco regime celebrated twenty-five years of rule with an elaborate publicity campaign and a referendum. The military and the party (Falange) uniforms were replaced by mufti. Within this process of opening Spain towards the anti-communist world generally, the press repeated, with ever greater frequency, whatever scientific news was drawing the attention of that world. It was at this time that magazines such as *Destino*, and publishers like Tecnos, which sought synchronization with the cultures of other countries, were founded.

Atomic energy formed part of this new spirit, and so it appeared frequently in the press, sometimes in a particularly picturesque way, as in an article which affirmed that Sodom and Gomorrah had been destroyed in a nuclear explosion.[63] In general, however, the articles referred to the beneficial role that atomic energy might play in the development of Spain and other more developed countries. These articles show that there was a definite interest in, or rather, insistence on, the peaceful uses of atomic energy. Particularly emphasized were its advantages (for industry) relative to other forms of energy, together with the fact that it could be used to benefit humanity in the fields of agriculture and medicine, where it was said to be effective in fighting cancer.[64] (It is, however, interesting to point out that despite this "pacifist" spirit, Spain never renounced the possibility of possessing its own arsenal of nuclear weapons; whilst Franco was alive, the United States never managed to persuade the Madrid Government to sign a treaty for the prevention of the proliferation of nuclear weapons.)

It would be reading too much into the "Atoms for Peace" campaign, and the way in which it was developed in Spain, to say that it was the principal reason for the emphasis on the peaceful aspects of atomic energy. However, it seems that it was indeed, for a time, an important element in the idea of a bright nuclear future which pervaded the press of the day – a press which was, in effect, controlled, given that there existed censorship prior to publication. In this sense, Weart's conclusion that "Atoms for Peace" "gave people everywhere the idea, which until then had seemed convincing only to a small elite, that atomic utopia could become a reality within their lifetimes" is confirmed in the Spanish case.[65]

Yet the Spanish case is also exceptional, precisely because of the close connection that the Falangist propagandists sustained between the peaceful atom and a pacified, if not altogether utopian, Spain. Thus, early in 1965 the leading Catalan newspaper linked the discovery of nuclear fission and Franco's victory in the civil war:[66]

While, in Spain, we are celebrating 25 years of well deserved peace, 25 years have also passed since the public announcement of the discovery which marked the beginning of the new atomic age.

Consistent with the 'all's well' rhetorical stance of the regime at the time,[67] only rarely was the nuclear holocaust mentioned.[68] Moreover, never during the long years of the Franco regime (it ended effectively with the dictator's death, in 1975) did there appear in Spain any significant opposition to the peaceful uses of nuclear energy, such as began to appear in more democratic milieus by 1957.

Let us look, therefore, at some of the principal examples of public promotion of "Atoms for Peace" in Spain during the sixties.

In order to spread knowledge and public support of Eisenhower's program of integration of atomic energy into civilian life, the United States Government prepared an exhibition, "Atoms in Action," which was sent throughout the world in the early sixties. It reached Spain in the spring of 1964, having already been on the road for two years and displayed in such countries as Japan, Pakistan, Lebanon, Greece, Thailand and Yugoslavia.

The Nuclear Energy Board (JEN), which had changed little in its internal structure and which still depended on Spain's militaristic state, was the organism chosen to collaborate with the US AEC in putting the exhibition on in Madrid. The Spanish Nuclear Board was not only interested in strengthening the ties which already linked it to the United States, but it too wanted to join the effort to show the human face of atomic energy and, in doing so, to help the Franco regime put off its military image.

The exhibition "Atoms in Action" was installed on a plot of land belonging to Madrid's University City, less than 200 metres from the JEN itself. The stainless steel pavilion had a floor area of 1000 m^2. Its purpose was to help spread the applications of atomic energy to various fields, including industry and health. According to the newspaper *Informaciones*, the exhibition joined together "three main aspects: a fully working scientific laboratory, a teaching institute and, finally, a varied and explicit demonstration for the general public, who would be able to see with their own eyes the various different ways in which atomic energy was being used at that time to benefit Humanity."[69]

Reports in the press seem to indicate that the exhibition managed to achieve some of its purposes, particularly that of promoting the image of the United States as the originator of a new energy source which it would generously share for the good of humanity:[70]

[this exhibition] not only serves to show the general public the most interesting aspects of the peaceful use of atomic studies, but it also serves to awaken and encourage the interest of school children and teachers in this important area of science. At the same time, it serves to improve the collaboration with the country being visited, in this case Spain, with regard to problems and questions on which the North American teams of scientists have made important steps forward.

"Atoms in action" was a success in the popularisation of science. According to the press,[71] approximately eighty thousand people visited the exhibition and more than eight thousand students followed intensive courses in nuclear science. School children from all over the country travelled to Madrid to take part in courses given by Spanish high school teachers who had been trained in the Institute of Nuclear Studies in Oak Ridge, Tennessee.[72]

The official propaganda, at the time of the "Atoms in Action" exhibition, was triumphant; thus, for instance, the Madrid newspaper *Pueblo* published an article entitled "Nuclear science will substitute humanistic education for a more technical training."[73] Together with this kind of analysis, one also finds articles commenting on the prospective applications of nuclear science.[74] The majority of newspaper articles, however, were concerned with the situation of atomic energy in other countries, such as France or Germany,[75] while also emphasising the necessity in the long term for countries throughout the world to find new forms of fuel. In fact, atomic energy was shown as already playing a significant role in world energy consumption,[76] and emphasis was placed on the fact that peaceful activities connected with nuclear science had begun to be the object of private initiatives in other countries, such as the United States.[77]

But information designed to persuade the public of the advantages of the new source of energy, was not restricted merely to journalistic rhetoric. Prestigious scientists also took part. Particularly noteworthy is Carlos Sánchez del Río, professor of Nuclear Physics at Madrid University, whom we earlier encountered as a beneficiary of EPALE.[78] Sánchez del Río was one of the best physicists of the time – a time, it is true, when mediocrity was the keynote. He was, and would continue until the eighties to be, a significant member of the JEN. In declarations published at the time of the aforementioned exhibition, he spoke, in an academic tone, about the conditions necessary for making the new technology viable:[79]

Nuclear technology is more complex than other methods of obtaining energy. This is not, in itself, an obstacle, for all new technologies tend to be more complex than their predecessors. And, yet, the most advanced technologies end up taking the place of those that have gone before, provided that one of the following three conditions is fulfilled: 1. That by the new technology the same product can be obtained for a lower cost. 2. That the new methods may be applied to previously inaccessible items. 3. That the raw materials or circumstances which led to the previous technology's use are exhausted.

There is no doubt that Sánchez del Río's comments were, in general, more astute than the majority of those which appeared at the time of the exhibition,[80] but it is also true that his presentation took it for granted that, for the world at large, atomic energy could only benefit humanity.

9. POLITICS AND NUCLEAR ENERGY

The information which appeared in the press in the early sixties, indicates that all the political forces of that time agreed that it was necessary to encourage the development of nuclear energy. Two such political forces were especially important: the Falangists and the new "technocrats". Although both agreed on the importance of nuclear energy for Spain, they were led to this opinion from opposite directions. The Falangists insisted that nuclear energy was necessary to enhance Spain's autonomy, whereas the technocrats considered it to be a way of opening up the country, of liberalising its economy, and of definitively breaking its isolation through active participation in the international arena. Indeed, credit for promoting nuclear energy became an issue in the political and ideological fight between Falangists and technocrats during the sixties.

Old members of the Falange, through their powerful information organs, insisted that the Spanish nuclear program had begun in 1955, the year it began its public life as an agency dedicated to research into nuclear energy, and thus long before the new economists had arrived on the scene. The technocrats for their part, tried to change the directions and regulations controlling nuclear energy in Spain to accord with their economic and political viewpoints. Thus, in 1963 the minister of Industry, Gregorio López Bravo, one of the leaders of the technocrats, proposed that surveys, research, and exploitation of radioactive minerals be liberalised, and not subjected to the strict control imposed, as we saw, in the fifties. The occasion for this proposal was the completion of a geological map of the Spanish uranium deposits, together with an evaluation of their potential. The JEN played a key role in carrying out the Ministry for Industry's plans: the map was drawn up by Francisco Pascual Martínez, General Technical Secretary of JEN.[81] Morever, a nuclear energy law was presented in Parliament on April 24, 1964. The idea behind this law was to establish liberal criteria for controlling the activities of the nuclear industry, with the intent of enhancing the nuclear activities of those industries related to agriculture and medicine.

10. THE ACTIVITIES OF THE JEN DURING THE SIXTIES: AN APPROXIMATION

A reader of Spanish newspapers in the sixties with a particular interest in the JEN would have found that the information those newspapers supply can be placed into four separate blocks: 1) the activity of the Board itself in its role as promoter of educational courses and international meetings; 2) its research activities directed towards the practical utilization of nuclear energy, especially the construction of nuclear electric power stations; 3) its role as spokesman in international nuclear research agreements; and, finally, 4) its participation in the exploitation of Spanish uranium deposits.

With regard to the first of these activities directed out towards Spanish society at large or towards specific but non-scientific persons or groups, it is interesting to note the significant increase in the numbers of courses (and in the participants in them) in nuclear science and engineering during the economic boom of the sixties. Initially, the courses were rather general meetings, like the Spanish-French Nuclear Conference of October, 1963, the purpose of which was "not to prepare a catastrophic bomb, but to see a way of applying atomic science to peaceful everyday life."[82] From this kind of meeting, the JEN moved on to interdisciplinary meetings in collaboration with university institutions, such as the Faculty of Law of the University of Madrid. Thus, for instance, on January 15, 1964, a course was begun on Nuclear Law, with collaboration between the Board and the University. The inaugural paper was read by José María Otero Navascués, in the presence of the Minister for Industry, López Bravo, and also of the Minister for Education, Manuel Lora Tamayo, who had earlier been a founding member of the JEN.[83]

Over the following years, the activities of the Nuclear Energy Board diversified still further. In 1967, the Board hosted a Symposium on Radioisotope Applications, one of the branches of the atomic world that had most important repercussions for industry. The same political figures present in the 1964 meeting participated also at the closing session of this Symposium. The Director of the Board and the Minister for Industry emphasized "the targets reached in the production of radioactive isotopes and in obtaining users for these products in the fields of science and technology" as evidenced in the fact that "in this country, there are 244 authorised radioactive installations, half of which are dedicated to medical applications, and the remainder to studies for the industrial and agricultural sectors."[84]

The JEN, however, dedicated its greatest efforts to the training of engineers specialised in the building and the operation of nuclear power stations. The periods of training lasted two years and were given to students from the University, Engineering Schools, and Officers from the Technical Corps of the Armed Forces (the Navy, the Army and the Air Force). López Bravo, in his closing speech for the 1967 course, emphasized[85] the importance that research and teaching have on the development of peaceful uses for atomic energy. That development requires specialised personnel to be adequately trained, and this can only be done if we have adequate means, which is precisely what the Board can offer its researchers.

As a matter of fact, the training and research activities of the Board had increased a great deal during that year. There were 56 grant holders who were working on their doctoral dissertations in the Board's laboratories, eight of whom were members of the atomic establishments of foreign countries. The Board had also signed fifteen different research contracts with universities and engineering Schools. In the closing session of the aforementioned course, Professor Quinteiro, Director of the JEN "Instituto de Estudios Nucleares" (Nuclear Studies Institute), pointed out that during the academic year, besides the basic activity of training specialised engineers, six short, but useful, courses had been offered on applications of radio isotopes.

An important goal of the JEN's training programs was the construction of new nuclear power stations. In fact, the Spanish nuclear power station program had began earlier, at least in 1963, when the building of a heavy-water reactor using 1.6% enriched uranium was proposed. The purpose of that reactor was to test fuels which were to be manufactured in Spain. It was then also expected that the plans for Spain's first three nuclear power stations would soon be completed.[86] To support these and other projects, the Nuclear Energy Board put forward an ambitious program "to strengthen the position of the JEN":[87]

A permanent research programme. A continuous plan for surveys and treatment of radioactive minerals. The methodical training of adequate technical personnel. A valuable and well equipped complex of laboratories. A uranium concentrate plant. A pilot plant for producing fuel. A group of accelerators. One experimental reactor and isotope plant. Two experimental reactors in two Engineering Schools.

A truly great effort was made to carry out this program, at least in part. The pilot plant for processing irradiated fuels, known as M–1, was put into service in 1966 as an adjunct to the operation of the Board's research reactor.[88]

Running contrary to this constant tendency towards self-sufficiency was the JEN's conviction that it had to broaden its collaboration with other countries; that the cooperation agreements with the United States were not sufficient.[89] The technocrats wished to open up the Spanish economy towards Europe, and, to this end, they made use of the JEN, which, again, served political purposes. Thus, Spain entered a partnership with France, which also jealously guarded its nuclear independence. The Spanish-French conference of 1963 has already been mentioned. In 1966, the JEN sent the Government, for its approval, a project for building of a new nuclear power station in Vandellós (Tarragona). The project required the collaboration of a European partner, which collaboration would, supposedly, help Spain in its integration into Europe, and decrease its dependence on the United States. This move implied a significant change in the technology (to the use of natural uranium as fuel), and a turnaround in the economic arrangements – as can be seen from declarations made by the Minister for Industry, López Bravo.[90] And in the attempt to find new partners, Spain contacted other countries, such as Brazil, that were interested in enhancing their own technology.[91]

11. CONCLUSIONS

In this paper we have studied how Spain integrated itself into the "nuclear energy world" which came onto the scene with the dropping of atomic bombs on Japan. As we have seen, this integration depended essentially on the diplomatic, military and economic relations between Spain and the United States – or did at least until the sixties, when Spain began to diversify its sources of nuclear energy information and materials. In this sense, the introduction of nuclear energy in Spain is as much an exercise in political and military history as in the history of science. Indeed, this connection with politics and the military was not only through the geopolitical interests of the United States, as exemplified by the uses it made of nuclear energy information and propaganda, but also appears within Spain, as shown, for instance, in the different views that Falangists and technocrats held of nuclear energy during the early sixties.

During the years of our study, Spain was far from being a democratic state. This fact turns out to be closely connected with the attitudes – unambiguously in favor of nuclear energy – that Spanish society sustained all throughout this period. To the Franco regime, nuclear energy was a link – no matter how slim – with the modernity (scientific and technological) that was then so scarce in Spanish society. It is not surprising, therefore, that Spanish newspapers – almost completely controlled by the Government – hailed the new source of energy as a dramatic step towards a new and wonderful world; a world to which Spain – the new Spain that had arisen from the Civil War – had gained access. Eisenhower's "Atoms for Peace" fitted perfectly into this opportunistic ideology.

Thus, from 1945 until the mid-sixties, the 'tale of nuclear energy' in Spain was a 'happy story'. Its main protagonists, the Spanish and North Americans *political* leadership, were happy with the (largely political and military) benefits they extracted from it. As to Spanish *society*, it was a passive and obedient instrument in the hands of its rulers.

Universidad Autónoma de Madrid

NOTES

[1] This isolation was broken only in certain exceptional cases, the fruit of individual initiative rather than official planning. One of these exceptions can be found in the field of aeronautics.

See Roca-Rosell, A. and Sánchez-Ron, J.M., 1991/1992. *Esteban Terradas. Ciencia y técnica en la España contemporánea* (Barcelona: INTA/Serbal) and *Ciencia y aeronáutica* (Madrid: INTA/Algaida).

[2] See, on this regard, Sánchez-Ron, J.M., 1992. "Política científica e ideología: Albareda y los primeros años del Consejo Superior de Investigaciones Científicas," *Boletín de la Institución Libre de Enseñanza*, n.° 14, 53–74. Also Santesmases, M.J. and Muñoz, E., 1993. "Las primeras décadas del Consejo Superior de Investigaciones Científicas: Una introducción a la política científica del régimen franquista," *Boletín de la Institución Libre de Enseñanza*, n.° 16, 73–94.

[3] Quoted in the *New York Times*, September 27, 1945, 1–2.

[4] Quoted in Lleonart y Amselem, A.J. and Castiella y Maiz, F.Mª., 1978. *España y ONU -I (1945–46)* (Madrid: Consejo Superior de Investigaciones Científicas), 30. It should be remembered that México unambiguously sided with the Republic during the Spanish Civil War. With the victory of Franco, many Spaniards (among them a large number of intellectuals) fled to México, were the majority settled for the rest of their lifes. Llorens, V. "La emigración republicana de 1939", in: Abellán, J.L. (ed), 1976. *Exilio español de 1939*, vol. I (Madrid: Taurus), 95–200.

[5] Lleonart y Amselem and Castiella y Maiz[4], 38.

[6] Acheson's letter is reproduced in Lleonart y Amselem, A.J., 1991. *España y ONU -IV (1950)* (Madrid: Consejo Superior de Investigaciones Científicas), 329–330.

[7] Indeed, in July, 1949, the United States had awarded Spain a loan of 25 million dollars through the Chase National Bank.

[8] Already towards the end of 1948 a North American military delegation had visited Spain.

[9] Lleonart[6], 56–66.

[10] Sánchez-Ron, J.M., 1988. "La Edad de Plata de la física española: la física de la Junta," in *1907–1987. La Junta para Ampliación de Estudios e Investigaciones Científicas 80 años después*, Sánchez-Ron, J.M., coord., vol. II (Madrid: Consejo Superior de Investigaciones Científicas), 259–280.

[11] On Catalán see Sánchez-Ron, J.M., 1994. *Miguel Catalán. Su vida y su mundo* (Madrid: Fundación Menéndez Pidal/Consejo Superior de Investigaciones Científicas).

[12] Sánchez-Ron[2].

[13] In the forties, the CSIC encouraged technology (construction, coal, steel) and applied sciences (in physics and chemistry, for instance, special attention was paid to optics, thermology, photochemics, electrochemistry and organic chemistry). About this "applied" dimension of the CSIC, see García, S.L., 1993. "Ciencia, tecnología e industria en España. Herencias institucionales y nueva política científica en la constitución del Patronato 'Juan de la Cierva'," *Documento de Trabajo* 9302, (Madrid: Fundación Empresa Pública).

[14] *Ibérica 1* (2nd period), no. 32 (1945), 180–182, 188.

[15] He was also professor at the Escuela Superior de Armas Navales (Naval Ordnance Academy).

[16] Martín Artajo, J.I., 1946. *La energía atómica. Sus características y su aplicación para fines militares* (Madrid: Dossat), 5 (the "Presentation" is dated January, 1946).

[17] The expression "cultured" was used here as a compliment, rather than to imply that only a small fraction – the cultured fraction – of the American people were the intended audience of Smyth's report.

[18] In 1946, the number of Spanish readers of the Smyth report increased when Espasa-Calpe Argentina published a Spanish translation. Smyth, H.D., 1946. *La energía atómica al servicio de la guerra* (Buenos Aires-Mexico: Espasa-Calpe Argentina).

[19] See Sánchez del Río, C., 1983. "José María Otero y la energía nuclear," in: *Homenaje al Excmo. Sr. José Mª Otero Navascués* (Madrid: Real Academia de Ciencias Exactas, Físicas y Naturales), 25–29.

[20] We are grateful to prof. A. Durán for providing us with a copy of this document.

[21] Here "thermonuclear pile" meant a heat-producing fission reactor.

[22] About the availability of uranium ore in Spain, see Arribas Moreno, A., "Yacimientos españoles de uranio," in: García Guinea, J. and Martínez Frías, J. (eds), 1992. *Recursos minerales de España* (Madrid: Consejo Superior de Investigaciones Científicas), 1403–1419.

[23] Indeed, the world sources of uranium ore known in the late 1940s and early 1950s were few. Moreover, it was known that at some time in the near future, the Shinkolobwe mine in the Belgian Congo would become exhausted. There was the fear that even if new sources were found, all available uranium would be consumed within a few years by the expanding production of fissionable materials. As a consequence, the US government was keenly interested in the problem of how to provide for its future requirements, but equally how to prevent other nations from gaining control of useable uranium deposits. Thus, in the memorandum prepared by the US Government to guide Bernard M. Baruch, its representative in negotiations for international control of nuclear weapons, it was stated (June 7, 1946) that "the Atomic Development Authority when set up should have as one of its earliest purposes to bring under its complete dominion world supplies of uranium and thorium." Williams, R.C. and Cantelon, P.L. (eds), 1984. *The American Atom. A Documentary History of Nuclear Policies from the Discovery of Fission to the Present, 1939–1984* (Philadelphia: University of Pennsylvania Press), 94. See also, Hewlett, R.G. and Duncan, F., 1990. *Atomic Shield* (Berkeley: University of California Press), 54, 426.

[24] As we shall see, this Board was soon transformed not into an industrial company, but into a Junta de Energía Nuclear (Nuclear Energy Board), dependent on the Ministry for Industry.

[25] Roca-Rosell and Sánchez-Ron, *Esteban Terradas* [1], 301–310.

[26] Some indications in this regard are made in Sánchez del Río, C., 1977. "La investigación nuclear en España," *Arbor*, no. 374, 87–90.

[27] As reproduced in the newspaper *Madrid*, November 30, 1950. The declarations were made to the Catalan physicist, Miguel Masriera.

[28] Otero's declarations were quite extensive and included a reference to the Board's efforts to send young scientists abroad so that they might begin to study "experimental techniques and theoretical disciplines." He also mentioned the fact that EPALE was a research group which came under the auspices of the CSIC.

[29] In the early 1950's, important deposits of uranium had been discovered in Spain, in Sierra Albarrana (Córdoba), between Hornachuelos and Fuenteovejuna. These deposits (of pitchblende) were first recognised in 1939; later, the company "Berilio y Radio Español S.A." carried out certain surface reconnaissance studies. There were also deposits in Monesterio (Badajoz). In 1946, the mining engineer, Antonio Carbonell, maintained that "with regard to world reserves, these uranium deposits may be considered as being the fifth most important in the world. They have been found to contain some 1000 tons of uranium oxide" (Carbonell, A., 1946. "Los yacimientos de uranio", *Ibérica 3*, n°. 57, 155–157. In 1954, two further deposits were discovered: in Santa María de la Cabeza (Jaén) and in Sierra Pedroches (Córdoba).

[30] Some information on the JEN can be found in Lorenzo, J. de, 1963. *España y el átomo* (Madrid: Publicaciones Españolas), and in an anonymous paper published in 1965: "La Junta de Energía Nuclear", *Las Ciencias 30*, 312–320.

[31] The fact that Spain was not a member of NATO (the European members of the military organisation did not accept the possibility of its admission), and consequently its territory unavailable to the forces of the treaty, should also be taken into account when understanding the US attitude.

[32] "Defense agreement between the United States of America and Spain", *U.S. Treaties and Other International Agreements* vol. 4, part 2 (Washington, D.C.: Government Printing Office, 1955), 1876–1894, 1896–1902.

[33] For "Atoms for Peace", see Hewlett, R.G. and Holl, J.M., 1989. *Atoms for Peace and War,*

1953–1961 (Berkeley: University of California Press); in particular, chapter 8. A shorter, but in many senses more perceptive vision of the origins of "Atoms for Peace", is that of Weart, S.R., 1988. *Nuclear Fear* (Cambridge, Mass.: Harvard University Press), chapter 8. Eisenhower's address is reproduced in *S.3323 and H.R.8862, to amend the Atomic Energy Act of 1946. Hearings before the Joint Committee on Atomic Energy, Congress of the United States*, second session, part I (Washington, D.C.: Government Printing Office, 1954), 4–13.

[34] The McMahon bill is reproduced in: *The American Atom, 1939–1984* [23], 79–92.

[35] The propagandistic aspect of "Atoms for Peace" is stressed also by Weart[33], 156, although he considers mainly the Eisenhower administration's conviction that the liberalization of their nuclear information policy would constitute an effective *"propaganda war"* against the Communists. As a consequence, he leaves aside some important propagandistic effects, *not necessarily connected with the "war against communism"*, that "Atoms for Peace" had in other countries.

[36] "Agreement for Cooperation between the Government of the United States of America and the Government of Spain concerning civil uses of Atomic Energy", *United States Treaties and other International Agreements* vol. 6, part 2 (Washington, D.C.: Government Printing Office, 1955), 2689–2694.

[37] The Agreement (Article X) made it clear what was meant by "research reactor": "a reactor which is designed for the production of neutrons and other radiations for general research and development purposes, medical therapy, or training in nuclear science and engineering." The term did not cover, therefore, power reactors, power demostration reactors, or reactors designed primarily for the production of special nuclear materials.

[38] It was also mandatory that American experts be permitted to inspect reactors in which leased fuel was used.

[39] Hewlett and Holl[33], 184. Some aspects of the industrial dimension of the Atomic Energy Act of 1954 have been considered by Balogh, B., 1991. *Chain Reaction. Expert debate and public participation in American commercial nuclear power, 1945–1975* (New York: Cambridge University Press), chapter 4. Just how unprecedented this role was for the AEC can be understood easily if one remembers that up to 1953 over 90 percent of reactor development funds had been poured into direct military applications (Balogh, 98).

[40] To appreciate the economic and industrial dimension of nuclear energy, we must take into account the ideas prevailing at the time in American industrial circles. Thus, a study dealing with the possibilities that the atomic market had for United States companies abroad, stated in reference to Western Europe: "There are strong indications that, in the coming years, the market there will be most promising". Mayer, K.M. "The atomic power market in the United States and overseas," in: Guéron, J. *et al.* (eds), 1956. *The Economics of Nuclear Power* (London: Pergamon Press), 194–211; on p. 210.

[41] Roca-Rosell and Sánchez-Ron, *Ciencia y aeronáutica* [1], 157–172.

[42] Carbon copy of letter to "Teddy", signed "Theodore", in: "Th. Von Kármán Collection," Robert A. Millikan Memorial Library, California Institute of Technology Archives, box 67.6. For a description of von Kármán Archives, see Goodstein, J.R. and Kopp, C. (eds), 1981. *The Theodore von Kármán Collection at the California Institute of Technology* (Pasadena, Cal.: California Institute of Technology).

[43] These treaties are reproduced in *United States Treaties and other International Agreements* [36].

[44] By 1961 the United States had negociated thirty-eight agreements with thirty-seven countries. Hewlett and Holl[33], 236.

[45] "Agreement for cooperation concerning the civil uses of atomic energy between the Government of the United States of America and the Government of Belgium," July 21, 1955, *United States Treaties* [36], 2551–2562; p. 2552.

[46] Gowing, M., 1974. *Independence and Deterrence: Britain and Atomic Energy 1945–1952*

(London: MacMillan).

[47] "Agreement for cooperation on the civil uses of atomic energy between the Government of the United States of America and the Government of the United Kingdom of Great Britain and Northern Ireland," *United States Treaties* [36], 2709–2719. Although this is not the theme of the present paper, it must be pointed out that the subject of the relationships between Americans and British as regards the civil uses of nuclear energy, is a complicated one. Thus, for example, in the late forties, Great Britain was viewed in America as the major threat to its dominance in the field of the development of electrical generating reactors, and it was considered that the leaders in this promising new technology would enjoy an export advantage and thereby improve considerably their balance of trade. Some comments on this point are made in Blumberg, S.A. and Panos, L.G., 1990. *Energy and Conflict: A Biography of Edward Teller* (New York: Charles Scribner's Sons), pp. 172–173.

[48] "Agreement between the Government of the United States of America and the Government of the United Kingdom of Great Britain and Northern Ireland for cooperation regarding atomic information for mutual defense purposes"[36], 2721–2724.

[49] For more information on the Anglo-American nuclear cooperation, see Twigge, S.R., 1993. *The Early Development of Guided Weapons in the United Kingdom, 1940–1960* (Chur: Harwood), 33–43.

[50] "Agreement for cooperation concerning civil uses of atomic energy between the Government of the United States of America and the Government of Canada"[36], 2595.

[51] Gowing, M., 1964. *Britain and Atomic Energy, 1939–1945* (New York: MacMillan), 394–398. The well-known M.A.U.D. Report dealt with the use of uranium for a bomb, and it was prepared in the summer of 1941 by a British Committee; see, on this regard, Peierls, R., 1985. *Bird of Passage* (Princeton: Princeton University Press), 156–169. The famous letter that Albert Einstein wrote (and Leo Szilard prepared) to President Roosevelt on August 2, 1939, also mentioned that "there is some good ore in Canada and the former Czechoslovakia, while the most important source of uranium is Belgian Congo." Reproduced in: Feld, B.T. and Szilard, G.W. (eds), 1972. *The Collected Works of Leo Szilard. Scientific Papers* (Cambridge, Mass.: The M.I.T. Press), 199–200.

[52] The story of the relationship between the US Government and Eldorado Gold Mines, Ltd. began in 1942, when Americans and British considered that the uranium controlled by Eldorado, then a private company, was essential to their atomic programs. The strategical importance of the possession forced the Canadian Government to finally take control of the company. See, Hewlett, R.G. and Anderson Jr., O.E., 1990. *The New World* (Berkeley, Cal.: University of California Press), 85.

[53] Especially, though not only, meetings related to the "European movement". Thus, for instance, the European Cultural Conference of December 1949, was held in Lausanne.

[54] One of the reasons for selecting Geneva was because it would be more economical to use existing United Nations buildings there rather than build new facilities elsewhere. Hewlett and Holl[33], 232–235.

[55] The reactor was assembled, a few weeks before the opening of the Conference, on the grounds of the Palais des Nations, where it was shown to the participants in the meeting. It was the first reactor (of the pool-type) ever built in Western Europe.

[56] On January 7, 1966 a U. S. B–52 bomber disintegrated over Palomares, a small village on the Spanish Mediterranean coast, releasing four hydrogen bombs. One was soon recovered; the conventional explosives in two others detonated, spewing plutonium over a wide area. A lengthy search went on for the fourth bomb, eventually located 760 meters below sea level five miles offshore. Not until 44 days after the accident did the U. S. admit that one of its H-bombs was missing.

[57] See *El átomo y sus aplicaciones pacíficas* (Madrid: Sindicato Nacional de Agua, Gas y Electricidad, 1958).

[58] The previous year the Union had also organised a series of conferences on Atomic Energy.

[59] "La energía nuclear y la estructura de la producción eléctrica nacional"[57], 297–315. See also, Caso Montaner, A., "Economía y energía nuclear", *ibid.*, 355–390.

[60] *Ibid.*, 306.

[61] "Aplicaciones de los isótopos radiactivos en la industria"[57], 45–72; on 46–47. Needless to say, the figures for 1953 and 1956 were already far too generous, and the projected savings a wild exaggeration.

[62] See Table 2 included in the text of Gamboa's paper, p. 70.

[63] *Destino* (Barcelona), March 7, 1964.

[64] See, for instance, an article ("Espectaculares servicios de la energía atómica para el bienestar del hombre") which appeared in *Ideal* (Granada), March 14, 1964.

[65] Weart[33], 163.

[66] *La Vanguardia* (Barcelona), January 26, 1965.

[67] One should not forget that the legitimacy of the Franco regime was due to a war.

[68] Normally, articles of this kind referred to reviews of books concerned with the atomic bomb holocaust. For example, in the *Diario Regional de Valladolid*, July 19, 1964, there is a review of Fernand Gigon's *Apocalypse de l'atome*.

[69] *Informaciones* (Madrid), April 22, 1964.

[70] *Ibid.*

[71] *Pueblo* (Madrid), June 27, 1964.

[72] The organiser of the courses was a North American, Robert Chambers.

[73] "La ciencia nuclear sustituirá la formación humanística por un contenido más técnico", *Pueblo* (Madrid), June 27, 1964. As a matter of fact, it can be argued that articles like this were giving a Falangist ideological twist to the American's idea of nuclear science in the service of mankind.

[74] For example, in *La Hoja del Lunes* (Madrid), July 13, 1964, an article commented the application of atomic energy in the desalinisation of sea water.

[75] *El Correo Gallego* (Santiago de Compostela), August 25, 1964.

[76] *Mundo* (Madrid), October 18, 1964.

[77] *Informaciones* (Madrid), January 6, 1965.

[78] Cf.[26]

[79] *Ya* (Madrid), May 1, 1964.

[80] For example, that published by *Mundo Hispánico* in June, 1964.

[81] In *Arriba* (Madrid), on December 29, 1963, Pascual Martínez evaluated the energy content of the Spanish uranium reserves.

[82] Pablo Corbalán in *Noticiero Universal* (Barcelona), October 24, 1963.

[83] The course was directed by Federico de Castro, professor and judge on the European Atomic Energy Tribunal, but it was given by Alfonso de los Santos, legal adviser to the JEN.

[84] *Ya* (Madrid), June 22, 1967.

[85] *Arriba* (Madrid), July 7, 1967.

[86] *España Semanal*, January 13, 1963.

[87] *Ibid.*

[88] We do not know if doubts were raised as to whether this action was consistent with the agreement with the United States.

[89] This was no obstacle for the ten-year period Agreement with the United States being extended to thirty years in 1966 (the new agreement authorised the sale of enriched uranium for very particular reactor programs).

[90] *Arriba* (Madrid), July 6, 1966.

[91] News given in *Ya* (Madrid), July 28, 1966.

BRUCE HEVLY

THE TOOLS OF SCIENCE: RADIO, ROCKETS, AND THE SCIENCE OF NAVAL WARFARE

INTRODUCTION

After World War I, the military establishment of the United States, and especially the U.S. Navy, was convinced of the efficacy of modern technology in warfare. On the Army's side, infantrymen came home from Europe pondering the meanings of the machine gun and chemical warfare for their profession, cavalrymen began to develop armored tactics, and fliers pressed their case for the power of the air arm. The Navy realized that in the twentieth century its strength would depend on new technology as well: on aviation, improvements in gunnery, a new naval architecture, the submarine and the torpedo, and advances in radio communications. While conservative forces in the service resisted some of these changes, it seemed clear that the Navy's potential adversaries would not allow America's navy, pressing into the front rank, to ignore them.[1] A new kind of research institution seemed necessary to advance the tools and weapons of warfare at sea, one which could exploit modern research into the physical sciences.

The Navy already had venerable institutions dedicated to developing ship designs and new types of ordnance. In aviation, the naval establishment could look to its well-established links to the aeronautics program at the Massachusetts Institute of Technology and to the newly-founded National Advisory Committee for Aeronautics.[2] But in radio research and anti-submarine warfare research (an area of study mainly concerned with underwater acoustics), the Navy had nowhere to turn. These subjects had only come under concentrated study during the course of the war, with the studies undertaken by *ad hoc* committees of volunteer scientists and engineers, as well as industrial researchers.[3] In the period after World War I, these studies were undertaken by scientists who devoted their careers to the development of advanced technology for the U.S. Navy, the character of whose research represented an important interwar scientific tradition, more industrial than academic. This essay is devoted not to analyzing a set of scientific experiments or results, but rather to examining this research tradition, particularly the way in which science and technology interacted within it, and to arguing for the significance of the tradition for twentieth-century science.

After supporting experts in radio and underwater sound in several locations for the half-decade after World War I, the Navy brought civilian researchers in these new areas together in a new institution, the Naval Research Lab-

P. Forman and J. M. Sánchez-Ron (eds.), National Military Establishments and the Advancement of Science and Technology. Studies in the 20th Century History, 215–232.
© 1996 Kluwer Academic Publishers. Printed in the Netherlands.

oratory (NRL), on the eastern bank of the Potomac River in Washington, D.C.[4] Interest in radio at NRL, and, in particular, research and development work conducted on the problems of shortwave radio, led to a more abstract topic for investigation: the physics of the earth's atmosphere and the sun's role in atmospheric processes. These were scientific questions with important technological implications. Over the next 35 years, up until the International Geophysical Year (1957–1958), a group of NRL scientists established themselves as leaders in experimental atmospheric and solar physics.

Their research, superficially considered, would seem to fit into the most commonly-held notion of how technology and science interact: that science defines a catalogue of natural facts which serves as the basis for advanced technologies.[5] But recent efforts to understand the historical intersections of science, technology, and the culture of organizations (both governmental and mercantile) contradict this simple view. Indeed, more closely considered, the interactions of science, technology, and institutional culture in the field of atmospheric and solar physics at NRL do not fit the old, standard pattern.

At the center of NRL's ionospheric research stood two instruments, each at once being employed for research, and developed for military operations. Researchers used experimental radio transmitters before World War II, and prototype guided missiles after the war, as tools for their studies, and it is the dual character of these central instruments – investigatory *and* operational – which holds the key to understanding the character of research at NRL and of the knowledge produced there. Scientists at the laboratory often published their results in scientific journals, but it would be incorrect to make the simple statement that scientific research laid the foundations for the advanced technologies of shortwave radio and guided missiles. Such statements assume a distinct chronology of discovery, in which "science" is established prior to the derivative creation, technology. Rather, because the objects of operational interest – radio sets and rockets – were used as experimental tools, the Navy's technological concerns were in effect built into the knowledge produced at NRL.

Such hybrid intellectual products were characteristic not just of NRL, of a single institution from the 1920s on, but of a new group of institutions which emerged around the turn of the century in government and industry, and within which science and technology were amalgamated. Contemporary social commentators recognized the emergence of such institutions throughout the spectrum of American enterprise; American intellectual historians such as Merle Curti and Charles Beard saw them as characteristic signs of a period in which the institutions of high culture began to be shaped by the power of the business mentality. Corporate interests, then, provided institutions not only for the production of "engineers and technicians for the new age of business enterprise," but also for the production of "new secrets" upon which their enterprises might be based.[6] Charles and Mary Beard approvingly described the 1920s, the period of NRL's establishment, as a time in which "the multi-plying institutions which furnished support for brain workers" were matched by "expanding markets for ideas which could be transmuted into commodities or technological improvements."[7]

These scholars, assuming that scientific ideas represent objective knowl-edge independent of the process of discovery, saw the new institutions for amalgamated research as efficient settings for the uncovering and application of useful knowledge. Considering the tools of science at NRL, we can affirm the significance of such institutions, but also begin to understand the process of knowledge production through sponsored research by bringing to bear the current understanding that scientific knowledge is not independent of the means used to produce it.

RADIO TRANSMITTERS AND THE CEILING OF THE SKY, 1924-1940

Naval expansion, modernization, and the adoption of new military technolo-gies between the Spanish-American War and World War I introduced a new set of organizational and tactical problems into the naval officer's world. At the higher levels of command in the U. S. Navy after 1918, a more centralized management structure, led by the Chief of Naval Operations, had to control the activities of fleets spread throughout the world.[8] At the local, tactical level, twentieth-century naval doctrine called for fleets to operate as inte-grated units, with scouting forces, cruisers, and battleships acting in concert and tied together by communications services which articulated the voice of central command.[9] Aircraft, in particular, represented a new technological el-ement to be incorporated within the naval structure, but one which presented a singularly difficult target for communicators – one difficult to reach by flag, blinker, or voice.

While the demands on naval communications had increased with modern-ization, radio technology also increased in proven reliability and promise. Radio had been only grudgingly admitted in the Navy before World War I; it represented in many ways an element of disorder to naval officers, and a threat to the perquisites of command.[10] But after the war, radio emerged as a proven technology, one so important that its use – even by civilians – had been strictly managed by the Navy during the conflict, and one bound to be so important after the war that the Navy intervened to establish the Radio Corporation of America in 1919, ensuring a reliable American supplier of communications gear.[11]

The Navy maintained three radio research groups in the immediate postwar period, and then brought all three activities (although not all the researchers) together at NRL in 1923. Leading NRL's Radio Branch was A. Hoyt Taylor, a Göttingen Ph.D. (1909) who left the physics faculty of the University of North Dakota for wartime service as a Navy communications researcher and never returned. Taylor was an enthusiastic amateur radio operator; commissioned as a naval reservist, during the war he carried out research on low-frequency antenna arrays and managed a transatlantic radio transmitter. Well acquainted with the problems of navy radio communications, after the war Taylor continued to develop radio technology for the Navy. His enthusiasm for radio became intertwined with his enthusiasm for the Navy; he was proud of his status as a reserve officer as well as a scientist.[12]

Taylor's group took up the problem of communicating with aircraft. Radio was the obvious general solution; any kind of visual signalling from the ground (and especially from the surface of the sea) to a relatively fast-moving aircraft was difficult at best. But the specific solution to using radio for ground to air communications was hedged around by the limitations of an aircraft's interior space and its load capacity. An airborne radio would have to be one that drew a small amount of power, with light, compact components, and a light-weight power supply. To fit aboard an airplane, its antenna would have to be small, compared to a ship's large radio mast.

Here Taylor the navy scientist and Taylor the radio amateur met. What an aircraft required was something more like an amateur's home sending set than the hulking communications gear designed for cruisers or the Navy's shore facilities. And amateur operators after World War I, restricted to the less-desirable, high-frequency region of the radio spectrum and using vacuum tube transmitters, were finding that they could achieve astonishing results with relatively little power.[13] For Taylor's purposes, shortwave gear presented advantages for antenna design as well. Since antennae grow in general with the wavelength they are designed to receive, shortwave signalling required shorter antennae, which fit in well with the limitations imposed by the size of aircraft. Working along these lines, Taylor and his collaborators had become well acquainted with shortwave techniques by the time they moved to NRL in 1923.[14]

At NRL, the radio researchers worked to extend shortwave techniques to serve the needs of the Navy's widespread fleets. While many in the Navy assumed that shortwave signals were suitable only for short-distance communication, Taylor was convinced that the results discovered by amateur operators – the ability to send signals great distances with a very low power output – could be exploited for long-distance communications by the Navy as well, allowing direct communication between Washington and ships around the world.[15] The problem would be to make shortwave reliable, by making it predictable. Taylor's task, then, was to provide a set of rules for the behavior of shortwave signals in their travels between elements of the U.S. Navy. In these efforts, he found a new collaborator in E.O. Hulburt.

Like Taylor, Hulburt had left a professorship in physics for military service as a technical specialist during World War I. A Baltimore native, Hulburt followed tradition throughout his education at Johns Hopkins: as an undergraduate, he was captain of the lacrosse team, and as a graduate student in physics he became a specialist in optics. Serving in the Army Signal Corps during the war, Hulburt worked in the Corps' Paris laboratory, and was mainly concerned with the operation of field telephones. In postwar research at Hopkins, while he looked for a new academic position, Hulburt added a working knowledge of radio to his research skills in optics and electricity.[16] Like Taylor, then, Hulburt offered the Navy a strong ability to undertake physics research mixed with experience developing new technological systems under wartime pressure.

Hulburt came to NRL in 1926 as head of a new division, Heat and Light (later renamed Physical Optics), which was created to devote itself to "real research." [17] Taylor's Radio Division and NRL's other original component, the Acoustics Division (which was concerned with the problems of antisubmarine warfare), were established with the expectation that they would answer quite well-defined technical problems of immediate concern to the Navy. Hulburt's responsibilities were more wide-ranging, and so less well-defined. His essential task was to attach problems in physics to the Navy's technological needs, a relationship which was not always obvious to naval officers in general.

Shortwave radio provided a set of questions appropriate to Hulburt's skills. The problems of reliable long-distance transmission was the problem of understanding the ways in which electromagnetic waves at high frequencies travelled through the atmosphere between senders and receivers. For Taylor and Hulburt, the most fundamental piece of data to be measured was the skip distance: the area passed over by a radio signal as it left the transmitter, bounced off the ionosphere overhead, and returned to a receiver on the earth's surface.[18] This was essentially a complex problem in Maxwellian optics: it was Hulburt's job to determine, as best he could, how the electromagnetic waves emitted by the transmitter were bent and rearranged as they passed through the air. In this case, the charged particles of the upper atmosphere bent radio waves, much as the electrons around the atoms in a sheet of glass reflect and refract visible light.[19]

Hulburt and Taylor, then, set out to formulate a science of radio which would allow the routine use of shortwave sets for naval communications. While the radio-reflecting layer, as it was sometimes called, could not be directly measured by means of instruments carried aloft in aircraft or balloons, the navy scientists planned to investigate its physical state by passing radio signals through it. Their research tool was NRL's shortwave transmitter, station NKF; their fundamental experiment was to use this device, a piece of technology under development for operational use, to measure the skip distance effect for messages sent around the world from NRL.

The skip distance depended on characteristics of the radio signal – its frequency, in particular – and on the elements of nature which affected the ionosphere, such as the time of day, the season of the year, and the level of solar activity (which follows an eleven-year cycle of its own). Under given conditions of the upper atmosphere, radio signals will be bent over to return to earth at different altitudes, depending upon each signal's particular frequency. Understanding the relationship between atmospheric conditions and shortwave frequency would allow NRL's scientists to provide navy radio operators with the knowledge to make shortwave communications work reliably. By understanding the conditions governing reflections off the upper atmosphere, Taylor and Hulburt could rationalize the process of finding out which frequency to use, to make any desired "bounce shot" and carom a signal off the ionosphere to a receiver in any specific location.

Because NKF was designed to hold a designated frequency with great precision, it was an ideal research tool to allow Taylor and Hulburt to map the ionosphere, using radio signals as probes. They announced NKF's broadcasting schedule in the wireless operator's journal, *QST*, and requested that those listening report back to NRL by post card whether or not they could hear the signal from Washington.[20] In addition, arrangements were made for navy operators and radio operators aboard research vessels sent out by the Carnegie Institution of Washington to keep a listening watch for signals for NRL. The International Polar Year of 1932–1933 provided an opportunity to put a listener in the Arctic, again in cooperation with the Carnegie.[21]

This empirical approach allowed Taylor and Hulburt to employ the shortwave transmitter, NKF, to compile a series of frequency charts giving the specifications for reliable radio operation in the shortwave band. Here certainly science and technology were, at the least, interpenetrating. Taylor noted in his memoirs, "we were able not only to get up good practical communication charts as guides to the use of high frequency, but to determine fairly closely the number of free electrons per cubic centimeter in the ionosphere under different seasonal and diurnal conditions."[22] The needs of practical communications and the physical conditions of the ionosphere were amalgamated at NRL; each defined the other to a remarkable degree, within the context of a military laboratory.

Taylor and Hulburt presented their results in reports to the U.S. Navy, and also through articles in engineering and scientific journals. Their point in these reports was two-fold: in addition to contributing to the growing field of atmospheric physics, they demonstrated the NRL could develop the knowledge necessary to make shortwave radio a reliable working system, not restricted to short-distance communications, but suitable for routine long-distance communications. Even publishing in the *Physical Review*, the final result presented was a lower limit for useful transmission frequency. "This is due to the mechanism of propagation entirely apart from the inherent technical difficulties in the apparatus," they noted.[23] To a navy hierarchy skeptical of shortwave radio, Taylor and Hulburt presented a set of ideas which made nature technologically comprehensible. With the assurance provided by NRL's improved technology and a firm vision of the ionosphere expressed in useful terms, the Navy adopted a new communications plan which made extensive use of shortwave radio in 1925, after a series of reports by Taylor and Hulburt emphasizing that high-frequency systems could be managed effectively, and could provide a sufficient number of channels to allow effective direction of America's administratively complex naval effort.

This research represented a successful effort to apply physics to the Navy's technological needs – or, more precisely, to create a system of ideas which used technologically-defined categories to express nature's regularities. From some distance in time, Hulburt and his subordinates congratulated themselves on finding a means of combining scientific research and technological development. The skip distance research was "not only of theoretical interest to theoretical people, but it was useful to the Navy," Hulburt recalled.[24] A close associate of Hulburt's in the Optics Division believed that the division's research followed a pattern defined in its early years: "a deliberate policy of looking for problems of general scientific interest which were not under thorough exploration elsewhere and which would lead to results of specific interest and usefulness to the Navy."[25]

Patterns were indeed established through radio propagation research in the early years at NRL, both for Hulburt personally[26] and for the research program of his division. It was a pattern in which research and development were combined in the very acts of theorizing and experimenting. In this process, questions – the central one being how did the radio-reflecting "ceiling of the sky"[27] come into existence – reflected technological necessity, and answers were produced by using operational technologies as scientific instruments. In this vein of research, the ionosphere appeared as a reflector for radio waves, and ionospheric science was the process of understanding how radio waves could be reflected in a regular and predictable fashion. But after examining an operational technology as research instrument – here the NKF shortwave transmitter – we cannot agree with the idea that the trick was to *find* areas that were at once technologically important and scientifically interesting. Rather, the trick at NRL and the other idea factories was to *create* them.

RESEARCH ROCKETS IN THE COLD WAR, 1945-1958

After World War II, Hulburt's successors in the Physical Optics Division were an ultraviolet spectroscopy specialist, Richard Tousey, and an x-ray physicist, Herbert Friedman. Hulburt advanced from leader of the optics researchers to the position of scientific director of the laboratory, where he served as a visible symbol of the Navy's postwar faith in "real research" as an important contributor to military strength.[28]

When A. Hoyt Taylor came to the newly-established NRL just after World War I, the scientific staff, as we have seen, was devoting itself to well-defined problems which had arisen in the course of the recent conflict. Hulburt's ability to attach his more esoteric research in physics to technological problems had secured support within the laboratory for his small group of optics researchers. But the U.S. Navy emerged from World War II with a much greater faith in scientific research, and also convinced that navy scientists needed to be concerned with predicting, even defining, and solving the problems of the next war rather than the last one.[29] Researchers would have to be given latitude to develop ideas which might, in the long run, have the importance of radar, missiles, or nuclear power.

Some well-established problems continued to be important. The military force of the United States had an even more extensive and complex world-wide reach after World War II; its need for reliable, abundant communications continued to grow. Thus the problems of ionospheric physics defined before World War II – what atmospheric conditions made communications via the ionosphere possible, and what led to interference with the process – continued to be of great interest. The Second World War also provided a new experimental tool for use in pursuit of these old queries: the research rocket, capable of lifting measurement instruments into the ionosphere for more direct observations of its state.

The Navy's interest in the rockets themselves added to the importance of ionospheric research, ensuring a greater effort in the area than communications alone could have inspired. World War II revealed a group of weapons, not decisive during 1939–1945 but of great importance for future conflicts. Submarine technology was greatly advanced, and during the postwar period a tremendous research effort was carried forward in anti-submarine warfare. The atomic bomb, of course, represented tremendous power, and tremendous danger to postwar strategists. But guided missiles, along with long-range, high-flying, fast bombers, threatened to make the Navy obsolete, and so research on the topic drew substantial support.[30]

The threat was made explicit by the aspirations of the U.S. Air Force, independent after 1948, which argued that America's future defense lay in an expanded frontier, the fortresses of which would be overseas strategic air bases threatening the Soviet Union.[31] As for the sea frontier, in which the oceans were conceived as a moat for America's protection and the U.S. Navy served as the nation's outmost guardian: new technology, the Air Force argued, rendered it meaningless. Guided missiles and strategic bombers could pass easily over the ocean, and so over the Navy, to threaten America. Prominent science advisors such as Vannevar Bush echoed the grave prediction that science-based technology had moved warfare to a new level of speed and distance.[32] In the new world seen by these military theorists, the Air Force's strategic bases would defend the United States. The Navy might convoy supplies overseas in support of the airmen, but it was no longer America's first line of defense.

The Navy, which traditionally had taken the position that "anything you can do, we can do better," quickly moved to develop its own guided missiles. Admiral of the Fleet Chester Nimitz returned from the Pacific to become Chief of Naval Operations, and publicly committed the Navy to joining the ranks of the rocketeers.[33] War-born laboratories at MIT, Johns Hopkins, and Inyokern were established on a continuing, postwar footing, and wartime specifications for devices such as the Lark guided missile were rewritten to make them suitable for continental defense – that is, able to intercept intercontinental missiles or bombers.[34] The Navy was eager to commit itself to the support of technical staff members engaged in work related to guided missile development, as a part of the effort to keep up with the pace of change in modern warfare.

The Navy's interest in guided missiles brought another central research tool into the work of NRL's optics researchers. Tousey and Friedman had inherited questions about the sun's interactions with the upper atmosphere which had gone unanswered before the war, when Hulburt had analyzed data gathered using the NKF transmitter. In the decades after the war, NRL's scientists would use research rockets, launched by the American military in order to gather data to support guided missile design, as platforms from which to gather information on the ionosphere. Tousey and Friedman conducted their research with less concern for the immediate application of their work, in part because of a new rhetoric surrounding scientific investigation which emerged after World War II from sources as disparate as the civilian Bush, Army Chief of Staff Dwight Eisenhower, and the Navy's coordinator of research and development.[35] But Friedman and Hulburt were both navy scientists by choice, who prided themselves on producing information which would ultimately benefit America's naval power, and they worked in an institution devoted to the task. Thus after World War II, as after World War I, their research instrument served to amalgamate scientific and technological concerns.

In the latter stages of the war, NRL's scientists knew of the German V–2 rockets from a variety of sources. Its general characteristics could be deduced from newspaper accounts of attacks on England, and in addition, authorized scientists at NRL read intelligence reports produced by Project ALSOS and the accompanying Navy Technical Mission which pursued Germany's technical secrets as the Allies advanced across Europe.[36] One of NRL's scientists, physicist Ernst Krause, helped to debrief German rocket scientists.

NRL had an interest in scientific problems related to rocket flight, and was determined to stake a claim for navy support in "basic research," a term which figured in discussions of postwar research policy around Washington, D.C. With a background in ionospheric physics and strong support for research related to guided missiles, groups of scientists at NRL began to plan to undertake rocket-based research after World War II. Krause led one group, which planned to investigate cosmic rays; Tousey led another, in a program to measure the Sun's extreme ultraviolet spectrum, a project of special interest to Hulburt. Research flights began in 1946.[37] Friedman, who had specialized during the war in applying x-rays and radioactivity to practical uses, began in 1948 to measure the sun's x-ray output and its impact on the upper atmosphere, his participation having been delayed by continuing applied research.

If rockets had not been available, NRL would have sought to build its own. But the U.S. Army made space available for instruments in the nose cones of captured German V–2s, which were being test flown at White Sands, New Mexico. While the Army began to produce American-built rockets of several types, including a miniature V–2 called the Hermes, NRL built its own Viking rocket. The Viking was generally modelled on the V–2, with some crucial design changes in both the body and the engine.[38] The Viking was designated a research rocket, but the term clearly had a double meaning: while it would carry scientific instruments to high altitudes, the changes incorporated in the design also represented an advance in the military rocket art. Simultaneously, then, rockets represented scientific instruments and the objects of an ongoing effort in military technical development. As shortwave radio transmitters had after World War I, rockets combined development and research in a single class of artifacts.

The ionosphere to Hulburt, before the war, had been important as a reflector of radio waves, and he had treated it in those terms. In the postwar world, from the point of view of the Navy's operational concerns, the ionosphere became a region of intense interest because guided missiles operated *within* it. They would be flown in it, detected in it, and, with luck, destroyed in it. The ionosphere's chemistry, its composition, and its dynamic interactions with the sun thus took on new importance, and this importance was reflected in the suite of experiments carried out aboard V–2s and Vikings.[39]

With rocket-borne detectors, Tousey and Friedman began to measure the sun's spectrum in detail, and to attach specific spectral components to particular atmospheric effects. Instruments carried aboard rockets gave a continuous profile of the ionosphere as the rocket rose to its maximum altitude. While compiling data about the sun's characteristics, these surveys also helped to answer well-established questions about the ionosphere's behavior and its impact on the Navy's technological systems. In 1953, NRL scientists reported that the weight of the evidence indicated that soft x-rays created the electron densities which reflected radio waves.[40] Their research then turned to the question of what disrupted the upper atmosphere, causing radio fade outs, and this question held their interest through the International Geophysical Year.[41]

By the time of the IGY in 1957–1958, the variety of rockets available to researchers had been augmented by the fruits of postwar development efforts. The Aerobee, a two-stage liquid/solid fueled rocket, became a workhorse for upper atmosphere research. Also available was a balloon-rocket combination dubbed the rockoon, in which the balloon served as the first stage, holding aloft for some time a solid-fueled rocket which could be fired off electrically whenever the sun showed incipient signs of interesting behavior.[42] Moreover, solid-fueled rockets designed for ground defense or air-to-air combat, such as the Nike and the Asp, were capable of carrying relatively light and rugged instruments – such as Friedman's x-ray detectors – into the ionosphere.[43] In a crucial experiment, these military rockets were used to loft instruments during a solar eclipse in the midst of intense solar activity. Friedman's detectors measured the arrival of an increased flux of solar x-rays, which caused a radio fade-out to occur.[44] This research, then, answered the second major question which Tousey and Friedman had inherited from Hulburt.

Just as had NKF before the war, in these postwar investigations rockets stood at the very heart of research on atmospheric physics at NRL. They made the research possible, not just by allowing access to the ionosphere in a new way, but by influencing the way in which scientific questions were framed, by helping to set the terms for a right answer – that is, one which responded to the Navy's technological needs – and by inspiring the institutions of the Navy, uniformed and civilian, to support the research. To imagine that the scientific knowledge exists somehow separate from this technological framework is to divorce it from human history.

While studies of bodies concerned essentially with research policy can tell us much about the nature and effects of military sponsorship for research, ultimately if we are interested in the development of knowledge within the context of military institutions we must turn to the practice of research within working environments.[45] To better understand at least one important way in which knowledge was created, one might consider the transmitter and the rocket as characteristic artifacts. That is, they shaped scientific ideas by instantiating utilitarian goals in the experimental process.[46]

CONCLUSION

Scientific research on the upper atmosphere was conditioned by the Navy's technological concerns: by the need for reliably operating equipment to meet the changing requirements of twentieth-century warfare. NRL's research rockets and shortwave radio transmitter stood as the physical manifestations of a system for research in which experimental tools were drawn from technologies being developed for very practical ends. Such technological artifacts served as organizing points, intellectually and organizationally, within institutions devoted to research and development. They crystallized the organizational and professional concerns of scientists devoted to producing new and useful knowledge.

Taking this view raises several general questions. First, did technological concerns – that is, the concerns of military sponsors – *define* or *determine* the content of science? Rather than making this strong statement, it seems more useful to consider the influences of sponsorship, institutional politics, and technological commitments as constraints which helped to shape scientific research.[47] Such factors set the conditions within which scientific knowledge was apprehended and defined.

Second, it is important to recognize that NRL represents a larger class of institutions, which began to appear in the late nineteenth century, devoted to the production of useful knowledge. These research centers were designed to conduct research in pursuit of national power, economic or military, or in pursuit of the advancement of corporate interests.[48] That is, research at these sites expressed the concerns of sponsors by defining the understanding of nature in terms of the systematic mastery of useful devices. Researchers then were able to restate technical obstacles in terms of usefully solvable "critical problems," to use the term advanced by Thomas P. Hughes.[49]

Industrial researchers involved in this process often found that academic training in physics or chemistry was not directly applicable to practical problems. Rather, they had to leave part of their training behind and work with the technologies before them, not the idealized experiments of the academic laboratory. These efforts required the researcher to imbue technological artifacts with scientific procedures and insights. Historians of technology and science have described cases of this transition in detail. Irving Langmuir and William Coolidge of the General Electric laboratories were successful as industrial researchers because they learned to *abandon* their academic studies and apply themselves to light bulbs and vacuum tubes as objects of experiment and as scientific instruments.[50] Not just patents, but even Nobel prizes could reward the ability to transform the interior of a vacuum tube into an experimental space, as Langmuir's history shows. Chemistry education at MIT, to take another example, was split between factions concerned with the theoretical bases of academic study and the elements of application, with the latter being institutionally successful.[51]

As workers in the institutions of applied research learned to shape scientific research around a series of important technological artifacts, they also integrated their work within the larger institutional cultures which had to accept the results and put them to use. Failure to do so at times led to technological failures. The copper-cooled automobile engine, for example, a product of the General Motors research laboratories, worked in the development shop but failed in production, almost certainly because the production engineers in the Chevrolet factory were estranged from the researchers and had no stake in making the engine a success.[52] Despite the failure of the copper-cooled engine, GM research director Charles F. Kettering did have a series of successful innovations, including the development of tetraethyl lead additives to prevent knocking in gasoline engines. This work required an analysis of fuel combustion in pistons: a series of experiments which transformed automobile components into test chambers for research.[53] At NRL, Taylor, Hulburt, Krause, Tousey, and Friedman were all successful in fastening on research tasks of both scientific and technological interest, creating appropriate research tools, and locating sponsors within the navy willing to support their work as ultimately utilitarian.

It is often convenient to think of science and technology as essentially separate things, with the former independently presenting a range of physical possibilities and the latter representing a choice of knowledge to be applied. While convenient, especially for defining the separate professional activities of history of science and history of technology, this understanding is in most cases incorrect. In the twentieth century, we have developed technical cultures which, by design, confound science and technology, binding them inextricably within a single context. Hence our difficulty, at times, in separating the two, even when we assume them to be two distinct entities.

In analyzing the cultures of science and technology in nineteenth-century America, Edwin Layton argued that the two spheres of knowledge were "mirror-image twins" which shared much common content but were conditioned by inverse sets of priorities and interests.[54] More recent work, especially that produced by scholars applying social constructivist approaches, has argued that the science/technology distinction is not only an artificial one, but also a detrimental one, because the utilitarian character of natural knowledge can be of vital importance in the project of organizing assent to a candidate "fact."[55] Such accounts have provided important insights into the spread of scientific knowledge, but not all scholars are as eager to fold the process of producing knowledge into the processes of persuading expanding circles of scientists to agree to it.

At NRL, the reliance on technologies then under development for naval operations, as experimental tools for the exploration of the upper atmosphere, ensured that research results would advance both scientific understanding and the Navy's technological capabilities. And in order to describe these 'Siamese twins' of science-based technology and technology-based science, at NRL and elsewhere, "invention" was replaced by "development", and "pure research" by "basic research" (a term implying that science would serve as the basis of something else). This represented a new organization of knowledge, one in which prototype weapons figured as appropriate experimental instruments.

Finally, perhaps the most general question is a moral one. On some fundamental level, should we be troubled by the intermixing of practical technical concerns with scientific knowledge? In industry, government, and academia, the twentieth century has witnessed the growth of research funding tied to the expected utility of science. Thus practical concerns have played a role in ordering science, and many research programs have crystallized around technologies under development. On the one hand, this could be perceived as the subversion of scientific autonomy by the forces of greed and gain. On the other, one could argue that science is never truly autonomous; rather, as a human endeavor, its practitioners fully participate in human society. Inevitably, then, "outside" influences will intrude as elements in the organization of scientific thought.

Charles and Mary Beard saw such perfusions of science through society as only beneficial, as they reflected on the decade which brought applied research institutions such as NRL into prominence. Among Americans, they wrote, "there was no doubt about the nature of the future... The most common note of assurance was belief in unlimited progress... a faith in the efficacy of that new and mysterious instrument of the modern mind, 'the invention of invention,' moving from one technological triumph to another."[56] In contrast, we now do doubt the underlying assumption that scientific knowledge exists separately from the history of its production. In the twentieth century, democracy, capitalism, and the state acted as agents of change through the creation of institutions and instruments which combined research and development into a single entity generating scientific knowledge as a malleable element perfused by society.

University of Washington

NOTES

[1] Rosen, P.T. "The Treaty Navy, 1919–1937," in: Hagen, K.J. (ed), 1978. *In Peace and War: Interpretations of American Naval History, 1775–1978* (Westport, Conn.: Greenwood Press). On the impact of new technologies on the social structures of naval life, see also Morison, E., 1966. *Men, Machines, and Modern Times* (Cambridge, Mass.: The M.I.T. Press), ch. 2 and 6; and Karsten, P., 1972. *The Naval Aristocracy: The Golden Age of Annapolis and the Emergence of Modern American Navalism* (New York: Free Press).

[2] Roland, A., 1985. *Model Research* (Washington, D.C.: NASA SP–4102), vol. I, ch. 4. See also NACA file 25–26, Box 152, Record Group 255, Federal Records Center, Suitland, MD., for documents on the relationship between the U.S. Navy and aeronautics researchers.

[3] Allison, D.K., 1981. *New Eye for the Navy: The Origin of Radar at the Naval Research Laboratory* (Washington, D.C.: Naval Research Laboratory, Report 8466), ch. 2, 3; Hughes, T.P., 1989. *American Genesis* (New York: Viking Penguin), ch. 3; Kevles, D.J., 1987. *The Physicists* (2nd edition; Cambridge, Mass: Harvard University Press), ch. 8, 9.

[4] For extensive descriptions of navy scientific research in the period after World War I, and the context of the establishment of NRL, see Allison, D.K., 1980. "The Origin of Radar at the Naval Research Laboratory," (Ph.D. dissertation, Princeton University), ch. 2 and 3; Allison, D.K., 1979. "The Origin of the Naval Research Laboratory," *U.S. Naval Institute Proceedings, 105*: 62–69; Hevly, B., 1987. "Basic Research Within a Military Context: The Naval Research Laboratory and the Foundations of Extreme Ultraviolet and X-Ray Astronomy, 1923–1960" (Ph.D. dissertation, The Johns Hopkins University), ch. 1; McBride, W.M., 1992. "The 'Great-

est Patron of Science'?: The Navy-Academia Alliance and U.S. Naval Research, 1896–1923," *The Journal of Military History, 56,* 7–33; Taylor, A.H., 1948. *Radio Reminiscences: A Half Century* (Washington, D.C.: Naval Research Laboratory); Howeth, L.S., 1963. *History of Communications-Electronics in the U.S. Navy* (Washington, D.C.: Government Printing Office).

[5] For a persuasive argument against this "assembly-line model" see Wise, G. "Science and Technology," *Osiris* (new series), *1*: 229–246.

[6] Curti, M., 1951. *The Growth of American Thought* (second edition; New York: Harper and Brothers), p. 518; see also, p. v.

[7] Beard, C.A. and Beard, M.R., 1930. *The Rise of American Civilization* (one volume edition; New York, Macmillan), II, p. 742.

[8] Morison, E.E., 1946. "Naval Administration in the United States," *U.S. Naval Institute Proceedings, 72*: 1303–1313. On the impact of changing naval organization on communications, see Gebhard, L.A., 1979. *Evolution of Naval Radio-Electronics and the Contributions of the Naval Research Laboratory* (Washington, DC: NRL Report 8300).

[9] Rosen, "The Treaty Navy,"[1]; Gebhard, *Contributions of NRL*[8], pp. 43–49.

[10] Douglas, S. "The Navy Adopts the Radio, 1899–1919," in: Smith, M.R. (ed), 1985. *Military Enterprise and Technological Change* (Cambridge, Mass: MIT Press).

[11] Douglas, S., 1985. *Inventing American Broadcasting, 1899–1922,* (Princeton: Princeton Press), ch. 4, 8; Aitken, H.G.J., 1985. *The Continuous Wave: Technology and American Radio, 1900–1932* (Princeton: Princeton Press), ch. 5–7. See also, Kent, H.D., 1921. "Fleet Radio Communications in War," *U.S. Naval Institute Proceedings, 47*, pp. 1059–1068 and Breckel, H.F., 1922. "The Vital Importance of Radio Communications in Modern Naval Warfare," *U.S. Naval Institute Proceedings, 48*, pp. 383–393.

[12] Louis Gebhard oral history interview, and Gebhard and Taylor biographical files, NRL historian's office. Taylor, *Radio Reminiscences*[4].

[13] On the technical advantages of shortwave, and the civilian inventors associated with it, see Aitken, *Continuous Wave*[11], especially ch. 4. On shortwave radio as a new social space for amateur operators, see Douglas, *Inventing American Broadcasting*[11], ch. 6 and 8.

[14] Gebhard, *Contributions of NRL*[8], pp. 21–24.

[15] E.G. Oberlin and A.H. Taylor to Chief of Naval Operations, "Radio Telegraph – Communications – Unusual Development In," 21 August 1924. History files, NRL. For the Fleet's low opinion of shortwave, see Gebhard, *Contributions of NRL*[8], pp. 43–47.

[16] E.O. Hulburt biographical file and oral history interview, NRL history office; "Edward Olson Hulburt: Frederick Ives Medalist for 1955," *Journal of the Optical Society of America, 46* (1955): 1; *Annual Report, War Department, Fiscal Year Ended June 30, 1919; Report of the Chief Signal Officer to the Secretary of War, 1919* (Washington, D.C.: Government Printing Office, 1919), pp. 287, 319.

[17] E.G. Oberlin to NRL director A.E. Krapf, June 6, 1963. "Personal Reflections" file, NRL historian's office.

[18] On the early history of ionospheric studies, see Gillmor, C.S., 1976. "The History of the Term 'Ionosphere'," *Nature, 262*: 347–348; Villard Jr., O.G., 1976. "The Ionospheric Sounder and its Place in the History of Radio Science," *Radio Science, 11*: 845–860; Hulburt, E.O., 1974. "Early Theory of the Ionosphere," *Journal of Atmospheric and Terrestrial Physics, 36*: 2137–2140.

[19] For a more detailed account of the research carried out by Hulburt and Taylor, see Hevly, "Basic Research in a Military Context"[4], pp. 30–35.

[20] See Taylor's note to amateur operators in *QST*, November 1925. Also, copies of responses from the volunteer listeners are held in "Radio – Taylor Correspondence" file, history office, NRL.

[21] NRL file A3–2(5)/GI, "Cooperation with Outside Agencies–Carnegie Institution," Box 4,

RG 19, National Archives.

[22] Taylor, A.H. *Radio Reminiscences*[4], p. 111.

[23] Hulburt, E.O. and Taylor, A.H., 1925. "The Propagation of Radio Waves Over the Earth," *Physical Review, 25*: 185.

[24] E.O. Hulburt oral history interview, NRL history office files, p. 18.

[25] Sanderson, J.A., 1967. "Optics at the Naval Research Laboratory," *Applied Optics, 6*: 2030.

[26] For some of Hulburt's later work on the ionosphere, see Hulburt, E.O., 1938/1947. "Photoelectric Ionization in the Ionosphere," *Physical Review, 53*: 344–351, and "The Upper Atmosphere of the Earth," *Journal of the Optical Society of America, 37*: 405.

[27] The term comes from one of NRL's commanding officers, Foley, P., 1925. "The Naval Research Laboratory," *U. S. Naval Institute Proceedings, 51*: 1927.

[28] On Hulburt's new position and its significance, see Hevly, "Basic Research in a Military Context"[4], ch. 6.

[29] See Hevly, "Basic Research in a Military Context"[4], ch. 4.

[30] Hevly, "Basic Research in a Military Context"[4], ch. 3; Schade, Commodore H.A., 1946. "German Wartime Technical Developments," *Society of Naval Architects and Marine Engineers, Transactions, 54*: 83–111; Millis, W. (ed), 1951. *The Forrestal Diaries* (New York: Viking Press), p. 244; Hewlett, R.P. and Duncan, F., 1974. *Nuclear Navy 1946–1962* (Chicago: University of Chicago Press). Much work was spurred by an examination of weapons for which Nazi Germany had plans, but had not deployed during the course of the war.

[31] Sherry, M.S., 1987. *The Rise of American Air Power* (New Haven: Yale Press), pp. 182–187; Yergin, D., 1977. *Shattered Peace* (Boston: Houghton Mifflin), pp. 208–211, 338–343.

[32] Bush, V., 1949. *Modern Arms and Free Men: A Discussion of the Role of Science in Preserving Democracy* (New York: Simon and Schuster), especially ch. 8 and 9.

[33] Declaring that the postwar world would bring a "new importance to seapower," Nimitz argued that "today our frontiers are the entire world." Hewlett and Duncan, *Nuclear Navy*[30], p.1. On the Navy's efforts to sponsor missile development in university-related laboratories, see Dennis M.A., 1991. "A Change of State: The Political Cultures of Technical Practice at the MIT Instrumentation Laboratory and the Johns Hopkins University Applied Physics Laboratory, 1930–1945" (Ph.D. dissertation, Johns Hopkins).

[34] Gebhard, *Contributions of the NRL*[8], p. 236–237.

[35] Bush, V., 1945. "Science–The Endless Frontier," (Washington, DC: Government Printing Office). Galambos, L. (ed), 1978. *The Papers of Dwight David Eisenhower*, VII: *The Chief of Staff* (Baltimore: Johns Hopkins University Press), pp. 1046–1050.

[36] The leader of the Navy Technical Mission was NRL's first postwar commanding officer. Schade, "German Technical Developments"[30]; Lasby, C., 1975. *Project Paperclip: German Scientists and the Cold War* (New York: Athenaeum); Kuiper, G., 1946. "German Astronomy During the War," *Popular Astronomy, 54*: 263–287; ALSOS report SAG/67, November 20, 1944, in National Archives record group 181, NRL file S-A9/Alsos. The larger process of the Americans' adoption of the V–2 for postwar research is described in: DeVorkin, D.H., 1992. *Science With a Vengeance: How the American Military Created the Space Sciences in the V–2 Era* (New York: Springer-Verlag), ch. 3, 4, and 5, and in his contribution to this volume: "The Military Origins of the Space Sciences in the American V–2 Era."

[37] For a summary of the early research work, see Durand, E., Oberly, J.J., and Tousey, R., 1949. "Analysis of the First Rocket Ultraviolet Solar Spectra," *Astrophysical Journal, 109*: 1–16.

[38] Rosen, M.W., 1955. *The Viking Rocket Story* (New York: Harper and Brothers). Early specifications and plans for the Viking are in National Archives record group 181, NRL file S78–1(119) #1.

[39] Navy Office of Research and Invention to NRL, "Army V–2 Firings: Data Desired From," record group 181, NRL file S78–1(119) #1. Krause, E.H., "Report on a proposed guided missile

program for the Navy Research Laboratory," in "Guided Missiles" file, NRL history office. DeVorkin, D.H., 1987. "Organizing for Space Research: The V–2 Rocket Panel," *Historical Studies in the Physical and Biological Sciences, 18*: 1–24.

[40] Byram, E.T., Chubb, T.A., and Friedman, H., 1953. "Contributions of Solar X-Rays to E-Layer Ionization," *Physical Review, 92*: 1066–1067.

[41] On general trends in ionospheric physics, see Gillmor, C.S., "Geospace and Its Uses: The Restructuring of Ionospheric Physics Following World War II," in: De Maria, M., Grilli, M., and Sebastiani, F., (eds), 1989. *The Restructuring of Physical Sciences in Europe and the United States, 1945–1960* (Singapore: World Scientific), 75–84.

[42] Van Allen, J., Fraser, L.W., and Floyd, J.F.R., 1948. "The Aerobee Sounding Rocket – A New Vehicle for Research in the Upper Atmosphere," *Science, 108*: 476–477; Hevly, "Basic Research in a Military Context"[4], ch. 5, 7, and 8; DeVorkin, *Science With a Vengeance*[36], ch. 10.

[43] Berkner, L.V. (ed), 1958. *Rockets and Satellites* (Annals of the IGY, 6; New York: Pergamon Press), pp. 52–58, 76–102.

[44] Friedman, H., 1977. *Reminiscences of 30 Years of Space Research* (Washington D.C.: NRL Report 8113), pp. 12–16.

[45] For an important study of the Navy's body for coordinating basic research, see Sapolsky, H., 1990. *Science and the Navy: The History of the Office of Naval Research* (Princeton, NJ: Princeton University Press).

[46] Similarly, Forman, P., 1987. "Behind Quantum Electronics: National Security as Basis for Physical Research in the United States, 1940–1960," *Historical Studies in the Physical and Biological Sciences, 18*: 149–229, and Forman's contribution to this volume, "Into Quantum Electronics: The Maser as 'Gadget' of Cold-War America." A compelling, although not air-tight, argument for a characteristic prewar American style of academic physics, which contributed to its utility, is made in: Schweber, S.S., 1986. "The Empiricist Temper Regnant: Theoretical Physics in the United States, 1920–1950," *Historical Studies in the Physical and Biological Sciences, 17*: 55–98.

[47] On the concept of constraints applied to the development of scientific knowledge, see Galison, P., Hevly, B. and Lowen, R.S. "Controlling the Monster: Stanford and the Growth of Physics Research, 1935–1962," in: Galison, P. and Hevly, B. (eds). *Big Science: The Growth of Large-Scale Research* (Stanford, CA: Stanford University Press), pp. 46–47 and 74–76.

[48] Among important works on this topic are: Hounshell, D.A. and Smith Jr., J.K., 1988. *Science and Corporate Strategy: DuPont R&D, 1902–1980* (New York: Cambridge University Press); Reich, L., 1985. *The Making of American Industrial Research: Science and Business at GE and Bell, 1876–1926* (New York: Cambridge University Press); Wise, G., 1985. *Willis R. Whitney, General Electric, and the Origins of U.S. Industrial Research* (New York: Columbia University Press); Cahan, D., 1989. *An Institute for an Empire: The Physikalisch-Technische Reichsanstalt, 1871–1918* (New York: Cambridge University Press).

[49] Hughes, T.P., 1983. *Networks of Power* (Baltimore: Johns Hopkins Press), pp. 14–15, 371–377; Forman, "Behind Quantum Electronics"[46], pp. 212–224.

[50] Reich, L., 1983. "Irving Langmuir and the Pursuit of Science and Technology in the Corporate Environment," *Technology and Culture, 24*: 179–221. Wise, G., 1983. "Ionists in Industry: Physical Chemistry at General Electric, 1900–1915," *Isis, 74*: 7–21.

[51] Servos, J., 1980. "The Industrial Relations of Science: Chemical Engineering at MIT, 1900–1939," *Isis, 71*: 531–549. Noble, D.F., 1977. *America By Design: Science, Technology, and the Rise of Corporate Capitalism* (New York: Alfred A. Knopf). D.F. Noble was with this among the first to draw attention to the fusion of science and corporate interests, especially in the process of educating professional engineers.

[52] Leslie, S.W., 1979. "Charles W. Kettering and the Copper-Cooled Engine," *Technology and Culture, 20*: 752–776.

[53] Leslie, S.W., 1983. *Boss Kettering* (New York: Columbia Press), ch. 7.

[54] Layton, E.T., 1971. "Mirror-Image Twins: The Communities of Science and Technology in 19th Century America," *Technology and Culture, 12*: 562–580.

[55] Among persuasive accounts which take this point of view, see Latour, B., 1987. *Science in Action* (Cambridge, Mass.: Harvard Univ. Pr.), especially ch. 2 and 3; and Bijker, W.E. "The Social Construction of Bakelite," and MacKenzie, D. "Missile Accuracy," among other essays included in Bijker, W.E., Hughes, T.P., and Pinch, T., (eds), 1987. *The Social Construction of Technological Systems* (Cambridge, Mass: MIT Press).

[56] Beard and Beard, *Rise of American Civilization*, II, p. 800.

DAVID H. DEVORKIN

THE MILITARY ORIGINS OF THE SPACE SCIENCES IN THE AMERICAN V–2 ERA

INTRODUCTION

The realm of the air – 'the atmosphere' – was one of the earliest foci of natural-philosophical speculation, and has remained such, whether as object of investigation or circumvention, for nearly every one of the physical sciences. Especially astronomers, meteorologists, and physicists have sought ways to get up into or above the atmosphere. Using whatever 'platforms' were accessible to them – mountain sites, balloons, kites, aircraft, and rockets – they studied how the atmosphere changes with altitude, what its circulation patterns are, and what exists beyond this obscuring and absorbing blanket over our heads. It was, however, only in the decade following World War II that the very highest stretches of the atmosphere, and the 'outer space' beyond it, were brought within the reach of physical scientists. How was access to these higher realms obtained? Who wanted this access, who provided it, and under what circumstances did they provide it?[1]

After the close of World War II, captured German V–2 missiles were used by all branches of the American military services as vehicles to launch scientific instruments to the edge of the Earth's atmosphere. The groups that formed within the American military laboratories and contract facilities to launch these 'payloads' and analyze the returned data focussed mainly on scientific questions, questions about the structure of the upper atmosphere which lay in the realms of solar physics, ionospheric physics, and cosmic-ray physics. But they were also the very questions that would logically come to mind when thinking about how to improve and to exploit a V–2 missile.[2]

Furthermore, when the V–2 rocket was applied to the scientific study of the upper atmosphere, a new way to conduct science emerged. Thus this case provides insight into how in postwar United States the military services promoted the alignment of scientific interests with their own. By making rockets available as launch vehicles for scientific packages, the military services created not only a new way to perform scientific research but also a new cohort of scientists. The scientists who packaged experiments for these rocket flights came to accept the vehicle that the military wished to develop as defining the boundaries and interests of their research.[3] More than by any intellectual problems or disciplinary orientations, the space sciences, at their inception, were defined by the military vehicles enabling them. Indeed, as late as 1980, the U.S. National Aeronautics and Space Administration's longtime

P. Forman and J. M. Sánchez-Ron (eds.), National Military Establishments and the Advancement of Science and Technology. Studies in the 20th Century History, 233–260.
© 1996 Kluwer Academic Publishers. Printed in the Netherlands.

Chief Scientist, Homer E. Newell, defined space science as science done in space: "those scientific investigations made possible or significantly aided by rockets, satellites, and space probes."[4]

This essay has two purposes: first, to describe how the U.S. Army Ordnance Rocket Development Division became responsible for initiating scientific research with rockets in the United States, and, second, to identify the types of individuals and groups who responded to its invitation. I will show not only that the institutional environments and infrastructures that were created fostered a set of scientific problems whose solution served the specific technological goals of rocket development, but also that these scientific problems and the interest in them were stimulated by those goals in the first place.

An allied interest here will be to describe what military program officers, on the one side, and their scientific colleagues, on the other side, wanted the world to know about themselves and their programs. How, specifically, did they paint the relationship between science and the military? Newell and other military scientists aligned their rhetoric with that of their sponsors in the Army and Navy. Both groups stated that captured German V–2 missiles were launched at White Sands to allow U.S. Army Ordnance technical crews and industry contractors experience in ballistic missile development. Even though the scientists also believed that much of the information they desired was indistinguishable from that desired by their military patrons to improve the missile itself, they nonetheless separated the military aspects of the work from the civilian. In the phrase of one commentator, they turned the dreaded V–2 'war head' into a benign 'peace head'.[5] Both groups also employed what Michael Smith has called the "display value" of science. Rocket development was given a higher meaning; science with the V–2 was fostered as a symbol of peaceful progress and vitality synonymous with strengthening the nation's military potential.[6]

But who here was using whom? Was it the case, as Newell would have it appear, that the V–2 and its successors were simply helping scientists to spin out an unbroken thread of scientific tradition? If so, why then had the first generation of scientists to perform research with rockets, men such as Newell, William G. Dow, Herbert Friedman, and James A. Van Allen, no prior professional experience in the scientific areas they would address with the V–2? Conversely, contrary to Newell's 1947 claim that "To the scientist, the advent of the rocket as a vehicle for high-altitude research is a welcome thing...,"[7] those scientists who had previously established career paths in these subject matter fields did not at first find the new vehicles a "welcome thing."[8] Rather, those who aligned themselves with rocket research possessed technical and practical training and experience in solid state physics, nuclear physics, electronics, optics, and radio telemetry, not in astronomy, meteorology or cosmic-ray physics. They also brought with them institutional agendas closely aligned with the interests of military agencies who were developing long-range ballistic missile systems. Thus questioning the reality of the unbroken thread highlights the importance of the military infrastructure in creating the space sciences in America. What distinguishes this case study is that rather than examining how scientific disciplines like quantum mechanics or solid state physics were realigned or re-oriented to address the needs of the national security state, I will show how a totally new way to do science emerged from a military infrastructure. First, however, I turn to examine the origins of that infrastructure.

"MR. MISSILE"

Research and development policy of the United States military was deeply influenced by the accomplishments during the course of World War II of the American Office of Scientific Research and Development (OSRD) and the Army's Manhattan Project, but also by the spectacular achievements of German scientists and engineers, especially in electronics and missile development.[9] In 1943, well before Germany's 14-meter, 12 800-kilogram, liquid-fueled A–4 missile, later called *Vergeltungswaffe Zwei*, or V–2, was operational, but as intelligence was collected of its potential, U.S. Army Ordnance contracted with the General Electric Company of Schenectady to form Project Hermes. Hermes was at first a research and development effort to build large-scale propulsion and guidance systems, but with the end of the war it also became part of Army Ordnance's evaluation of the V–2.[10] With the invasion of Germany, Ordnance competed, under Colonel Holger M. Toftoy's direction, with its counterparts in the British and Soviet military to capture what it could of the V–2. By April 1945 a mountain of captured V–2 parts and documentation gathered from the underground Mittelwerk V–2 production plant were under Army Ordnance control.[11]

Toftoy's top priority at the end of World War II, as chief officer in the Rocket Development Division of the Office of the Chief of Ordnance, was to gain support for a sustained long-range ballistic missile research and development program within the rapidly growing Army ballistic missile effort. Toftoy and other OCO staff were responsible for maintaining and upgrading traditional facilities within the Army's arsenal system, as well as building new facilities such as the White Sands Proving Ground, where the captured German V–2 missiles were to be tested in 1946. They also sought to maintain the activity of the wartime missile contractors such as the Jet Propulsion Laboratory, General Electric, and Bell Telephone Laboratories. These were all challenging tasks, especially since less than one-quarter of the annual postwar $50 million research and development budget of Army Ordnance went to missile development, of which long-range missiles formed only a part.[12]

Toftoy had a major selling job to do, both within the Army and beyond. The Army Ground Forces Equipment Review Board believed that long-range ballistic missiles were too far in the future to warrant serious attention at the moment:

While great progress has been made by our research facilities, during recent years, there is much yet to be learned relative to aerodynamics and controls for guided missiles. For this reason, it is felt that until satisfactory designs and controls are evolved for short range guided missiles and their capabilities determined, it is not advisable to undertake the development of a very large, very long range strategic guided missile.[13]

Toftoy lobbied for increased support for facilities and for domestic research, and was a chief architect of the plan to exploit the German scientists who had designed Hitler's "vengeance weapons."[14] He recalls meeting resistance from within: "I could tell some anecdotes about, '45, 6 and 7; those years in the Pentagon we were all considered batty."[15] His mission was to overcome entrenched skepticism and indifference he perceived within Army Ordnance. But he knew as well that he was in a race. All the services had champions for ballistic missile systems. Army Air Force General Hap Arnold, the ardent champion of what Michael Sherry has called "a permanent alliance of officers and scientists," told Jet Propulsion Laboratory's Theodore von Kármán in late 1944 that "I see a manless Air Force" composed of missiles that would also require less in manpower than planes to produce.[16] Arnold was specific: the V–2 demonstrated that missiles were "ideally suited to deliver atomic explosives." He was captivated by "a dehumanized technology of war." Even when air-power doctrine remained wedded to the piloted long-range bomber, Arnold and other Air Staff sympathetic to strategic missiles were able, by October 1945, to solicit proposals from American industry for design studies of missile systems with ranges up to 5000 miles.[17]

In the face of internal skepticism and external competition, Toftoy's trump card was possession of the V–2 missile parts – parts sufficient to construct dozens of V–2 missiles.[18] He had been responsible for capturing and carting them off to America in the summer and autumn of 1945. How Toftoy, known to his admirers as 'Mr. Missile,'[19] used the V–2 to gain political support for his programs is the subject of the first half of this paper. I will focus specifically on how Toftoy exploited the growing perception of military dependence upon scientific talent to create a body of civilian scientific researchers who defined their research in terms of the needs of missile development. In the process, they became a new advocacy group for the Army's missile program and, ultimately, the embryonic nucleus of the space sciences in America.

ARMY R&D POLICY

In the absence of a national strategy for control of postwar defense research, and in an effort to survive the shake out of postwar reconversion, entrepreneurial officers within the research and development services of the various branches of the military moved rapidly to perpetuate the OSRD model.

Thus they contracted with civilian-based industrial and academic laboratories for a wide range of services. Army Ordnance was no exception.[20]

At war's end Army Ordnance was already heavily involved with industrial contracting for ballistic missiles. During the war, Ordnance's natural goal was to rush to production anything that might be an effective weapon. But by October 1945, it modified its policy into a longer-term research and development posture. Reflecting this shift, Toftoy repeatedly emphasized the developmental aspects of Ordnance programs, justifying ongoing research and development contracts with General Electric, Bell Laboratories, and the Jet Propulsion Laboratory for the Hermes, Nike, and Corporal missile programs. His bench mark for a developmental posture was the V–2, which by the spring of 1945 had become part of Hermes.

Ordnance's testing of the captured German V–2 at White Sands was Toftoy's way of learning the lessons of the V–2; not satisfied with simply building an American version, he wanted to build bigger missiles. In October 1945, at the first of a series of interservice conferences on ballistic missile development sponsored by the Joint Chiefs of Staff, Toftoy's vision for progress was firmly set on the V–2 as a symbol of what was possible. It was his centerpiece for a multi-point rationale with practical interests in all aspects of missile fire control. It was also a plan for gaining new fundamental knowledge about missile behavior and control.[21] Toftoy's push beyond the V–2 strengthened in 1946, when he argued before that same JCS-sponsored group that Ordnance should not go hastily into production with a too naive or limited design:

More progress will result from stressing basic research and component development for the next 5 years or so, than by hurrying into large-scale production improved existing missiles or early models of promising new weapons.[22]

Toftoy interpreted Army R&D policy to fit his perception of the needs of guided missile development. He wanted to maintain a developmental research atmosphere within Army Ordnance, and found that he could take advantage of a new Joint Chiefs of Staff policy that stressed interservice coordination and cooperation. The Joint Chiefs wanted to overcome intense interservice competition to relieve the scarcity of trained and experienced talent for building new weapons. Toftoy both shared and exploited these concerns.[23]

Early in 1946, Toftoy created a mechanism that anticipated elements of the Joint Chief's 'Joint Research and Development Board'. He organized a body of civilian scientists, representing agencies within the services, who would aid in the development of missiles by using the V–2 in a coordinated visible program to conduct scientific research. As he stated at that same June 1946 interservice conference:

The primary aim of the Ordnance rocket program is to produce free rockets and controlled missiles that meet military requirements, but the field is so new we are placing emphasis on fundamental and basic research. Our program has been planned not only to provide the necessary basic knowledge from all fields of science and apply this knowledge to missile design, but also to conserve personnel and funds and to prevent unwarranted duplication.[24]

CONNECTING THE CIVILIAN TO THE ROCKET

Amidst the divisive turbulence of postwar reorganization, Toftoy exploited a commonly held belief in the importance of expanding the pool of scientific personnel available to help design and build missiles. This stance was both prudent and calculated, for calls were coming from all quarters to preserve and expand the scientific manpower pool that had helped win the war. So, for example, in the Spring of 1947, the first article of the first issue of the *Air University Quarterly Review* claimed, in line with Hap Arnold's prognostications, that future pushbutton wars were thinkable only because "... a new element has been added to air warfare that has revolutionized our thinking. The new element is the *decisive contribution of organized science to effective modern warfare*. Never before have so many scientific workers been united in the application of science to military purposes."[25] The author, a colonel in the Army Air Forces, warned that military planners could "... no longer wait for science to present them with new weapons. Rather, by acquiring a knowledge of the capabilities of science today, they must anticipate tomorrow's weapons."[26]

What better way to "anticipate tomorrow's weapons" than to capture and exploit Germany's so-called 'rocket team' that had built the V–2. Toftoy did just that, in a manner both celebrated and condemned.[27] Von Braun and his Germans were invaluable short cuts for sorting out and interpreting the mountains of documentation on the V–2, and for advising Army personnel and Hermes contractors on how to launch reconstructed V–2s at White Sands. Less than a year into the program, however, in October 1946, Toftoy told a superior that "We can no longer justify the enlarged project as work on the V–2's."[28] By then his Germans were at work designing a larger missile, and were aiding in an Ordnance effort to propose a delivery system for atomic warheads, dubbed "Comet."[29]

For Toftoy, tomorrow's weapons, his "enlarged project," required not only German expertise, but links to domestic scientific manpower aligned with the Army's needs. With the V–2 as bait, in late 1945 Toftoy cast his line beyond the boundaries of Army Ordnance, well into Navy, Signal Corps, and Army Air Forces territory, looking for groups of scientific workers whose attention might be diverted to his missile aspirations. He knew well that sympathetic program officers and civilian scientists looked to the V–2 as the vehicle to meet and supersede.[30] To these groups he offered space in the warheads of V–2s that would be flying at White Sands. In return, Toftoy asked that those who got involved adhere to Army Ordnance and Project Hermes guidelines, and form themselves into a visible, coherent group to organize the research that would be done with the V–2.[31] What I will show is that in return, Toftoy gained an advocacy group that on several occasions helped him maintain the V–2 program itself.

In January 1946, Toftoy and an assistant, James Bain, had little trouble finding activist entrepreneurs within both military and civilian laboratories who wanted to get involved. The V–2 was ready to fly at White Sands, and in the mind of at least one of the scientists who was doing the recruiting: "It's a screwy program, but here are the V–2's and we should do something about them."[32] Indeed, as NRL member F.S. Johnson recently recalled, "... the interest developed there because of the opportunity."[33]

Toftoy found a powerful and energetic ally to manage the effort, thus allowing him to distance himself from it, and attend to his higher priorities. Ernst H. Krause was head of a recently formed Rocket-Sonde Research Section within the Naval Research Laboratory. As a civilian PhD physicist, Krause was wholly fit, in both form and substance, to lead the group of scientists who made up what was called at first the V–2 Panel. Each member represented groups within the services who would prepare instruments for V–2 flights, and each provided Toftoy with just the rationale he was looking for: arguments couched in the jargons of cosmic-ray physics, ionospheric physics, atmospheric physics, and solar physics justifying the scientific uses of rockets like the V–2.

At the June 1946 Joint Chief's meeting, Toftoy proudly highlighted both Army Ordnance's technical progress, and its provision of missile berths to the Navy, Air Corps, and Signal Corps through the agency of the V–2 Panel. Although he felt that "So far we have just scratched the surface," Toftoy believed that an important step had been taken that would make maximum exploratory use of the desert firings of the V–2.[34] Both here and in his frequent public addresses, Toftoy linked the V–2 program to the Army's need to improve its pool of scientific personnel. He said as much to a radio audience in September 1946, and repeated the message in following years:

You are living in the beginning of a new and fascinating age. You have read in the newspapers and magazines about rockets, atomic power, and cosmic rays – all of which a few years ago were called imaginary and fantastic. The Army Ordnance today has rockets which speed along at thousands of miles an hour, and go higher than anything has ever gone. Now we are working to make them better.[35]

He also told the National Rifle Association in February 1947 that much basic research still had to be done by "experts in electronics, aerodynamics, physics, meteorology..." to perfect ballistic missile systems. This work could not be done by military ordnance laboratories alone, so Army Ordnance had enlisted "the most capable civilian and industrial organizations, so that the best talent in the country is engaged in this work."[36]

Connecting the civilian to the rocket was meant to enhance the public image of strategic missile development in the first years of the Cold War. The

V–2 attracted attention from the media, and civilian science gave it a new meaning. "Big Ben Works at Peace" declared the June 1946 issue of *Science Illustrated*, just after the launchings started at White Sands. Similarly, *Life* wanted to reassure its readers that the V–2 firings were not merely to prepare for "future atomic wars." They were "to be more than a development of new weapons and defenses. Later rockets [will] carry aloft instruments which send back information about temperature, gases and cosmic rays in the earth's little-known upper atmosphere."[37] *Fortune* showcased Baird Associates, Inc., of Cambridge, Massachusetts, highlighting Baird's participation in building the spectrograph that obtained the first ultraviolet solar spectrum from the V–2.[38]

Public relations officers in Army Ordnance wanted the world to know that the Army was relying upon "civilian scientists [to] build up the fundamental research" with the V–2, whereas a June 1946 Navy press release simply wanted the world to know that the "V–2 Works for Science: Captured War Weapon Now Probes Universe."[39] In preparation for the first night firing in December 1946, the War Department invited news agencies to attend what it predicted would be a "spectacular shoot."[40] The journalists there found many astronomers gathered in the desert, and excitedly reported on their plans to study artificial meteor trails using the V–2. *Science News* retained the astronomer Fred Whipple, who led the Harvard contingent, to prepare a major report on the firing. But when the 'shot' failed, a tired and disappointed Whipple wired the journal early the next morning: "Long feature article on meteor experiment scarcely justified. I need sleep."[41]

A host of visionaries and popularizing entrepreneurs ranging from Arthur C. Clarke and J.B.S. Haldane to Willy Ley and Edward Pendray linked the rocket indelibly to science and space exploration.[42] Many of those who were participants in the scientific experimentation, including Ernst H. Krause, the head of the NRL group, along with Newell, Van Allen, Charles V. Strain and Joseph Siry, wrote popular science articles that separated the science from its military underpinnings, making it appear to be a distinct goal. In June 1947 Newell wanted his readers to know that his group at the NRL consisted of "professional men engaged solely in upper atmosphere studies" even though they worked within "an intensive program of upper-atmosphere research" which was "one phase of a three-purpose program" including the "handling of large rockets" and the production of "ballistic and aspect data for the missile."[43] Outlining the history of rocket research in 1950, Joseph Siry identified "upper-air studies" as "one of the principal objectives of American postwar rocket research" which led to the Army funding the WAC Corporal and the Navy the Aerobee. "Even the martial V–2" Siry added, "was harnessed for peaceful purposes."[44] Gilbert Perlow of NRL noted in 1949 that "Rocket research in cosmic rays and in other fields was advanced several years by the capture of a quantity of German V–2 rockets."[45] Dorrit Hoffleit, a Harvard astronomer at Aberdeen Proving Grounds during the war, wrote about the "Peaceful V–2 Rockets" for *Sky & Telescope*, which also entertained similar articles from James Van Allen.[46]

All of these efforts left the impression that science was being conducted apart from military interests, even though from author affiliations and acknowledgements the military was hardly invisible. Waldemar Kaempffert of the *New York Times* followed Van Allen's cosmic-ray studies, and observed that "the rockets that are being sent up at White Sands, N.M. [were] for the benefit of scientists as well as of Army and Navy officers."[47]

THE BENEFITS OF THE CONNECTION

Within the first year of firings from White Sands, Toftoy found the V–2 Panel to be a useful lobby for continuing his program. In late October 1946, Ernst Krause, representing the V–2 Panel, joined with Toftoy to testify before oversight hearings at the Pentagon. Krause brought along Navy officers and in concert they all called for the continued firings of V–2s at White Sands. Krause, in fact, became an ardent Toftoy booster during these hearings; at an early point he testified that "in at least three fields more information had been collected from the program than in all the previous ten or twenty years of work... [on] cosmic radiation, solar spectroscopy and temperature and pressure measurements of the upper atmosphere." He argued that the firings were "dirt cheap" and that there was "nothing better than the V–2 for securing the high altitude data."[48] No one challenged Krause's testimony; it lent scientific imprimatur to the program.

Krause secured a confirming endorsement of the panel of scientists he represented, and gave the October 1946 oversight panel what it wanted: a rationale that went beyond the partisan interests of Army Ordnance. Toftoy also took a cue from many of his newly recruited scientific colleagues and promoted the potential applications of science from rockets themselves. To the same NRA audience noted above, Toftoy added that basic research with the V–2 has already "contributed greatly to man's knowledge of the atmosphere, including its composition, temperature and density ..." and that "It is also greatly advancing cosmic ray research which has the possibilities of energy greater than that produced by nuclear fission – and which, as yet, has not been duplicated in laboratories."[49]

Toftoy's public statements in following years strengthened the link between peaceful research and national security. In May 1953, a year after the last V–2 was launched from White Sands and when the Army's Corporal program was in full swing, he stressed that "In today's Armed Forces it is impossible for military science to exist separately from the broader fields of science." He also believed that "In order to retain high quality research personnel, it is important that their morale be maintained at a high level... what is even more important to the scientist is appropriate freedom of action, and recognition of professional accomplishments."[50]

Toftoy did what he could to allow the members of the V–2 Panel "appropriate freedom of action." What he got in return was not only an advocacy group, but a new breed of scientist who made the rocket the central ingredient in their research. In effect, Toftoy reflected what was then a growing military tendency to permanently mobilize science. He did so by making possible a new way to conduct research that depended upon an infrastructure that was wholly military in character.[51] It is only within this context that one can gain a full appreciation for the world that gave birth to the space sciences in America, a world where the alliance between science and the military was fostered equally by both partners, as well as by the popular media which tended to favor an image of the V–2 as a peaceful weapon for science.[52] As the rhetoric of war blended into that of peace, the boundary between the two was clouded. Navy officers at White Sands, according to a *Time* magazine reporter in 1949, wanted to state "For the record," that "the Navy maintains that its rockets are strictly for peaceful research."[53] At the time of the oversight hearings, Toftoy's own house organ *Army Ordnance* boasted how the Army was turning the evil Nazi warhead into a productive American "'peace head' filled with scientific paraphernalia for exploring the upper atmosphere and evaluating the performance of the... rocket."[54]

For Toftoy, as well as his many counterparts in the Air Force, Navy, and Signal Corps, the needs of the ballistic missile defined the science that was done with it. The question remains, to what extent was this constraint reflected in the agendas of the members of the V–2 Panel?

DEFINING SCIENCE WITH A VENGEANCE

First, who got involved? With but few exceptions, the people who entered upper air research with rockets were young scientists who experienced organized research for the first time during the war. Among them, the leaders who proposed, designed, and built instruments, were mainly experimental physicists with experience in optics, electronics, radio and mechanical engineering.[55] With the exception of a group from Princeton, none of them had prior experience in the areas of research they embarked upon with the V–2. Most were electrical or electronics engineers, or experimental physicists with training in electronics and nuclear physics, who had worked in military laboratories during the war. F.S. Johnson, who worked for Richard Tousey's solar ultraviolet spectroscopy group at NRL, recalls that his responsibilities were a direct application of his training and talent, even though he lacked any previous professional interest in the subject. He also felt that "[Herbert] Friedman's work typifies that best of all. He was working with various types of counters, and he had the ability and the people who could assemble such things and put them in rockets... It wasn't that he had been waiting a long time for an opportunity to make these measurements on the sun, or on astronomical objects."[56] Accordingly, Newell and his contemporaries recall having to undergo a period of "intensive self-education" to gain both the technical tools and the scientific insight they would need to do science with the V–2.[57]

Among the leaders in 1946, NRL's Krause, APL's Van Allen, and Princeton's Myron H. Nichols were all in their early thirties. Marcus O'Day, of the Air Corps' Cambridge Field Station, was 49. He was a contract and project officer who had taught radio physics and electronics at Reed College before the war, and had been at Harvard's Radio Research Laboratory since 1941 in charge of the Special Beacon Group, specializing in radio navigation and guidance systems. William G. Dow, professor of electrical engineering at Michigan, was the oldest group leader at 51. He was a senior vacuum-tube development engineer who had worked mainly at the MIT Radiation Laboratory during the war. Excluding Dow himself and one senior technician, the average age of the Michigan group was 28, similar to the average ages in the other groups. Krause was trained in nuclear physics but his career had been at the Naval Research Laboratory in the area of guided missile countermeasures and fire control. Van Allen had similar training in Iowa and worked for a year under Merle Tuve in nuclear physics before he followed Tuve to the Applied Physics Laboratory, where they concentrated on the development of an aerial proximity fuze. Nichols was an assistant professor of physics and specialist in thermionic techniques whose experience centered on radio telemetry. During the war he acted as technical liaison to APL for a Princeton telemetry subcontract.

The one common characteristic these people possessed was that they had talents and experience building devices that somehow dealt with command, communications, or control. They knew how to design and build instruments that would work under the harsh conditions of rocket flight, and were anxious at war's end to apply their acquired experience and talents to a broad range of activities. Save for Dow and a few others, they all were young and without compelling career demands established prior to the war.

That there was a cohort of young, enthusiastic, and technically proficient scientists ready to be tapped can best be appreciated by examining the observations and reflections of more senior, established members of the scientific elite as they entered war research. Gerard Kuiper was an established astronomer who found himself at Harvard's Radio Research Laboratory in mid–1943. His first impressions were negative. He saw in warwork

... a danger: lots of people work in this field without a full or even a half-understanding of its principles. The growth is so rapid that some of it is wild, and needs trimming. But such half-understood science is probably common in all fields of engineering. The danger is if youngsters think that this is the way science should always be.[58]

Yet after working at the Radio Research Laboratory for just six weeks, Kuiper had become fully co-opted. He soon declared that in the interests of national security, some of this work had to continue after the war "keeping employed some of the best young men in each field."[59]

Van Allen, Krause, Nichols, and many of the others who entered research with rockets represented just those "youngsters" Kuiper observed as they experienced large-scale research for the first time in their professional lives – research that bore little resemblance to what went on before. And indeed, they also fit nicely into the "best young men" category required to keep the nation secure. Few if any had compelling research agendas to return to; most, if not all of the new rocketry group leaders hoped to somehow exploit their wartime knowledge and experience.

The Naval Research Laboratory was the first to sign up with Toftoy, but close behind were the Johns Hopkins University's Applied Physics Laboratory (a contract laboratory of the Navy Bureau of Ordnance), the University of Michigan's Department of Electrical Engineering, Army Ordnance's Ballistic Research Laboratory at Aberdeen, the Evans Signal Laboratory of the Army Signal Corps' Fort Monmouth facility, and Princeton University's Physics Department. The Army Air Forces entered a few months later through the Cambridge Field Station (later the Air Force Cambridge Research Laboratory, AFCRL) of the Air Materiel Command. The CFS was created to maintain some of the research started by MIT's Radiation Laboratory and Harvard's Radio Research Laboratory; it also contracted with several universities for scientific rocketry. Here we limit discussion to major players on the V–2 Panel who best represent the first generation of space scientists.

THE NAVAL RESEARCH LABORATORY ROCKET SONDE RESEARCH SECTION

During the war, Krause had been head of NRL's Communications Security Section, twice touring Europe to gather intelligence on guided missile countermeasures. After the war he campaigned for an NRL leadership role in developing guided missile technology for the Navy.[60] During one cycle of interrogations of German scientists, Krause met R.W. Porter of Project Hermes at Garmisch-Partenkirchen, and learned much from him about German missile research. Porter gave him a sense of what future wars would be like and what had to be done to bring the Navy up to speed in guided missile research.

Upon his return from Europe, Krause found an ally in the new NRL director, Commander Henry A. Schade, who, like Krause was fresh from scientific and technical intelligence adventures in Europe. Early in December 1945, Schade developed a plan to demonstrate to NRL's new patron, the Office of Research and Invention (ORI), and to the older Navy bureaus, that NRL was a "research laboratory covering all fields of science and technology" and, as such, had to be thought of as a civilian institution in which militarily relevant research would be pursued in an atmosphere of open enquiry.[61]

With Schade's backing, Krause turned his Communications Security Section into a guided missiles research and development unit called the Rocket-Sonde Research Section. Although Krause initially had hoped to turn a large fraction of NRL's manpower to guided missile development, what he succeeded in accomplishing was more in line with Schade's image of NRL – a place where basic scientific research furthered institutional goals. Indeed, Krause's first formal proposal was for research into the upper atmosphere with a new Navy missile built in the image of the V-2.

The basic research Krause linked to guided missile development included studies of the ionosphere, atmospheric physics, cosmic rays, and solar physics. Each he justified by how it informed the development of guided missiles. Cosmic rays, Krause asserted, might reveal "staggering knowledge of how mass is converted into energy," whereas atmospheric physics observations would gather knowledge about the "medium through which missiles of the future are sure to go."[62] Performing these studies with rockets, Krause added, would provide practical experience in launching, guiding, and tracking rockets through the ionosphere:

Such knowledge could be as dangerous as it is useful. It could very conceivably lead to the destruction of the earth... Such awesome military applications [are] imposing enough to warrant a vigorous program.[63]

Krause, in concert with Schade, used rhetoric similar to Toftoy's to argue that if military laboratories like NRL wanted to attract the highest quality research scientists to their staffs, they had to make it worthwhile. Scientists had to be allowed to pursue activities interesting to them, with access to facilities not available elsewhere. Guided missile research, broadly construed, was an area which, Krause contended, would lift his best men out of the drudgery they all experienced during the war:

I am perfectly willing to sacrifice myself on such a bureaucratic altar during wartime, but I feel that my value to the Navy and to the nation will deteriorate very rapidly unless I gain more experience in the laboratory. This is a feeling which I share with many people... at NRL. The proposed project can provide us with the necessary experience.[64]

The NRL Rocket-Sonde Research Section was established within days of, and possibly in anticipation of, Toftoy's official invitation to the Navy to participate in the V-2 flights. In January 1946, the Section was defined by Schade's office as a unit that would "investigate the physical phenomena in and the properties of the upper atmosphere with a view to supplying knowledge which will influence the course of future military operations." Their task would be "basic research and [would] include development of the necessary techniques, instrumentation and devices required to carry out the function."[65]

THE APPLIED PHYSICS LABORATORY AND PRINCETON

In January 1945, the Applied Physics Laboratory succeeded in becoming a Navy Bureau of Ordnance (BuOrd) contract laboratory, after spending the war years building a successful proximity fuze. Under the code name "Bumblebee" they would perform a wide range of guided missile research.[66] The young institution thus defined itself and its postwar future in terms of guided missile development, and by the end of the war its senior staff were prominent members of the Guided Missile Committee of the Joint Committee on New Weapons and Equipment of the Joint Chiefs of Staff.[67] Thus APL's founding director Merle Tuve well knew what Army Ordnance was planning at White Sands.

In January 1946, APL's R.B. Roberts, looking for ways APL might diversify to avoid postwar retrenchment, suggested to Tuve that they find new projects related to their primary mission. Either "V–2 tests for upper air research including cosmic rays" or the "instrumentation of atomic bomb tests" might be appropriate. Roberts preferred the V–2 work, because it was less expensive, would not seriously affect the Bumblebee project, and because their subcontractor, the Princeton Physics Department, was "already active in this program."[68]

Indeed, Princeton physicists were already making plans. Urged on by Rudolf Ladenburg, Myron Nichols organized a full program of research, centered on cosmic-ray studies, that would utilize their telemetry system for the Navy's Lark missile, developed under a Bureau of Ordnance contract managed by APL. Ladenburg, an emigré physicist who had spent the war at Aberdeen, enjoyed an international reputation in many fields; at the time he was a member of the International Subcommittee on Ozone for the Royal Society's Gassiot Committee, and he claimed that his colleagues on the committee (Sydney Chapman and G.M.B. Dobson), would be "most interested... in the contemplated high altitude studies by use of rockets."[69]

APL's Tuve and Roberts approved of a modification of Princeton's BuOrd contract to include upper atmosphere research, but by February 1946, decided that they would also get in the business directly. Roberts carved out a small group with twin goals identical to those at NRL: to take immediate advantage of the V–2 through establishing a high-altitude research program, and extend it with a home-grown launch vehicle. The development of a new vehicle complemented their patron's requirement that the APL should acquire experience with liquid-fueled rockets in addition to the air-breathing ram jets that powered Bumblebee.[70] He put James Van Allen in charge.

Van Allen's background and training in nuclear physics, as well as his exposure to geophysical issues during his postdoctoral year with Tuve at the Carnegie Institution of Washington's Department of Terrestrial Magnetism, made cosmic-ray research appealing. In the first months of 1946, as soon as he was out of Navy uniform, Van Allen worked feverishly to build his group, stake out a place on the V–2 Panel, and design APL's launch vehicle, the Aerobee. By the end of February, Van Allen could boast to Nichols that the 15 members of his "group represents considerable experience in nuclear physics, spectroscopy, and geophysics."[71]

Although Van Allen had a strong scientific agenda in mind, like Krause he also was sensitive to applications. His group developed detector arrays for cosmic-ray studies, but they also used the same detectors in prototype cosmic-ray altimeters, which in 1947 were intended to guide "long range supersonic missiles" at altitudes between 12 and 32 km.[72] At Princeton, Myron H. Nichols and John A. Wheeler matched their military laboratory counterparts, pointing out how their scientific activities would also advance military technology by serving to test their Lark telemetry system, which was a major Princeton missile product.[73]

Princeton, however, had only a short-lived romance with the rocket. After two flight failures in late 1946, John Wheeler, who took direct control of the project after Nichols' departure for the Michigan College of Engineering, decided that they would drop out. At the same time they asked for, and had no trouble receiving, continued support from ORI (reformed as the Office of Naval Research) for cosmic-ray studies in support of nuclear research. As Wheeler noted to himself when he embarked on the project "if better results some other way[,] do it the other way."[74] Princeton's departure removed the only active rocketry group that had possessed a strong pre-existing scientific agenda.

Those who remained in the program, either in the Navy, or as contractor groups to the Air Force or the Signal Corps, could not afford the luxury of Wheeler's philosophy. Absent the momentum of a pre-existing research program, and keenly aware of their institutional priorities, remaining scientists defined their science in terms of a vehicle for research. Overall, as S. Fred Singer, a theoretical physicist enlisted with APL's group, clearly recalls, "We really were constrained to only do work that could be done on rockets."[75] Indeed, it was his strong impression that all the surviving groups thought in terms of the rocket itself.

THE UNIVERSITY OF MICHIGAN

Those defining research at the University of Michigan thought in terms of the needs of the rocket. William Gould Dow, professor of electrical engineering at Michigan, was not even remotely interested in the physics of the upper atmosphere. His involvement came directly as a result of Project Wizard, an AAF antiballistic missile program created and headed by Emerson W. Conlon. Conlon was chairman of Michigan's Aeronautical Engineering Department, and enjoyed close contacts within the Air Technical Services Command at

Wright Field. Michigan's Department of Engineering Research had developed an "effective pattern for sponsored research involvement" which coordinated campus talent to meet outside interests.[76] Dow became involved because of his expertise in the development and application of vacuum tubes; he came to view the rocket, however, not as a means of studying the high atmosphere, but as a means of studying the electrical behavior of the rocket in flight. The rocket thus became, as had Dow's vacuum tubes, objects of research as well as vehicles for research.[77]

In December 1945 the stated goal of Project Wizard was a pilotless aircraft system "capable of intercepting the V–2 type of missile."[78] The first phase of the project included a 'requirement study' of the performance characteristics of such a missile. Diagnostic instruments had to be devised to obtain data from the upper atmosphere that could reveal design constraints on the missile itself.[79] As Conlon stated to Wright Field officers, in terms similar to those used by Barnes, Toftoy, and Krause:

The region above 100 000 feet is essentially uninvestigated and in order to determine accurately the flight characteristics of long range rockets and their interceptors, densities, temperatures and composition need to be known. In order to shed more light on radio propagation through and reflections from the ionosphere, more data on the ionization in the upper atmosphere are required.[80]

Dow agreed to get involved mainly because it gave him support to attract some of his best graduates back to campus. He quickly formed a small group to apply vacuum tube technology to measure the phenomena identified by Conlon. He easily rationalized his participation in military activities, noting in October 1945 that "continued military competence is important to all of us, and it is part of the University's job, where it can, to serve the public interest by technical guidance as well as in education; also, military vacuum tubes offer special technical problems that are interesting professionally."[81] Dow did indeed bring special skills to the enterprise, as well as a competitive spirit that established Michigan as a leading player on the V–2 Panel. As an expert experimentalist "in the behavior of charged particles in a vacuum," Dow later rationalized placing his laboratory on a rocket as a study of an "outer-atmosphere vacuum."[82]

Dow was always more interested in experimentation with vacuum tubes than collecting data about the upper atmosphere; Nelson W. Spencer, an early associate, recalls that his old teacher regarded the upper atmosphere as "the greatest vacuum of all."[83] Dow's vision of his V–2 work therefore focussed upon building new devices. In 1950, he introduced Belfast atmospheric physicist David R. Bates to their work at Michigan:

Primarily what we are trying to do is to stimulate attention to a new type of instrumentation. The ionosphere exhibits many of the properties of the plasma of an ionized gas, with the exception of a very large scale factor and marked changes as to the origin of the sustaining energy. We should like to stimulate study of the ionosphere by various methods that have been extremely successful in the study of gas discharges.[84]

Dow's preoccupation with instrument development was matched at NRL by Krause's rationale for radio propagation and cosmic-ray studies. Krause recently recalled that NRL's cosmic-ray effort was really experimentation. Since electronics development was one of NRL's primary reasons for existence, his group built complex cosmic-ray packages that would help to develop and test sophisticated electronics systems that would work on rockets, and test the telemetry devices they had been developing for some years.[85] Indeed, the wording of the original December laboratory order which gave life to Krause's plan highlighted the need to develop "... the necessary techniques, instrumentation and devices required to carry out the function."[86]

Building instruments was a critical military interest. When Dow began attending V–2 Panel meetings, he renewed his contact with an old wartime friend, Lieutenant Colonel Harold A. Zahl, who was the newly appointed director of research at the Signal Corps' Fort Monmouth Laboratories (SCEL). Zahl was very interested in supporting a microwave-frequency vacuum-tube laboratory at Michigan, as long as the University of Michigan agreed in turn to build instruments for the Signal Corps for upper atmospheric research. Dow happily obliged; as he recently described himself, he always "wanted to do research on something that somebody wants."[87]

FRED WHIPPLE AND THE JOINT RESEARCH AND DEVELOPMENT BOARD

In a previous section I noted that Princeton's departure removed the only active academic group with an established research agenda. The Harvard astronomer Fred Whipple, however, was also a charter member of the V–2 Panel, although in deference to the wishes of his observatory director, Harlow Shapley, Whipple never actually prepared experiments for flight. Shapley was wary of getting too close to military activities, but also had other priorities; as he told Donald Menzel in January 1946:

For us there is hardly time to do anything about it [the Army's invitation to use the V–2] because they expect the cooperating universities to devise and build up their equipment, electronics & all, and have it ready by early summer. We are all too much tied up...[88]

Shapley wanted to rebuild the Harvard College Observatory, not build new gadgets for the Army. Whipple's role on the V–2 Panel was to act as adviser, which allowed him to remain in touch with an area of potentially great interest to his own research program, the study of the interaction of meteorites with the Earth's upper atmosphere. Whipple's participation was important to the V–2 Panel and its sponsors; at least one elite scientist recognized that 'science with a vengeance' was worth his attention.

From the 1930s, Whipple had been a student of meteor orbits and origins, and was keenly aware that there was a serious lack of knowledge of conditions in the upper atmosphere. Whipple's seminal 1943 study, "Meteors

and the Earth's Upper Atmosphere," helped him gain patronage after the war
from the Navy's Bureau of Ordnance for meteor studies that might inform
hyperballistic re-entry research.[89] As Shapley sarcastically put it, the military
funded Whipple's meteor research "because the shooting stars perform in
that same level of the earth's atmosphere where the shooting rockets and the
rocket ships of the future are planning to operate."[90]

Whipple, however, was a networker. In his multiple capacity as V–2 Panel
member, member of the Joint Research and Development Board (JRDB), and
of many geophysical panels, Whipple not only gave scientific rocketry cred-
itability, but acted as a critical interlocutor for upper atmosphere research.
Whipple's own research was needed by the Panel. It included determinations
of upper atmospheric conditions, seasonal variations in the density of the
upper atmosphere, aeroballistic heat transfer, and winds in the upper atmosphere. His role on the JRDB's Panel on the Upper Atmosphere bears special
mention because it will illuminate how members of the V–2 Panel, and their
counterparts on the JRDB, portrayed their activities to their patrons.

The Joint Research and Development Board was formed in June 1946 by
the Joint Chiefs of Staff out of the wartime Office of Scientific Research and
Development and the Joint New Weapons and Equipment Board. The JRDB
was to "coordinate all research and development activities of joint interest to
the War and Navy Departments."[91] It did so by forming a maze of committees and working panels staffed by civilian scientists and military program
officers. In July 1947 the JRDB Panel on the Upper Atmosphere, which reported to the JRDB's Geophysics Committee but had liaison responsibilities
to its Electronics and Guided Missiles committees, asked Whipple to chair a
"Working Group on the Upper Atmosphere Research Program for FY 1949."
The working group included another member of Toftoy's V–2 Panel, SCEL's
Michael Ference, NRL's Homer E. Newell, who was soon to replace Krause
on the V–2 Panel, and Aberdeen's T.H. Johnson, who was very friendly to the
V–2 Panel as chairman of the JRDB Panel on the Upper Atmosphere. They
joined L. M. Slack of ONR and other project officers from the services to
review the nation's active upper atmosphere research projects that employed
aircraft, balloons and missiles, under the charge of identifying "reasonable
objectives in upper atmosphere research... considering military requirements
and available resources."[92]

The Working Group met once, on 22 July 1947, and Whipple then drafted
its report, which began with the declaration that:

The Working Group is of the unanimous opinion that all basic upper atmospheric research is
fundamentally and inherently of value to the general problems of national defense.[93]

The Working Group report listed scores of observational projects possible
with rockets, balloons, and aircraft as vehicles, as well as supportive ground-
based research. It called for multi-year funding and a 5-year plan, both for
instrument development and for long-range statistical studies of slowly vary-

ing natural phenomena. And, consistent with the promise of its premise, it recommended that the overall effort should be "planned with military requirements conceived in the broadest possible construction of this term."[94] Although this blanket claim was apparently acceptable to the military members of the JRDB panel, they were more concerned with "tilting basic research toward the development of new technical capabilities."[95] Thus they criticized the draft for its lack of specifics, or more seriously, for its unwillingness to establish priorities.

Taking note of the criticisms of the military members of the panel, the JRDB ultimately endorsed this statement, confident that, left largely to themselves, scientists involved in scientific rocketry were sufficiently aware of military need to design their research programs accordingly. Clearly Whipple nourished this confidence by endorsing accomplishments of his associates on the V–2 Panel. For his part, Whipple knew that he had to keep his ear to the wall to remain *au courant* with military needs; his membership on numerous boards, committees, and panels in postwar Washington provided him with intelligence of value both to himself and to his institution.

The importance of intelligence gathering was not lost at the University of Michigan. The chief task of the JRDB (reformed in late 1947 as the Research and Development Board) was to identify the needs of, and resolve differences between, the services. Its tactic was to collect information, stimulate communication, and push for compromise. The RDB would typically turn to its committees and panels for advice when petitioned by manufacturers or service representatives. Membership on an RDB panel or committee, therefore, was not only useful politically, but was important for home institutions as a means to gather information about trends and needs in the military. When A.P. Fontaine, director of the Aeronautical Research Center at the University of Michigan's Willow Run facility, was asked to become a member of the Countermeasures Panel of the Guided Missiles Committee, his invitation from Frederick L. Hovde, and his acceptance, were distributed to Michigan's Technical Advisory Board of the School of Engineering with the admonishment that other members of Michigan's faculty should push for membership on these panels too. The University administration felt that "It will be a good thing for the University of Michigan to be represented on some of these panels."[96]

On one occasion in 1949, the University of Michigan's Aeronautical Research Committee was deliberating on how to reconcile faculty research initiatives with military trends, and learned from an RDB meeting representative that the military services were now looking for a "balance in emphasis" in guided missile research between hardware and pure research:

Army & Navy emphasize the need for hardware to serve as a focus for projects. Pure research projects [are] not looked upon with favor by Services because they are difficult to sell. On the other hand, too much hardware is being severely criticized.[97]

This vigilance paid off.[98] By the late 1940s, institutions like Michigan had a vested interest in maintaining their large grants and contracts, and needed more than the nine muses to plot their futures.[99] They had to play this game very carefully, reconciling faculty interests as best they could with what was being bought by the military. One way they did this was by staying in touch with the deliberations of the RDB committees and panels. Of course, the importance of intelligence was not overlooked by the faculty. By keeping in touch with Fred Whipple, some of his astronomical colleagues kept close tabs on what was transpiring at the RDB meetings.[100]

ALIGNMENTS

The constraints placed upon research with rockets identified by Fred Singer were in fulfillment of Toftoy's plan to align science with Army Ordnance needs. This was one way in which scientific talent became permanently mobilized in the wake of World War II. Under the umbrella of rocket research, fostered in the first instance by R&D project officers and technical managers like Toftoy, Schade, Zahl, O'Day, and Krause, one could imagine that a young scientist might feel relatively free to define a personal agenda even though it required an alignment with the rocket as a new way to conduct scientific research.[101]

Such alignments were hardly visible and generally not problematic to many who came of age in wartime laboratories. As Gerard Kuiper and his senior colleagues had predicted when they observed the attitudes and actions of junior colleagues at the Radio Research Laboratory and Aberdeen in 1943, the laboratory atmosphere fostered a new form of science that was intent on technical problem solving, where the principles of the science itself were less a concern. Indeed, Van Allen described the condition best in 1947 when he told a colleague what was needed to succeed in the new field: a "tough skin and a strong back are the chief requisites of a rocket experimentalist."[102]

Indeed, when individuals within NRL found that rewards were less than they desired after several years of flights of cosmic-ray instruments, they did not stick with cosmic-ray research using non-rocket techniques, but moved into more fruitful areas, mainly in geophysics, using rockets. That is to say, they chose to *break* the thread of scientific tradition and follow the rocket. Van Allen at APL was similarly able to choose what he did with rockets, but the rocket itself remained the one unshakable rock. Once he moved to Iowa in 1950/1951, as APL retrenched back to a wartime posture, Van Allen had acquired sufficient technical equity in rocketry to feel that it was his strongest suit. Though he had other options, Van Allen chose to remain in geophysical research using a highly efficient balloon/rocket combination.[103] Van Allen's initial alignment mapped well onto his newly found academic environment; his migration from the laboratory to the rocket was complete.

Persistence in the use of a vehicle for research distinguished a new way to conduct science. Unlike the acquisition of new technologies in familiar institutional frameworks, the scientific study of the upper atmosphere with rockets centered around a social institution whose existence and motives lay outside the traditional bounds of science, yet were critical to the conduct of the science itself. Superficially similar to Darwin's relationship to the *Beagle*, that of this new scientific cohort to the rocket was essentially different in that the vehicle defined the science, even though many who persisted in the effort eventually identified themselves with the fields of study they acquired. Even so, along with his colleagues on the V–2 Panel, Van Allen promoted the use of rockets as vehicles for research, and, when the time seemed right during the planning years of the IGY, actively campaigned for the scientific use of earth satellites. As a result, members of the V–2 Panel, such as Van Allen and Newell, eventually became the designers of the scientific rocketry programs for the IGY.[104] They saw the advantages of continuing a relationship with government and military agencies capable of supporting the infrastructure necessary for performing global atmospheric and geophysical research.

Van Allen, Newell, and colleagues from every group that formed to conduct research with rockets, ultimately carried this alignment into the academic, technical, and administrative infrastructure that made space science a possibility in the first years of NASA. Indeed, as NASA's highest ranking scientist upon his retirement, Newell had spent decades faced with the survival of science within NRL and then NASA. Schooled by more than a decade at NRL, working, as Krause's successor, with Toftoy, the V–2 Panel and the RDB, Newell was highly adept at defining scientific activities in terms of NASA's mission. Since NASA was in the business of building rockets that would launch satellites and space probes, Newell defined the science that NASA would support as those areas that could be studied using instruments built to fly on these vehicles. Thus it is not surprising to find that Newell's definition of space science in 1980, quoted at the outset of this paper, is so reminiscent of the original military rationale for such activity: space science was really *limited* to "those scientific investigations made possible or significantly aided by rockets, satellites, and space probes."[105]

To a large extent then, true to its origins in Toftoy's mind in late 1945/1946, space science was, and is, defined by the use of a vehicle for research, a vehicle available initially from only one source. In a very real sense, therefore, the space sciences in the United States, though conducted as civilian programs, were in fact created as extensions of the interests of the various branches of the armed services concerned with the development of an infrastructure capable of building and operating ballistic missile systems.

Smithsonian Institution, Washington

NOTES

[1] This paper is a distillation of material presented in DeVorkin, D.H., 1992. *Science with a Vengeance: How the American Military Created the Space Sciences in the V–2 Era* (New York: Springer-Verlag), and is a considerable revision of an early study, DeVorkin, 1987. "Organizing for Space Research: The V–2 Panel," *Historical Studies in the Physical and Biological Sciences, 18*, pp. 1–24. The author wishes to thank the archives cited in this paper, including the National Archives and Records Administration (NARA), the Alabama Space and Rocket Center (ASRC), the Jet Propulsion Laboratory Library; the Library of Congress (LC), the California Institute of Technology Archives (CIT), the NASA History Office (NHO), the Naval Research Laboratory History Office (HONRL), the Princeton University Library Manuscripts Division (P), the Bentley Historical Library of the University of Michigan (BHLUM), Harvard University Archives (HUA), University of Colorado Archives (UC), University of Chicago Archives, Chandrasekhar Papers (SC/UC), Smithsonian Institution Archives (SIA), Princeton University Physics Department records in the Princeton University Archives (PP/P), Johns Hopkins University Applied Physics Laboratory records relating particularly to James Van Allen (JVA/APL), of which copies are available at the National Air and Space Museum, and the Space Astronomy Oral History Project files of the National Air and Space Museum (SAOHP/NASM). Karl Hufbauer, Michael Neufeld, and the editors of this volume provided helpful criticism.

[2] This historical development at one U.S. military laboratory, and its import for our conceptions of 20th century science, has been studied by Hevly, B., 1987. *Basic Research within a Military Context: The Naval Research Laboratory and the Foundations of Extreme Ultraviolet and X-Ray Astronomy*, Ph.D. diss., Johns Hopkins University (Ann Arbor: University Microfilms). See, further, Hevly's contribution to this volume, "The Tools of Science: Radio, Rockets, and the Science of Naval Warfare," pp. 215–232. For the V–2 itself, see Neufeld, M.J., 1995. *The Rocket and the Reich: Peenemünde and the Coming of the Ballistic Missile Era* (New York: Free Press).

[3] This interpretation benefits from the work of Michael Aaron Dennis, who redefined the post World War II civilian in "'Our First Line of Defense:' University Laboratories and the Making of the Postwar American State," *Isis, 85* (1994), pp. 427–455, and in *A Change of State: The Political Cultures of Technical Practice at the MIT Instrumentation Laboratory and the Johns Hopkins University Applied Physics Laboratory, 1930–1945*. (Ph.D. dissertation, Johns Hopkins University, 1990; University Microfilms, 1991). Relevant literature on how sponsors' needs have influenced space science activities includes Needell, A.A., 1987. "Preparing for the Space Age: University-Based Research, 1946–1957," *Historical Studies in the Physical and Biological Sciences, 18*, pp. 89–109; DeVorkin, D.H., 1989. *Race to the Stratosphere: Manned Scientific Ballooning in America* (New York: Springer Verlag).

[4] Newell, H.E., 1980. *Beyond the Atmosphere: Early Years of Space Science* (NASA, SP–4211; Washington, D.C.), p. 11.

[5] Tangerman, E.J., 1946. "Nazi Vengeance," *Army Ordnance* (Sept./Oct.), p. 153.

[6] Smith, M.L. "Selling the Moon," in: Fox, R.W. and Jackson Lears, T.J. (eds), 1983. *The Culture of Consumption* (New York), pp. 177–209. Similar issues and connections are considered by J.M. Sanchez-Ron and J. Ordoñez in their discussion of "Atoms for Peace" in Spain, elsewhere in this volume.

[7] Newell, H.E., 1947. "Exploration of the Upper Atmosphere by Means of Rockets," *Scientific Monthly, 64*, p. 453.

[8] This has been argued before by Smith, R.W. "Selling the Space Telescope: The Interpenetration of Science, Technology, and Politics," in: Collins, M.J. and Fries, S.D. (eds), 1991. *A Spacefaring Nation: Perspectives on American Space History and Policy* (Washington, D.C.: Smithsonian), pp. 31–32; and more extensively by DeVorkin, 1992. Ch. 11, 14, 18.

[9] I treat this influence as one given at this time, but the question remains open as to how this influence was fostered, and by whom. See the various articles and essays by Forman, Schweber, Dennis and others, both in this volume and elsewhere.

[10] Porter, R.W., 1965. "Prologue – 1945–1954," *Challenge 4*, no. 1 and 2, General Electric Missile and Space Division.

[11] On Toftoy's role in capturing V–2 technology and it creators, see Lasby, C.G., 1975. *Project Paperclip* (New York: Athenaeum); Ordway III, F.I. and Sharpe, M.R., 1979. *The Rocket Team* (New York: Crowell); and McGovern, J., 1964. *Crossbow and Overcast* (New York: Paperback).

[12] During the period FY 1946–1948, OCO spent a total of $130 million on ordnance research and development, of which $29 million went to long-range guided missile development. Koppes, C., 1982. *JPL and the American Space Program* (New Haven: Yale), p. 34. The Army Ordnance contracting budget included $3 million for Hermes from its inception through 1946. Haviland, R.P., 1946 Jan. 7. "Informal Conference between representatives of Army Ordnance and BuAer on Rocket Development Program" Meeting held 21 December 1945. BuAer engineering division intelligence memorandum F41(1) A19(8) Aer-E-313-RPH. Jet Propulsion Laboratory History files, JPL-5-494.

[13] "Review of Report of Army Ground Force Equipment Review Board and Comments Thereupon by Theaters," December 7, 1945, p. 7 in: "Brief of Testimony of Army Ground Force Representative on Guided Missiles before War Department Equipment Review Board," OCO/NARA RG 156, B 911 J861.

[14] Lasby, 1975. Ch. 5; Bower, T., 1987. *The Paperclip Conspiracy* (London: Michael Joseph).

[15] "Leaders of Space," Committee Meeting, 16 December 1959, page 7, HT/ASRC.

[16] Sherry, M.S., 1987. *The Rise of American Air Power* (New Haven: Yale), pp. 186–187.

[17] Arnold, H. Quoted in: Sherry, M.S., 1977. *Preparing for the Next War* (New Haven: Yale), p. 229. See also Sherry, M.S., 1987. pp. 40–52, 187; Neufeld, J., 1989. *Ballistic Missiles in the United States Air Force 1945–1960* (Office of Air Force History), pp. 24–25, Chart 1–1, and pp. 44–45. For contrast, see Possony, S.T., 1949. *Strategic Air Power: The Pattern of Dynamic Security* (Washington, D.C.: Infantry Journal Press), ch. 11.

[18] See, for instance: McGovern, 1964; Bower, 1987; Lasby, 1975; Ordway and Sharpe, 1979.

[19] See, for instance, Bergaust, E., 1976. *Werner von Braun* (Washington DC: National Space Institute), p. 218.

[20] Army Ordnance policy was articulated in [Gladeon M. Barnes], "Section I: Long-Range Rocket-Powered Guided Missiles," n.d., circa October 1945, pp. 10–16, on p. 10. Box A763, "History, Planning for Demobilization Period." See also [Research and Development Office, OCO], "Outline of Ordnance Department Plans for Postwar Research and Development," 1 November 1945, p. 10. Box 763, RG 156, OCO/NARA. On the OSRD as a model for the Navy, see Sapolsky, H. "Academic Science and the Military," in: Reingold, N. (ed), 1979. *The Sciences in the American Context: New Perspectives* (Washington D.C.: Smithsonian), pp. 379–99, on p. 381.

[21] Transcript of meeting, "Joint Army-Navy Meeting on Army Ordnance Research and Development, the Pentagon, October 1, 1945," p. 40. OCO/NARA, RG156, Box 767A.

[22] Transcript of meeting, "Joint Army-Navy Meeting on Army Ordnance Research and Development, the Pentagon, June 26, 1946." p. 43. OCO/NARA, RG156, Box 767A.

[23] His diaries during this period record these concerns. See, for instance, Toftoy, H. "Diary," entry for February 20, 1946, HT/ASRC.

[24] Transcript of meeting, "Joint Army-Navy Meeting on Army Ordnance Research and Development, the Pentagon, June 26, 1946." p. 37. OCO/NARA, RG156, Box 767A.

[25] Emphasis in original. Glantzberg, F.E., 1947. "The New Air Force and Science," *Air University Quarterly Review, 1* No. 1, pp. 3–16, on p. 4.

[26] Ibid., p. 5.

[27] McGovern, 1964; Lasby, 1975; Jones, R.V., 1978. *Most Secret War* (London: Hamish Hamilton), chap. 45; Ordway and Sharpe, 1979; Hunt, L., 1985. "U. S. Coverup of Nazi Scientists," *Bulletin of the Atomic Scientists, 41*, April, pp. 16–24; Gimbel, J., 1990. "German Scientists, United States Denazification Policy, and the '*Paperclip* Conspiracy,'" *International History Review, 12*, pp. 441–65, on p. 448.

[28] Holger Toftoy Diary entry, October 2, 1946. ASRC

[29] "Use of Atomic Warheads in Projected Missiles," April 12, 1946. HQ, Research & Dev Sv Sub-Office (Rocket), Fort Bliss, TX, to Chief, Rocket Development Division, OCO [Toftoy]. Alabama Space and Rocket Center.

[30] Glantzberg, "The New Air Force and Science," p. 6. See also Haviland[12], and "First Meeting, Space Rocket Committee," JPL History Files 5–370; Haviland, R.P., 1945 Aug. 10. "Rockets – Comments Concerning Use of Development of", JPL History Files 5–381a–1, p. 6. See also Hall, R.C., 1963. "Early U.S. Satellite Proposals," *Technology and Culture, 4*, no. 4, pp. 410–34, esp. pp. 426–430.

[31] See Ladenburg, R., 1946 Jan. 8. "Notes on Meeting...", Tuve Papers, LC; "List of Attendance V–2 Conference," in: H.A. Schade to J.G. Bain, "Summary of Minutes of V–2 Meeting Held January 16, 1946 – Forwarding of," February 6, 1946, RSRS Box 32 R320–35/46 RF42–1/84 (320:EHK), NRL/NARA; see also Nichols, M.H., 1946 Jan. 23. "Notes on the January 16 meeting at the Naval Research Laboratory for further discussion on the use of the Army's V–2 missiles," Box 118, Tuve Papers, LC. Both documents contain a summary of a formative meeting on January 16. See also James A. Van Allen Interview #3, draft transcript, pp. 160–161, SAOHP/NASM.

[32] J. Allen Hynek to Jesse Greenstein, April 24, 1946. JG/CIT.

[33] F. S. Johnson Oral History Interview, transcript, pp. 27–28, SAOHP/NASM.

[34] Transcript of meeting, "Joint Army-Navy Meeting on Army Ordnance Research and Development, the Pentagon, June 26, 1946." p. 37. OCO/NARA, RG156, Box 767A.

[35] "Radio Script," Sept. 30, 1946, Toftoy Collection ASRC.

[36] "Rockets and Guided Missiles," speech before the National Rifle Association, February 1947. Holger Toftoy Collection, ASRC, pp. 6–7.

[37] 'Big Ben' was the British code name for the V–2. "Big Ben Works at Peace," *Science Illustrated* (June 1946), pp. 41–43; "U.S. Tests Rockets in New Mexico," *Life Magazine, 20*, no. 21 (May 27, 1946), pp. 31–35, on p. 31.

[38] "Instrument Makers of Cambridge," *Fortune*, December 1948, pp. 136–140.

[39] Tangerman E.J., 1946. p. 153. [NRL], "V–2 Works for Science: Captured War Weapon Now Probes Universe," n.d. [June 1946]; also in enclosure A to G. L. Webb to ORI, May 31, 1946, pp. 1–2. Box 98, folder 2, NRL/NARA; copy in V–2 folder, HONRL.

[40] [War Department Special Staff, Public Relations] to Watson Davis, December 6, 1946. Science Service files, NASM.

[41] Whipple to Watson Davis, December 18, 1946, from El Paso. Science Service files, NASM.

[42] DeVorkin, D.H., 1992. "War Heads into Peace Heads: Holger M. Toftoy and the Public Image of the V–2 as a Vehicle for Upper Atmospheric Research in the United States," *Journal of the British Interplanetary Society, 45*, pp. 439–444.

[43] Newell Jr., H.E., 1947. "Exploration of the Upper Atmosphere by Means of Rockets," *The Scientific Monthly, 64*, #6, pp. 453–463; quotes from 463, 453–4. David Edgerton elsewhere in this volume also discusses how scientists and historians have tended to separate developments in pure science and those fostered by the military.

[44] Siry, J.W., 1950. "Rocket Research in the Twentieth Century," *The Scientific Monthly, 71*, pp. 408–421, on p. 418.

[45] Perlow, G.J., 1949. "Cosmic Ray Measurements in Rockets," *Scientific Monthly, 69*, pp. 382–385, on p. 382.

[46] Hoffleit, D., 1946. "News Notes: Peaceful V–2 Rockets," *Sky & Telescope, 5*, no. 8, June,

p. 10. Van Allen, J.A., 1948. "Exploratory Cosmic Ray Observations at High Altitudes by Means of Rockets," *Sky & Telescope, 7*, no. 7, May, pp. 171–175.

[47] Kaempffert, W., 1948. "Science in Review: Some New Facts About the Upper Atmosphere Are Acquired from German V–2 Rockets," *New York Times*, Oct. 10, p. E9.

[48] Green, C.F., 1946 Oct. 30. "Meeting of Advisory Committee for V–2 Firings". Meeting date, October 25, 1946. G. E. Aeronautics and Marine Engineering Division Report no. 145. Army/G.E. Project Hermes (V–2). USMR(1946–1951) File/NHO, p. 2.

[49] Toftoy[36]

[50] Armed Forces Day Speech, May 15, 1953, Muskegon, Michigan; "Ordnance Research and Development," February 15, 1952. Toftoy Collection, ASRC.

[51] On science mobilization, see, for instance, Sapolsky, 1979; and Seidel, R.W., 1983. "Accelerating Science: The Postwar Transformation of the Lawrence Radiation Laboratory," *Historical Studies in the Physical Sciences, 13*, pp. 375–400.

[52] The "attitude of awe and admiration" for science and scientists held by the American press is explored in Nelkin, D., 1987. *Selling Science: How the Press Covers Science and Technology* (Freeman). See p. 107.

[53] "X Marks the Minute," *Time* (19 August 1949), n.p. The term 'warhead' persisted for a number of years in internal documents.

[54] Tangerman, E.J., 1946. p. 153.

[55] Detailed group profiles are in DeVorkin, 1992. Ch. 6, from which the review in this section is taken.

[56] Johnson, F.S., OHI, pp. 27–28; see also Tousey, R., OHI, p. 70. SAOHP/NASM. The first cohort of American 'space scientists' is in these respects similar to the first cohort of 'radio astronomers' in Australia and Britain. See Forman, P. 1995. "'Swords into Ploughshares': Breaking New Ground with Radar Hardware and Technique in Physical Research after World War II," *Reviews of Modern Physics, 67*, pp. 397–455, and the references cited there.

[57] Newell, 1980. p. 33. C.Y. Johnson, F.S. Johnson and E.H. Krause oral histories, SAOHP/NASM.

[58] Gerard Kuiper to S. Chandrasekhar, August 5, 1943. Box 19.21, S. Chandrasekhar Papers, University of Chicago Library. Similar concerns were expressed by Bridgman, P.W., 1955. *Reflections of a Physicist* (New York: Philosophical Library), p. 437, based upon his essay "Science and Freedom, Reflections of a Physicist' in 1947.

[59] Kuiper to Chandrasekhar, September 19, 1943. Box 19.21, SC/UC. Although he exploited his military connections to gain access to infra-red detector technology, Kuiper apparently did not possess the same degree of "heightened expectations" exhibited by Paul Forman's physicists, discussed elsewhere in this volume, although many of his contemporaries, such as Donald Menzel and Fred Whipple at Harvard did. See DeVorkin, "Back to the Future: American Astronomers' Response to the prospect of Federal Patronage, 1947–1955: The origins of the ONR and NSF programs in astronomy," in: Keuren, D. van and Reingold, N. (eds), *Science and the Federal Patron* (forthcoming).

[60] Ernst Krause Oral History Interview, draft transcript, pp. 29–34, SAOHP/NASM. On Krause's wartime missile countermeasures work, see Hevly,[2] and DeVorkin, 1992. Ch. 4 and 6.

[61] Excerpted from J. H. Garvin to "Distribution," December 13, 1945, "NRL Policy," A3–2(8) (200–106/45). HONRL. How different this plan was to NRL's research agenda prior to the war has been discussed by Bruce Hevly (1987). Hevly also argues elsewhere in this volume that NRL really only shifted its focus from developing radio technology to developing the missile as simultaneously a tool for research and as an object of research.

[62] See Krause, E., 1945 Dec. 3. "Report on a Proposed Guided Missile Program for the Naval Research Laboratory," p. 10. HONRL; copies in EHK/NASM.

[63] Ibid., p. 7.

[64] Ibid., p. 8. Krause's rhetoric must be considered in context of his goals. He was then lobbying to develop a group far larger than the one he administered during wartime. Clearly it was the management of large-scale research and development programs, distinct from the traditional testing and evaluation mission of pre-war NRL, that Krause found more desirable.

[65] "Function of the Rocket-Sonde Research Subdivision," Laboratory Order no. 46–45 (December 17, 1945). HONRL, copies in EHK/NASM. This laboratory order was put into effect in January 1946.

[66] G. F. Hussey, Chief BuOrd to Coordinator, Research and Development, January 11, 1945. JVA/APL. Dennis (1990; 1994) provides the most detailed analysis of the postwar development of the Applied Physics Laboratory, and his perspective is adopted here.

[67] Merle Tuve, APL's founder-director, was chairman of the Committee's Propulsion Panel; others were headed by Clark B. Millikan of JPL/Caltech (Aerodynamics and Ballistics); W.A. MacNair of Bell Laboratories (Guiding and Control); R.E. Gibson of APL (Launching); and R.H. Kent of Aberdeen (Warheads and Fuses). See "Announcements," Guided Missiles Technical Panels. Announcements of December 8, 1945 and January 5, 1946. JCNWE, 911 J859, OCO/NARA. On Tuve's prominence in postwar planning, see Dennis, 1990, and Kevles, D., 1975. "Scientists, the Military, and the Control of Post War Defense Research: The Case of the Research Board for National Security," Technology and Culture, 16, pp. 20–47.

[68] R. B. Roberts to Tuve, January 28, 1946, p. 1. Box 119, "Blue Notebook," MAT/LC.

[69] Ladenburg, R., 1945 Nov. 1. "Memorandum Concerning Study of the Upper Atmosphere," appended to Nichols to Tuve, November 23, 1945, p. 3. MAT/LC.

[70] See R. B. Roberts to Tuve, January 28, 1946, p. 1. Box 119, "Blue Notebook," MAT/LC; and Smith, C.H., 1945 Feb. 13. "Post Conference Notes," Record of Consultative Services, p. 4. Box 32, NRL/NARA.

[71] Van Allen to Nichols, February 25, 1946. NRL/NARA.

[72] In 1947 they built a 10-counter double-coincidence circuit based on a Rossi design, tested the instrument against barometric readings in high-altitude aircraft, and predicted that they could get instantaneous altitude readings accurate to +/- 150 meters when above 3600 meters. Fraser, L.W. and Siegler, E.H., 1948. "High Altitude Research Using the V–2 Rocket, March 1946–April 1947," APL Bumblebee Report, no. 81, p. 68.

[73] Wheeler, J.A., 1946 Aug. 28. "General Survey of the Princeton Project Program–Cosmic Rays and Telemetering," appendix IV to Smyth, H.D. "Annual Report NOrd 7920," n.d. PP/P.

[74] Wheeler, J.A. Fragmentary notes from a July 29, 1946 meeting with Gibson, Tatel, and Hafstad. A–475 folder (Wheeler), PP/P.

[75] Singer Oral History Interview no. 2, draft transcript, p. 25, SAOHP/NASM.

[76] Quotation from William G. Dow to the author, October 25, 1986, p. 2. See also Weeks, R.P., 1964. "The First Fifty Years" (typescript draft of history of Department of Aeronautical and Astronautical Engineering), p. 19. Box 15, Michigan, College of Engineering Collection, BHLUM.

[77] Hevly in this volume discusses how scientists at NRL, following counterparts in industrial laboratories, focus on specific objects as the goals of experimental research as well as the instruments for scientific research. He argues that not only did rockets make research possible by allowing in situ access to the ionosphere, but their needs also determined what questions were asked about the ionosphere.

[78] Emerson Conlon to Commanding General, ATSC, "Proposal for Ground-to-Air Pilotless Aircraft Research Program," January 26, 1946. WGD/BHLUM.

[79] Michigan would also study supersonic aerodynamics, radio, and radar methods of target location, and infrared detection devices. Ibid.

[80] Conlon to Commanding General ATSC, Wright Field: re "Proposed Contract for Development of Instrumentation for Pilotless Aircraft. Draft proposal to be used for discussion during meetings at Wright Field on 21–22 January," January 19, 1946. WGD/BHLUM.

[81] Dow to Major Walter N. Brown (Special Projects Lab., Wright Field), October 11, 1945. Box 4, folder 2, WGD/BHLUM.

[82] Dow to the author, October 25, 1986.

[83] Nelson Spencer Oral History Interview, pp. 35–36, SAOHP/NASM. In 1953, when planning to attend summer meetings at Oxford on the upper atmosphere, Dow spent most of his time and energies planning for tours of British vacuum-tube facilities. See Dow to Guy Suits (GE), May 18, 1953, and Dow to H. C. Steiner (GE), May 18, 1953; Dow to A. L. Samuel (RDB Panel On Electron Tubes), February 25, 1953, A. L. Samuel to Dow, April 7, 1953; and Dow to A. W. Schrader, June 30, 1953. Box 4, WGD/BHLUM.

[84] Dow to D. R. Bates, c/o William H. Pickering, March 21, 1950. M824 files, "technical," WGD/BHLUM.

[85] Krause Oral History Interview, p. 90.

[86] "Function of the Rocket-Sonde Research Subdivision," Laboratory Order no. 46–45 (December 17, 1945). HONRL, copies in EHK/NASM.

[87] Dow to the author, October 25, 1986, p. 2; August 8, 1990, pp. 35–38. Harold A. Zahl had been in the Signal Corps since obtaining his Ph.D. in physics from Iowa in 1931 and was an important patron throughout the early years of upper air rocket research. See Zahl, H.A., 1952. "Physics in the Signal Corps." Physics Today, Febr., pp. 16–19. Zahl, 1968. Electrons Away, or Tales of a Government Scientist (New York: Vantage Press). The Signal Corps eventually funded Myron Nichols after he joined Michigan's Department of Engineering Research. Dow to the author, August 8, 1990, pp. 36–38; quotations from p. 38. On Zahl's proprietary attitude, see Forman, P., 1987. "Behind Quantum Electronics: National Security as Basis for Physical Research in the United States, 1940–1960." Historical Studies in the Physical and Biological Sciences, 18, pp. 149–229, on pp. 205–06.

[88] Shapley to Menzel, January 9, 1946. HUG 4567.5.2, DHM/HUA. See also Shapley to Walter Orr Roberts, July 25, 1947, p. 1; August 8, 1947; Roberts to Shapley, August 13, 1947. WOR/UC. Various published sources identify Shapley's fear of dependence upon military and governmental connections. See Kidwell, P.A. "Harvard Astronomers and World War II – Disruption and Opportunity," pp. 285–302 in: Rossiter, M.W. and Elliott, C.A. (eds), 1992. Science at Harvard University (Bethlehem, PA: Lehigh Univ. Pr.); DeVorkin, 1992. Ch. 6; and DeVorkin. "Back to the Future" (forthcoming)

[89] Whipple, F.L., 1938. "Photographic Meteor Studies, I." Proceedings of the American Philosophical Society 79, pp. 499–548. Whipple, F.L., 1943. "Meteors and the Earth's Upper Atmosphere," Reviews of Modern Physics, 15, pp. 246–264. On early meteor studies and their importance for determining the conditions in the upper atmosphere, see Doel, R., 1990. Unpacking a Myth: Interdisciplinary Research and the Growth of Solar System Astronomy, 1920–1958. Ph.D. diss., Princeton (University Microfilms, Ann Arbor), Ch. 1 and 2. See also McKinley, D.W.R., 1961. Meteor Science and Engineering (New York: McGraw Hill).

[90] Shapley H., 1946 Aug. "Tenth Informal Memorandum from Harlow Shapley," p. 6. Box 4, HCO Council folder 1946–48, FLW/SIA.

[91] "Charter of the Joint Research and Development Board, Secretaries of War and the Navy, June 6, 1946, amended July 3, 1946," excerpted from: [Science Policy Research Division, Congressional Research Service, Library of Congress] United States Civilian Space Programs 1958–1978 (U.S. Gov. Printing Office: January 1981), p. 36. On the origins of the JRDB, see Rearden, S.L., 1984. The Formative Years, 1947–1950. Vol. 1 of Goldberg, A. (ed), History of the Office of the Secretary of Defense. Historical Office, Office of the Secretary of Defense, pts. 1 and 2, pp. 96ff; Needell, (forthcoming). "Science and Defense: Institutionalizing the Partnership 1945–1949;" and DeVorkin, 1992. Ch. 9, 10.

[92] Whipple, F.L., 1947 Aug. 19. To Panel on the Upper Atmosphere. Box 240 UAT 30/2.1 JRDB/NARA. This report went through several drafts after its July 22, 1947 meeting.

[93] ... Panel on the Upper Atmosphere, August 19, 1947, p. 1. Attached to Whipple.[92]

[94] *Ibid.*, p. 3.

[95] Forman, P. "Into Quantum Electronics...". This volume, p. 271.

[96] J. Ormondroyd to Technical Advisory Board, March 24, 1948, with attachments, including A. P. Fontaine (Director, Aeronautical Research Center UM) to Frederick L. Hovde, February 3, 1948. Box 4, Folder 3, WGD/BHLUM.

[97] "Meeting of the Aeronautical Research Committee," March 14, 1949. Box 5, "1949–1950," WGD/BHLUM.

[98] By 1951 Michigan had the largest number of faculty researchers in physics of any university in the United States and more than 90 percent of them were sponsored by defense contracts. Forman[87], pp. 195–197.

[99] Stuart W. Leslie shows how MIT appreciated the value of intelligence in *The Cold War and American Science* (Cambridge: MIT Press, 1993), pp. 24–29, 191–192. See as well the essays by S.S. Schweber and others in: Galison, P. and Hevly, B. (eds), 1992. *Big Science: The Growth of Large-Scale Research* (Palo Alto: Stanford).

[100] See, for instance, G. P. Kuiper to Whipple, October 27, 1947. Box 3, 1940–1950 K-New Mexico Meteor folder, HUG4876.810. FLW/HUA. Academic intelligence gathering has been explored nicely by Leslie, S.W., 1987. "Playing the Education Game to Win: The Military and Interdisciplinary Research at Stanford," *Historical Studies in the Physical and Biological Sciences, 18*, pp. 55–88.

[101] In the opinion of a DoD contract monitor, military support definitely shaped the type of laser research that went on at Stanford, even though the scientists themselves did not see it that way. See Leslie[99], pp. 180–181.

[102] Van Allen to Greenstein, June 4, 1947, p. 2. JG/CIT.

[103] Van Allen, J.A., 1983. *Origins of Magnetospheric Physics* (Washington, D.C.: Smithsonian Institution Press). On APL's retrenchment, see Dennis[3].

[104] DeVorkin, 1992. Ch. 17.

[105] Newell, 1980. p. 11.

PAUL FORMAN

INTO QUANTUM ELECTRONICS: THE MASER AS 'GADGET' OF COLD-WAR AMERICA

In a paper published several years ago, under the title "*Behind* quantum electronics," I offered an estimate of the funds made available in the United States for basic physical research, 1930–1960, and I argued that the enormously increased resources provided in the latter half of this period, i.e., in the fifteen years following the Second World War, were primarily intended not to increase the physicists' knowledge, but to increase the physical security of the United States. Further, I suggested that these ample resources supplied to American physicists in order to promote national security effected a reorientation of physical research in the United States toward technique. And I pointed to the maser – the historical, conceptual, and nominal progenitor of the laser, the paradigmatic quantum-electronic device – as exemplary for this reorientation.[1]

The present paper is a step toward a close examination of the origination of the maser, as is necessary to give definition and substance to this claim.[2] Here I try to get more nearly inside the protagonists' heads than I did in the earlier study, focused, as it was, on social forces rather than individual actors. Thus the present paper begins by revisiting the theme of rising levels of financial support – now, however, emphasizing its association with elevated levels of expectation among American physicists. I then turn to the particular institutional setting – the Radiation Laboratory at Columbia University – wherein, with the exceptional resources there provided, the maser was created from the conception and under the direction of Charles H. Townes.[3]

The contingence of the novel is amply exemplified by the account here presented of the emergence of the maser from this institutional milieu. And yet, the litany of contingencies notwithstanding – indeed, through that litany – one comes finally to feel that the result, although not inevitable, was certainly overdetermined by a web of priorities and incentives that, integrating academic physics with the national security state, oriented academic physicists toward technique.

The U.S. military's interest in hardware and technique – and the American physicists' recognition of, and compliance with, this circumstance – I have here made vivid through an inquiry into 'gadget' as a term of art among physicists and, eventually, within the 'security community' generally. This etymological excursus is the more appropriate as in this paper I treat the maser as a 'gadget' – rather than, say, as an 'artifact.' With this distinction I mean

P. Forman and J. M. Sánchez-Ron (eds.), National Military Establishments and the Advancement of Science and Technology. Studies in the 20th Century History, 261–326.
© *1996 Kluwer Academic Publishers. Printed in the Netherlands.*

to say that I have here largely avoided describing and analyzing the physical-technical presuppositions of the maser's design, construction, and operation.[4] Whereas the term 'artifact' calls for close attention to physical structure and function in relation to the producing culture, 'gadget' is a thingamajig term, a form of reference employed specifically to evade specificity, to avoid saying just what the device so referred to does and just how it does it. And evasion of specificity, I suggest below, was especially important among those features of the word 'gadget,' as used in common parlance, that led American physicists to adopt the term into their own argot – to adopt it as code for such devices as they created to meet the requirements of the military.

1. THE REVOLUTION OF RISING EXPECTATIONS IN AMERICAN PHYSICS

A striking result of Spencer Weart's "statistical reconnaissance" of the physics enterprise in the United States between the two world wars is the evidence of a very rapid rise in the levels of research expenditure by academic physics departments in the later 1930s. Thus it would be incorrect to attribute the expectations for a much elevated standard of scientific living, with which American physicists returned to their university laboratories from the Second World War, entirely to their experiences in that war.[5] Nonetheless, their participation in military research, and the sense of their own importance to the maintenance of national security with which they emerged from it, remains the most important factor, responsible for perhaps 90% of the change in American physicists' level of expectation of funding. A striking illustration of this circumstance – and a basis for this quantitative estimate – is provided by the prospectus for Heilbron and Seidel's history of the Radiation [i.e., nuclear physics] Laboratory created and dominated by Ernest O. Lawrence at the University of Califonia, Berkeley. "We have," the authors observe:

a measure of the effect of the success of the Manhattan Engineer District [in producing workable nuclear weapons] in the form of plans for the future that Lawrence drew up in [January] 1944 and revised in [March] 1945. In the earlier plan he assumed that the Laboratory would continue as a division of the University's physics department... He expected to have a small permanent staff of scientists and technicians, and to reimplement the frugal policy of allowing visitors and students to do much of the work. As for expenses, he thought to make do with $85,000 a year and some war surplus: "a considerable proportion of the [wartime] laboratory equipment, particularly supplies and machine tools, [might] be kept available either through gift from the Federal government or purchase by the University."

By... [March] of 1945, however, Lawrence knew that science would be both honorably discharged and held in ready reserve for the national defense and welfare... [Thus] Lawrence wrote the MED offering to accept $7 to $10 million for the Laboratory's first year of postwar operation, a hundred-fold increase over the budget he had estimated the year before. After Trinity [the July, 1945, test explosion of a plutonium bomb detonated by implosion], which confirmed his confidence in dealing with [Leslie] Groves [the commanding general of the Manhattan Engineer District] as it did [U.S. President] Truman's in negotiating with Stalin, he set the postwar laboratory staff at 239, including 66 scientists.[6]

Berkeley in Lawrence's day remained unequaled in grandiosity. Nonetheless, that the greatly heightened expectation of material support for their researches was arising among American physicists as a general phenomenon, and not merely as a distemper of this specific locale (commonly known as 'Berkeley-itis'), is certainly suggested by the lyrics of Arthur Roberts' "How nice to be a physicist." This satirical song was composed for the reunion party of the Massachusetts Institute of Technology Radiation [i.e., radar] Laboratory, held during the spring 1947 meeting of the American Physical Society in Washington, D.C. Here, as the first refrain, his colleagues are ironically enjoined:

Oh,/ Research is long,/ And time is short/Fill the shelves with new equipment,/Order it by carload shipment,/Never give/ A second thought/You can have whatever can be bought.[7]

Some ten years later these circumstances were more prosaically, but more fully, described by Paul Klopsteg, a senior and widely experienced physicist who, following a successful career in the design and manufacture of scientific instruments for teaching and research, was serving as Associate Director of the National Science Foundation. From this vantage point, but emphatically not as spokesman for this agency, Klopsteg observed of the demobilized physicists that:

... they had become accustomed to large funds, easily acquired. Many now felt unable to pick up where they had left off, except with sums of money a few magnitudes greater than they had ever before had for their own work. And they felt the need of paid assistants, because the military jobs furnished them paid assistants. Many were convinced that team research was the new order, with management and organization set up on industrial lines, and that they should, as "principal investigator," manage the teams and the funds.

In their feeling of need for financial aid and their desire to initiate team work, they did not discriminate between the military developments on which they had worked and their own prewar basic research. The new pattern of large operations had been extended, in their minds, to include their own work. They would be greatly handicapped unless they had a great deal of money and many assistants. In many instances their university administrations were easily persuaded to adopt the same view. Institutions participating in war work had greatly, and in unaccustomed ways, expanded their operations but apparently found this to their liking and were loath to see reduction in activities and budgets.[8]

An instance of the phenomenon Klopsteg describes that is particularly pertinent to the subject of this paper is offered in a letter written by I.I. Rabi in September, 1944, to the administrators of his home university, Columbia, specifying quite exactly his desiderata for the support of his post-war research. Compared to Lawrence's, Rabi's demands were very modest. (The announcement of *his* Nobel prize was still two months in the future.) But even so, Rabi is quite definite that he wants and expects, *inter alia*, "to employ two technicians at about $2000 per year in the general capacity of all-around handy men within the laboratory."

Experience of the last few years has shown their great usefulness in physical research. For example, it would make it possible for me to get out an experimental paper under my own name in a reasonable time. A Ph.D. research assistant provides no continuity. One always has an undefined responsibility to find him a job, to share the credit for a piece of work. Furthermore, it takes more than a year to train him properly in the molecular beam techniques. By the time he is trained he is gone, or should be. This step I recommend as one of the most important to increase the efficiency of University physical research.[9]

And just as Lawrence's post-war research support would come from the Army Corps of Engineers – and, after 1946, its successor organization bearing responsibility for 'atomic secrets', the Atomic Energy Commission (AEC) – so Rabi's desiderata would be more than met by funds from the Army Signal Corps and the Navy's Office of Research and Inventions – after 1946, Office of Naval Research (ONR) – and eventually from the AEC as well.

The rise in expectations for financial support that the war had won for America's academic physicists was paralleled, indeed exceeded, by the rise that the war and its immediate aftermath had brought in the career expectations of that considerable fraction of America's physicists who were of Jewish origin – typically, first or second generation immigrants from Eastern Europe. Here included was Isador Isaac Rabi (whose surname was, understandably, occasionally miswritten as 'Rabbi').[10] Having gotten a big boost in celebrity and authority by the award of a Nobel prize late in 1944 – only the sixth in physics to an American – Rabi was permitted to become chairman ("Executive Officer") of Columbia's Physics Department in the summer of 1945, the earliest possible date for a Jew to clamber into this academic driver's seat.[11]

With Enrico Fermi and Harold Urey lost as spoils of war to the University of Chicago, Rabi was then the most celebrated scientist – and only Nobel laureate – at Columbia. He was also to be by far the most influential. A member of the General Advisory Committee of the Atomic Energy Commission from its formation late in 1946, he succeeded J. Robert Oppenheimer as its chairman in 1952, and sat as well on numerous other high-level advisory committees. Dwight D. Eisenhower, who served as Columbia's President from 1948 to 1951, quickly came to "lean most heavily" upon Rabi, finding that "His brilliant mind is matched by his sound common sense and his unswerving devotion to America."[12] Rabi retained this exceptional confidence, and the access it implies, following Eisenhower's election as President of the United States in November 1952 – retained it by, *inter alia*, correcting his typically Jewish leftist leanings and allying himself with his President's party, the Republicans.[13]

2. THE COLUMBIA RADIATION LABORATORY: QUID PRO QUO

Much of the scientific research during World War II was performed not merely by university scientists, but – here was the novelty – on university campuses. Thus when these scientists turned their minds to "reconversion" their first thoughts were to keep hold, for their post-war work, of such instrumentalities as had been created for, or in consequence of, their wartime researches. Such, as we saw, made the greater part of E.O. Lawrence's first, still relatively modest, plan for a post-war Rad Lab at Berkeley. At MIT, physicists and electrical engineers on its faculty – John C. Slater and Julius Stratton, in particular – who could find no comfortable role in the "Radiation Laboratory" that had been installed on their campus in the autumn of 1940, to carry forward the development of microwave radar, were anxious to ensure that if they were not to be the operators, they should at least be the heirs of this unique facility. The result was the preservation of a rump of the MIT Radiation Laboratory as an inter-departmental entity, between physics and electrical engineering. Funded by the "Armavy" (Harlow Shapley's term of derision and apprehension) through a contract with the Army Signal Corps, MIT's Research Laboratory of Electronics (RLE) had almost a million dollars a year. A staff of about 30 machinists, technicians, etc., supported the researches of some 25 faculty and nearly 100 students.[14]

Similar in origin to MIT's RLE, though on a much smaller scale and less liberal plan, was the Columbia Radiation Laboratory. Rabi, who had been helping to direct the MIT Radiation Laboratory since its inception, wanted his home institution, Columbia University, to obtain a piece of this physical-technical action – and especially so after it was decided, early in 1942, that research on nuclear fission chain reactions, in which Columbia had had the leading position, would be consolidated at the University of Chicago.[15] By this date the MIT Rad Lab had, after a year and a half of extraordinarily rapid growth, aroused much envy among the staffs of other academic institutions and was already under pressure from various sides to downsize. Thus as it then came to taking the next step up in frequency of the microwaves employed, Rabi was able to arrange for a satellite laboratory of advanced research to be established at Columbia, in its Physics Department. The remnant of Rabi's molecular beam team served as nucleus, and Rabi himself served as non-resident Director. This adjunct to the MIT Radiation Laboratory, occupying all of the eleventh and much of the tenth floors of Pupin Laboratory, Columbia University's physics building, was charged to develop miniature "magnetron" oscillator tubes for radar systems projecting (and detecting) electromagnetic radiation of 1–2 cm wavelength.[16]

As the war was ending, and the on-campus laboratories dissolving, the Army Signal Corps was anxious to see the Columbia Radiation Laboratory continue its R&D on still higher frequency magnetrons and related microwave components. Rabi hammered out a deal with Harold Zahl, Director of Research in the Signal Corps' principal R&D laboratories at Fort Monmouth, New Jersey. The Signal Corps would provide $250 000 per year to support a technical staff of about twenty, the salaries of nearly as many graduate student research assistants, and part of the academic year salary, as well as all of the summer salary, of four to six faculty members.[17]

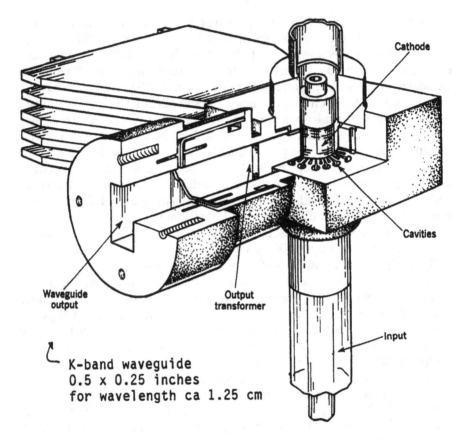

Fig. 1. Columbia Radiation Laboratory K-band magnetron[16].

The contractually agreed-on use of these resources and operatives was to:

maintain the tube and circuit laboratory established [during the war]... and conduct experimental investigations and fundamental research in connection with, but not necessarily limited to:

   (i) Generation of mm radiation [i.e., microwaves with wavelengths in the millimeter range] – pulsed and C.W. [continuous wave];

   (ii) Theoretical work on electronics;

   (iii) Electronic frequency control;

   (iv) Absorption of cm and mm radiation;

   (v) High power magnetrons.

Of these five 'tasks' only the fourth – investigations of the absorption of microwaves whose wavelengths lay in the centimeter and millimeter range, investigations which also had direct pertinence to the operational capabilities of advanced radar systems – could be construed as authorizing a program of fundamental physical research, viz., microwave spectroscopy, particularly of gases.[18] However, until the arrival of Charles Townes, on the first of January, 1948, molecular microwave spectroscopy of gases languished at Columbia, a stepchild to Willis Lamb's theoretically extremely important, but practically unimportant, microwave spectroscopic investigations of the finer differences between the energy levels of the hydrogen atom.[19]

Rabi and his molecular beams group, while not formally part of the Columbia Radiation Laboratory, were located in premises interspersed with those of CRL, and it was understood that they were to enjoy a significant share of the resources made available by the Army Signal Corps, both in salaries and in technical services. The bulk of CRL's resources, however – the work of its technical staff, especially – was devoted to the design, fabrication, and testing of magnetrons and associated microwave hardware.[20]

From this arrangement Rabi gained much. But it also cost him something – some degree of hypocrisy, and, presumably, some frowns from others in his Department and elsewhere in the University who were not so ready to compromise those principles of academic purity that had long been standard operating ideology at Columbia. Rabi and his fellow physicists were members of Columbia's Graduate Faculty of Pure Science, formed in the late 19th century in self-conscious emulation of the German university – and of its idealized, 'mandarin' conception of the university professor. The object of those 19th century academic reformers was "the full establishment in America of the pursuit of science for its own sake, as a controlling university principle."[21]

Where previously the natural sciences had been taught at Columbia chiefly within its eminent School of Mines, by 1905 advanced instruction in engineering subjects was under the aegis of the Faculty of Pure Science.[22] And indeed by the end of the first decade of the 20th century the three Graduate Faculties – "Philosophy," "Political Science," and "Pure Science" – had become "the center point and the controlling influence in the life of the entire institution."[23]

The members of Columbia's Graduate Faculties "would be, the founders expected, 'pure' scholars and 'pure' scientists, with no other professional thought than the increase of human knowledge." Such was the appraisal of John Herman Randall, Jr., in the early 1950s, writing as historian of these Faculties. And while he saw this expectation as unrealistic, his history is of its repeated reassertion.[24] Even as Randall was preparing to write that history, the Pure Science Subcommittee (including Rabi) of the President's Committee on the State of the University was stressing once again that "The function of the Pure Science Faculty" – including the Departments of Civil, Electrical, and Mechanical Engineering, Mining, and Metallurgy, which were distinguished only as being of lesser distinction than "the strictly pure science

departments" – "is to develop and disseminate a more complete understanding
of fundamental science."[25] And in 1957, at the end of the period with which
we are here concerned, a distinguished panel representing the three Graduate
Faculties reasserted Columbia's obligation to pure = fundamental = basic
research:

the position taken in the present discussion is that a graduate school should be an institution for
the development and the cultivation of those creative abilities and specialized skills which are
involved in basic research. Membership in the faculty of a graduate school should therefore be
reserved to those with proved capacity for this activity. And the highest degree which a graduate
school confers should be granted only to candidates with an adequately tested competence for
participating in such activity.[26]

It is thus not surprising to find that Rabi too espoused those principles –
when wearing his academic cap. In September, 1946, at the first of Princeton
University's year-long series of Bicentennial Conferences – just a year after
Rabi had struck his deal with the Signal Corps – he introduced a discussion
of "The relation of research in universities to government and commercial
laboratories" by asserting that the university and its research are essentially
different in kind from that performed by other institutions. Furthermore, said
Rabi on this occasion of academic affirmation, "A university professor is
not an ordinary individual because he does not follow an ordinary calling."
But wherein, exactly, does university research differ from that in industry and
government? In two regards: first, its "impracticality"; second, its "frankness."
It is on this latter point – the lack of communicativeness by commercial and
government laboratories – that Rabi is passionate, for here, as he sees, "*Vis-
à-vis* the universities, there are great difficulties." In such laboratories:

there must of necessity exist a strong form of commercial security... It exists to receive and
retain information but not to transmit it... When this is coupled with military security we get
what amounts to a system surrounded by a semi-permeable membrane. Knowledge can diffuse
inward, it is true, but with some difficulty; and nothing comes out...

This same Rabi who has incorporated into his own university department
what is, in effect, a governmental laboratory, now raises the question:

Should the universities become more like the commercial laboratories and the government
laboratories? I think not. Our problems are getting to be so serious, so practical, and so
regimented that we should not endanger the existence of the little oasis that is free and not
under the direct necessity of continually justifying itself in the material sense.

And to the ensuing discussion Rabi contributed, as concluding word,

My views are rather conservative; I think that science is infinite – we are not going to exhaust
it by gathering more facts faster. I think it is a way of life; the way one goes at the task
has important cultural and aesthetic values which are at least as important as the quantity of
scientific results.[27]

Such was Rabi's parole in public. In private, he was "conservative" in a rather different sense: concerned much more with "the quantity of... results" than with "cultural and aesthetic values."[28] And in the wake of the war, Columbia's Physics Department – a department that in the course of the 1930s had built itself up to one of the nation's best – badly needed that "quantity of... results."[29]

Unquestionably, had Rabi not been in charge in those immediate postwar years Columbia's Physics Department would not have done nearly so well in rebuilding itself despite the financial weakness of the University. Columbia, more than other Ivy League universities, was hurt by the prewar depression and the postwar inflation.[30] From 1944 to 1950, the university's real income from sources other than tuition and fees paid by students was a flat 60% of what it had been in 1931, and it rose only slightly from 1950 to 1955. Fully 10% of that non-fee income was Federal funds that Rabi, largely, had brought in for the Physics Department to spend. Totaling about a million dollars a year, these government funds (nearly all from agencies with military missions) were three times what the University was budgeting annually from its own funds for that Department.[31]

3. THE GADGET AS OBJECT OF MILITARY SPONSORSHIP

Before going any further in describing the conjunction of historical circumstances out of which the maser emerged from Columbia University's Physics Department in the early 1950s, two rather lengthy excursions are required: first, in the present section, to address more directly the question of the military's motive in funding research in academic institutions after World War II, and, second, in the following section, to describe how the term 'gadget' came to express the physicists' self-*dis*respect at being valued not as the creators of an intellectual culture, but of a practical culture of militarily useful devices.

Why were the consequences, for physics and physicists, of World War II so different than had been those of the First World War?[32] American physicists had been mobilized to a considerable extent for military research during World War I, as well. With the conclusion of those hostilities, however, the military services showed far less interest in maintaining close relationships with academic physicists, and none in supporting their researches in academic institutions.[33] What then had changed? Obviously *much* had changed in America between 1918 and 1945, but three factors seem especially important: 1) the extent of the surplus resources in American society at large and in the military services particularly; 2) the perceived likelihood of America's involvement in major military conflict and, especially, of its vulnerability to aggression; and 3) the breadth of the concept of 'weapons' and the expectations regarding the rate of succession of existing weapons by new ones.

The general level of America's material-technical 'civilization' is probably the most important factor differentiating the early twenties from the late forties. In that quarter century the United States economy grew enormously in size, complexity, and surplus resources.[34] The 'peace-time' military budget was nearly ten times larger after World War II, and a much smaller fraction of that larger budget went to feeding and paying soldiers and sailors. So much more, relatively and absolutely, was spent after the Second World War equipping the military services with technologically advanced weapons. Thus, when pay and allowances of military personnel are excluded, one finds that in 1940 the Federal government's expenditure, in constant dollars, on its military services was, per 'man,' just what it had been in 1930 and in 1910. But by 1950 it had nearly doubled; by 1960 nearly tripled.[35]

The high level of the military budget in the late 1940s – as insufficient as it seemed to the military services – reflected both the general sense of America's vulnerability to aggression at the close of a war that had begun with the debacle of Pearl Harbor, and the expectation, widespread in the nation's military and political leadership, that this Second World War, far from ending all wars, was but the prelude to another.[36] As Julius Stratton – who would become the founding Director of MIT's Research Laboratory of Electronics – wrote in October 1944 to MIT's Vice President, James Killian:

Twenty-five years ago everyone talked about the end of the war. Today we talk about World War III, and the Navy and the Air Force, at least, are making serious plans to prepare for it.[37]

In short, it was taken for granted that the defeat of Germany and Japan would be followed by a far more protracted and problematic geopolitical confrontation with the Soviet Union. Indeed, even during the war, U.S. military intelligence was far more concerned about Soviet than German scientific espionage.[38]

Third, and most pertinent to this paper, the Second World War experience, "the technological revolution in warfare" from Blitzkrieg to A-bomb, had quite altered earlier conceptions of what were to be counted as weapons, as well as earlier expectations about rates of obsolescence and of renovation: "We're in a research and development race – a tough one," was the way Clifford Furnas, as Assistant Secretary of Defense for Research and Development (1955–57), characterized the continuing military confrontation with Soviet Russia.[39]

But what exactly did 'research' mean to the Defense Department? Customarily it is said that the U.S. military emerged from World War II persuaded that basic research leads to new military capabilities.[40] But this proposition, like the more general proposition that basic research leads to new technologies, is a mystification in the service of an ideology of basic research. Rather we should see the proposition that basic research leads to new military capabilities as the conclusion of a syllogism, whose major premise – a premise that indeed the U.S. military came largely to accept in and after World War II – is that every new *technical* capability is, under the conditions of modern warfare, potentially a military capability. Basic research enters, however, only in the minor premise, and only in so far as it leads to new technical capabilities.[41] Thus those doling military dollars for research were perfectly clear that not every sort of basic research is in the interest of increased military capability. "To truly remain technologically ahead of our potential enemies," Harold Zahl, the Signal Corps' chief doler in the postwar decade, admonished those who, in 1951, might be thinking of getting on his dole:

research must produce and hold world leadership in our scientific knowledge, which at the same time must also be geared to the direct military effort, for the criterion of research success, in a military sense, lies in the performance of military weapons in the hands of our troops.[42]

Or, as one of Zahl's subordinates put it a few years later, when budgets had again become tighter, the Signal Corps must restrict itself "to sponsoring of work which promises technically sound end items rather than purely theoretical results."[43]

In the post-war decade the military's interest and the interest of the military lay in tilting basic research toward the development of new technical capabilities. As recalled by Emanuel ("Manny") Piore, one of the principal managers of the system of contracts let by the Office of Naval Research (ONR) to support physics research in universities after the war,

We were trying to make sure that the brightest [of the physicists engaged in military research projects during the war] could go back to research. Our judgement [at ONR as to who were the brightest] was reinforced by how well they performed during the war in designing equipment ... no peer review or anything.[44]

And such a tilt has indeed been characteristic of physics, preeminently, and other sciences increasingly, since that time.[45]

But just how, then, was this tilt brought about? Again, Piore, responding to an interviewer, felt quite sure that, "No, it isn't a question of doing research, even. It's a question of conversation, the controls, the influence. It's a very subtle thing."[46] And in order to appreciate just how subtle, it is important to add that Piore was widely perceived and esteemed by university-based physicists as a strong advocate of their interests. As one of them recalled decades later of ONR's timely support of his neophytic efforts,

... the guy at ONR was a man named Manny Piore... and he invented the whole idea of federal support of basic research in this country. I give Manny a lot of credit, he sold it to the admirals... and the idea that you find good young scientists and just give them money...[47]

Is it then any wonder that this revolution in the conditions and the goals of scientific life in America seemed to Lawrence's and Rabi's European colleagues an "enslavement of American science to the military forces" – an enslavement no less effective for being perceived domestically as generosity.[48]

4. 'GADGET' IN THE PARLANCE OF PHYSICISTS

Indicative of this reorientation of research toward "designing equipment" (Piore) – and indicative, moreover, that the physicists recognized, if only latently, in this reorientation a derogation from their traditional stance of "impracticality" (Rabi), i.e., of concern only with creating transcendent knowledge – is the postwar shift in the parlance of physicists of the connotation and application of the word 'gadget' (and its derivatives, 'gadgetry,' 'gadgeteer,' and 'gadgeteering').

From uncertain colloquial origins in the late 19th century, 'gadget' had become by the eve of World War II a common, unpejorative term for any convenient device or appliance.[49] Thus Arthur H. Compton, distinguished physicist, commenting publicly in September 1941 regarding the recent 50th anniversary celebrations of his own University of Chicago – and implicitly taking a stand against a frequently heard depression-era deprecation of technology as destructive of human values – drew the attention of his radio audience to the fact that,

In one of the lectures of last week's impressive symposia, Dr. Hu Shih, the Chinese Ambassador, remarked upon the humanizing effect of *gadgets*. It is a point that has impressed itself strongly upon me.[50]

Notwithstanding that Arthur Compton was already deeply involved in what would become the Manhattan Project – the committed national effort to build atomic bombs – the connotations of the term 'gadget' were evidently then still, in his mind, without ominous overtones.

Possibly such overtones were then also still wholly absent from the mind of Arthur's more worldly older brother, Karl Compton, President of MIT and host to its Radiation Laboratory – that cornucopia of microwave radar gadgets. Yet what we find in the course of the following year, 1942, is the use of 'gadget,' down in the ranks, at the lab bench, there at brother Karl's Rad Lab. Namely, when, in 1942, the MIT Rad Lab was reorganized into "divisions," Division 7, Special Systems – headed by the exceptionally inventive Luis Alvarez – was referred to informally as "Luie's Gadgets."[51] And, at brother Arthur's Manhattan Project Metallurgical Laboratory, we find, to my knowledge, not 'gadget,' but various other playful monikers for equipment developed and deployed. Thus, again in the course of 1942, Met Lab physicists chose whimsical names for their radiation detection instruments: Pooh, Heffalump, Roo, Pluto, Sneezy, Fishpole, Cutie-Pie, etc., etc.[52]

Such christenings foreshadow the postwar shift in attitude of American physicists – to a considerable extent these very same physicists – toward the naming of their scientific constructs. From the earnest, even pedantic, seriousness that since the early nineteenth century had characterized their nomenclatorial coinages, American physicists swung quickly over to an affected frivolousness – "affected," because during the war and in the postwar

Fig. 2. Cover of pamphlet published by the U.S. Office for Emergency Management in 1943 describing the military materiel equivalent to each of a wide range of consumer products – with the intent to persuade the civilian population to accept drastic reductions in the production of consumer goods. Evidently, for the authors of this pamphlet, as for physicist A.H. Compton – and presumably for the American public – 'gadget' still connoted innocuous consumer goods, and could thus then appropriately be contrasted with weaponry.

decades such frivolousness served, rather, to disguise a deep, and not altogether unconscious, ambivalence. Frivolity distracted attention – the physicists' own attention – not from the importance of their work, but from the seriousness of its consequences.[53]

It was perhaps Harold D. Lasswell who first furnished 'gadget' with ominous overtones. Late in 1939 this brilliant, prolific, polymathic political scientist had been called by Archibald MacLeish, recently appointed Librarian of Congress, to head a new "Division of War Communications Research" at the Library.[54] MacLeish and Lasswell had many affinities, and here in particular they shared a sense of solidarity with European high culture and a conviction of the necessity for active opposition to European fascism.[55] In January 1941, in a special issue of *The American Journal of Sociology* devoted to perspectives on war by distinguished representatives of a range of scholarly disciplines, Lasswell put forward the concept of "the garrison state," the form of social-political organization towards which the twentieth century seemed inexorably tending.[56]

The elite of the garrison state, as Lasswell conceived it, "will have a professional interest in multiplying gadgets specialized to acts of violence."

As long as modern technical society endures, there will be an enormous body of specialists whose focus of attention is entirely given over to the discovery of novel ways of utilizing nature. Above all, these are physical scientists and engineers. They are able to demonstrate by rather impersonal procedures the efficiency of many of their suggestions for the improvement of fighting effectiveness.[57]

Given Lasswell's prominence on the Washington scene, and more particularly as a leader – along with Vannevar Bush and James B. Conant – of the effort among the *wissenschaftliche* elite to prepare the United States to join the European war, he was surely well known to these mobilizers of science. It seems likely that J. Robert Oppenheimer, whether through his mentor Conant or otherwise, soon gained some awareness, dim or direct, of Lasswell's essay and of the menacing role in it of 'gadget.'[58]

Be that as it may, it seems more than likely that 'gadget' was Oppenheimer's gift to Los Alamos, and that its subsequent significance in the physicists' parlance is thus also largely his legacy: "I started lecturing," Robert Serber recalled in introducing the recent publication of *The Los Alamos Primer*, the course of lectures he had delivered in early April 1943 to the first cohort to arrive at that secret laboratory. "Oppie" was among the auditors at these lectures that were to acquaint colleagues with the essentials of fission bomb physics, as worked out in Oppenheimer's group at Berkeley in the previous two years. Edward U. Condon, his Deputy Director at that time, himself a senior and distinguished theoretical physicist, was taking the notes that would then become this *Primer* read by subsequent cohorts as well.

I started talking about the "bomb." After a couple of minutes Oppie sent John Manley up to tell me not to use that word. Too many workmen around, Manley said. They were worried about security. I should use "gadget" instead. In the *Primer* Condon wrote it down both ways. But around Los Alamos after that we called the bomb we were building the "gadget."[59]

'Gadget' became so firmly fixed in the lingo of the lab, that, when Los Alamos was reorganized in August 1944 to meet the challenge of constructing a bomb "assembled" by implosion – truly the premier performance of the entire war effort "in designing equipment" – the unit created to concentrate on this problem received the official designation "G Division (G for gadget, code for weapon)."[60] Indeed, the term became so much a matter of course, so taken-for-granted in our Cold-War culture, that among all the many histories of the atomic bomb, official and unofficial, not one – so far as my acquaintance extends – has taken 'gadget' to be in any way problematic, or even deserving of remark, let alone investigation.[61]

In the post-war years "gadget" and "gadgeteering" became the accepted self-denigrating term for the physicists' technical work, especially as weaponeers. "Must we always be gadgeteers?" theorist Philip Morse asked his fellow physicists late in 1950, contributing to the lively discussion of what should be done with and for science in the 'emergency' following the outbreak of war in Korea. Morse urged that, rather than being "gadgeteers," physicists instead involve themselves, as had he, in the analysis of military operations and weapons systems. This modest proposal for elevating the physicists's labors – ten years' experience of operations research had persuaded Morse that physicists could not aspire to be policy-*makers* – had necessarily to remain of very limited application. [62]

Thus in 1954, Lee DuBridge, who had directed the MIT Rad Lab throughout the war, and afterwards became President of Caltech and constant advisor to the Federal government on matters military-scientific, wondered before a wider audience:

Why is it that there has grown up the widespread impression that science and gadgeteering are the same thing?... [Industry and government] people have often heard that universities need money. So they naturally conclude that by buying a gadget from a university, they kill two birds with one stone... Possibly we have over-emphasized the gadgets – including the weapons... [63]

Certainly novelist and sometime physicist C.P. Snow thought so. His Godkin lectures at Harvard in 1960, published the following year as *Science and government*, are a diatribe against "the euphoria of gadgets" as the principal source of distortion of scientific judgement, and more particularly of the illusion

that the West as a whole can expect a permanent and decisive lead in military technology over the East as a whole. That expectation is a typical piece of gadgeteers' thinking. It has done the West more harm than any other kind of thinking. [64]

Snow effected no change in that "kind of thinking," but succeeded in fixing in the parlance of all those concerned with science and government that connotation (as well as that denotation) of 'gadget,' and of its derivative 'gadgeteer,' which in the previous two decades had become fixed in the parlance of physicists. [65]

But it took the war in Vietnam to propagate the ominous overtones of 'gadget' beyond those esoteric circles into wider spheres of American Cold-War culture. Thus we find *Gadget warfare* as the title and subject of one of the volumes by paratrooper F. Clifton Berry, Jr., in Bantam Books' multivolume "Illustrated history of the Vietnam war." (His other volume is *Sky soldiers*.) "Until Vietnam," Berry explains, "we had come to think of gadgets as any small, ingenious device or dodge for making an existing task easier." There they learned to think differently – more, to *fantasize* differently.

We wanted gadgets to solve our problems. Somehow the miraculous gadgets would eliminate ambiguities and frustrations, and let us go home... we got gadgets – by the score. As the war progressed, the flow of gadgets turned into a flood.

In Washington and Saigon, and in countless laboratories, the search for gadgets was pressed hard during the Vietnam War years. To many in high levels in Washington, faith in gadgets and numbers replaced reality. By concentrating on gadgets they overlooked or ignored the need to face fundamental truths about fighting a strange war in a far-off place... [66]

Fig. 3. Front cover (original in color) of *Gadget warfare: the illustrated history of the Vietnam war*, by F. Clifton Berry Jr. Copyright (c) 1988 by Rufus Publications, Inc. Used by permission of Bantam books, a division of Bantam Doubleday Dell Publishing Group, Inc.

In that flood of gadgets, there were many – including some of the most fantastic – that derived from Charles Townes' discovery/invention of the maser at Columbia University fifteen years earlier.

5. TOWNES, MICROWAVE SPECTROSCOPY AND ATOMIC CLOCKS

In the summer of 1947 – and rather precipitately – Rabi chose Townes to fill an opening in the Physics Department and Radiation Laboratory created by the departure (for Harvard) of Norman Ramsey, Rabi's ablest protege. Townes was Rabi's third choice to fill this gap, into which he failed to draw, first, Robert Dicke, and, then, Donald Hamilton, both then at Princeton, both alumni of the MIT Rad Lab. The first had particularly distinguished himself "during the war in designing equipment," and the second had been Rabi's student, continuing in the field of molecular beams. Twice Rabi's much prized intuition had led him to choose a good man who was not the right man. Both men, wisely, turned him down – in part because of the precarious finances of the university and its Radiation Laboratory. Finally Rabi turned to Townes, who in order to get into a first class university – he was then thirty-two and in his eighth year at Bell Telephone Laboratories – was prepared to make substantial sacrifices.[67]

Quickly, Townes proved himself to be an excellent choice, the right man for the position at Rabi's disposition. And all the more so as the extraordinarily wide scope of activity that Townes developed at Columbia was to a great extent the effective exploitation of opportunities created by Rabi's presence and influence – opportunities that could never have arisen prior to the Second World War.

Townes had come to Bell Telephone Laboratories in September 1939 – just as war was breaking out in Europe – straight from graduate studies at the California Institute of Technology.[68] In February of 1941, barely settled into a research problem of his own – one closely connected, however, with technical questions of importance to his employer – Townes was reassigned to war work, participating in the design, development and testing of radar bombing systems. As he recalled circa 1961, responding consciously or unconsciously to C.P. Snow's deprecation of gadgeteering, "I wasn't hot for gadgeteering, but there was a war on."[69] With these tasks, relative to successively shorter wavelength radars, Townes was occupied throughout the war – in close association with Dean Wooldridge, later of Ramo-Wooldridge fame – and, much against his personal inclination, still for some months after war's end. Where Wooldridge went on to make his fortune in systems engineering for the Air Force, Townes fought his way out of systems engineering, and back into research.[70]

In the winter of 1944/45, when war-working physicists were giving much thought to their "pure" post-war research activities, Townes too was determined to catch the crest of this wave. The research field he chose – he had considered radio astronomy very seriously – was microwave spectroscopy of gases. This field, for which only the most rudimentary experimental means had been available before the war, had been endowed, through the development of microwave radar, with both refined instrumentation and an urgent problem to investigate: the absorption of microwaves by atmospheric gases. Thus wherever the development of radar was being pushed into previously unexplored frequency domains – as at the Columbia Radiation Laboratory – there also we find approaches to the problem of microwave spectroscopic measurement, especially of oxygen and water vapor.[71]

Although still heavily involved with the testing of radar bombing systems, Townes, in his spare time, made, as was his wont, a careful analysis of the physics of such microwave-spectroscopic measurements on gases. The results of that analysis were surprising and promising, indeed, seemingly too good to be true. In May, 1945, seeking to persuade his superiors to reassign him to research in this field, Townes incorporated these results in a lengthy memorandum stressing, in title as in text, "Applications of microwave spectroscopy." Townes opened this memorandum by sketching his vision of a new sort of physical electronics, one which the steady advance to higher frequencies made both necessary and feasible:

Microwave radio has now been extended to such short wavelengths that it has overlapped a region rich in molecular resonances, where quantum mechanical theory and spectroscopic techniques can provide aids to radio engineering. Resonant molecules may furnish a number of circuit elements of future systems using electromagnetic waves shorter than one centimeter. Many difficulties in manufacture of conventional circuit elements for very short wavelengths can be obviated by using molecules, resonant elements provided by nature in great variety and with the reproducibility inherent in molecular structure.[72]

The principal application Townes here explored, and to it the bulk of the memo was devoted, was the use of molecular microwave resonances – in particular, the so-called inversion transition of the ammonia molecule at a wavelength of about 1.25 cm – for an unprecedentedly accurate control of the frequency of microwave oscillators. Such an apparatus, as Townes briefly pointed out, was, in principle, an "atomic clock," requiring only that its ultra-high frequency oscillations be integrated to measure elapsed time.[73]

Released to research early in 1946, Townes, as he had promised, constructed such a microwave frequency stabilization system as an adjunct to his microwave spectroscopic apparatus, and, in 1947/48, cooperated with Bell Labs in preparing two patent applications.[74] There Townes would in all probability have left this subject when he stepped over into Columbia's "pure" research atmosphere on the first of January, 1948. But Townes' patent applications had barely been sent off by Bell Labs' lawyers when, at the American Physical Society's annual spring meeting in Washington, D.C., Harold Lyons recruited Townes as a consultant to the atomic clock program that Lyons was launching at the National Bureau of Standards.[75]

Lyons was the chief of the Microwave Standards Section of the Bureau's Central Radio Propagation Laboratory, a continuation into peacetime of the Interservice Radio Propagation Laboratory created during the war to provide the military essential communications data. Lyons and those above him in the Bureau were very well aware of the important role of frequency stabilization in military communication systems with their premium on freedom from overhearing and from jamming, as also of the importance of highly stabilized oscillators for radio guidance and position location of ships and airplanes – and eventually of missiles.[76] The National Bureau of Standards was also responsible, in some generalized hyper-civic sense, for the maintenance and improvement of "the national standard of frequency." Thus with a national responsibility covering a situation in which the military was the principal – though generally unacknowledged – beneficiary, Lyons promoted within the Bureau of Standards, and publicized outside of it, an ambitious, comprehensive program for the development of "atomic clocks."[77].

Townes was drawn into this program as expert in microwave spectroscopy, but not because of his earlier conception and partial realization of just such applications of microwave spectroscopy to the construction of frequency standards as Lyons was then pursuing. Of this anticipation Lyons was initially unaware – and he was not altogether pleased to learn of it. Presumably, Townes' recruitment came about in consequence of his connection with Rabi. Lyons looked to Rabi because, early in 1945, Rabi had taken advantage of the attention turned to him by the award of the Nobel prize, to propose publicly the construction of "atomic clocks." The apparatus Rabi envisaged was based not on the resonant absorption of short electromagnetic waves by a gas – Townes' technique, and the one that Lyons chose for NBS's first atomic clock – but upon the absorption of electromagnetic waves by a beam of molecules in the kind of experimental apparatus for which he had been awarded his Nobel prize.[78]

Lyons soon added to his ambitious program the development of a Rabi-type atomic clock as well. But Lyons also had his own notions of what would constitute an authentic "atomic oscillator." He contrasted the 'passive' devices of Townes and Rabi with his own concept of an 'active' device: in the former, the 'oscillations' of the atoms or molecules were not themselves the source of the output signal, but served only to control the frequency of an entirely separate, electronic, microwave oscillator; in the latter, as Lyons envisioned it, the contained molecules themselves would be the source of fixed-frequency microwaves. Into the realization of such an "atomic oscillator," as he called it, Lyons threw much of the resources of his program – vainly, as should have been clear much sooner. And yet, Townes' association with Lyons could not have failed to keep the ideal of an "atomic oscillator" before his mind.[79]

6. MILLIMETER WAVES AND THE MILITARY

Townes' field of physical research, molecular microwave spectroscopy of gases, owed its very existence to the development during World War II of microwave radar – a debt then often commented upon, for it was seen as a reversal of the accepted and expected relation of science to technology.[80] The scientific value of this technology increased markedly with increase in frequency, i.e., decrease in wavelength. This was by no means equally true of the military value of microwave radar. Apart from the mounting technical obstacles as wavelengths were decreased below a few centimeters, radar systems encountered operational limitations arising from just those same facts of nature that made short waves so valuable to the physicist. Molecular microwave absorption lines, still few and far between around ten centimeters wavelength, come thickly enough to make microwave spectroscopy a fruitful field only at about one centimeter, and become luxuriant only in the millimeter range and below. Thus the microwave hardware for the "K-band" (1–2 cm), with whose development the Columbia Radiation Laboratory had been charged, was far more valuable to the spectroscopist than "S-band" (10 cm) and "X-band" (3 cm) equipment, developed earlier at the MIT Rad Lab. It was to this well recognized circumstance that Townes was alluding in opening his May, 1945, memorandum by reminding his readers that, "Microwave radio has now been extended to such short wavelengths that it has overlapped a region rich in molecular resonances..."[81]

The scientific interest in pushing the frontier of microwave generation technology from one centimeter – where it lay at war's end – down to one millimeter, and beyond to sub-millimeter waves, was manifest. Indeed, Townes arrived at Columbia with a "burning interest" in getting to shorter wavelengths, confiding not long after to his first student, as they stood conversing in a lunch line, "whoever can open up that region of the spectrum will get a Nobel prize."[82] But it was not the burning interest of any, or many, spectroscopists that brought substantial material and social resources to be applied to extending proven techniques and devising new techniques for the generation of shorter microwaves. All this occurred because those responsible for advancing another societal enterprise, enjoying far higher priority – namely, "national security" – saw some reason to believe that the capability of generating millimeter and sub-millimeter waves would prove advantageous in their program.[83]

Columbia's Radiation Laboratory was being maintained, as we have seen, by the Army Signal Corps in anticipation of future military applications of shorter microwaves. The addition of the Navy, in 1947, and then the Air Force, in 1953, as sponsors served further to broaden the range of thinkable applications.[84] Those charged with thinking up such applications, and preparing a tempting technological smorgasbord for the armed services, were confident that:

A general exploration of techniques, properties of materials, and atmospheric transmission in the millimeter range should be quite helpful in future design and application of radar or communication systems of very short wave length.[85]

The advance to shorter wavelength brought higher resolution, smaller size, lighter weight, and, if desired, greater directivity – all characteristics of significance for military operations. And with national security enjoying so high a priority in 1950, even the operational limitations of higher frequency radars – e.g., their restricted range – were reconceived as advantages under special circumstances.[86]

Until the early 1950s the Columbia Radiation Laboratory continued to hold its wartime lead in the development of higher frequency magnetrons, pushing the technology toward shorter wavelengths by further reducing the diminutive dimensions of existing tube types. In this push to shorter wavelengths, magnetrons were not without some advantages over the other principal types of microwave generators, klystrons and traveling wave tubes. In particular, magnetrons lent themselves to pulsing. Through intermittent operation it was possible to mitigate the principal limitation encountered in every attempt to reduce the wavelength by scaling down the dimensions of the microwave tube, namely, the rapid fall in the amount of heat that could be dissipated in/by the tube, and consequently in the amount of microwave power that it could generate.[87] By the end of 1949 CRL had pulsed magnetron oscillators generating, at peak power, 1–2 kW of 3–4 mm microwaves. The Army Signal Corps was anxious to have a quantity of them in order to measure the still largely unknown microwave propagation characteristics of the atmosphere at those frequencies.[88]

Millimeter magnetron development activities were in full tide when Townes came into the Columbia Radiation Laboratory early in 1948, and it was not long before he began to exploit this unique capability, both for the sake of microwave spectroscopy and for the challenge of setting new records for the shortest electronically generated microwaves. Townes threw himself into this competitive enterprise with the more gusto as Walter Gordy, with whom Townes was developing an intense rivalry, had, for a year or two already, been establishing such records at Duke University – with Air Force backing. Gordy's personal best was 2 mm, achieved by employing silicon crystal diodes to generate harmonics of the rather pure note put out by K-band klystrons. Townes would surpass Gordy by operating CRL's pulsed magnetrons in such a way as to enrich their output in harmonics, and then sort out the higher harmonics from this cacophonous mixture by means of diffraction gratings and tapered waveguides.[89]

Pressing the local technology into service, Townes and company quickly edged past Gordy. But further progress was slow. After nearly three years of experiments that "have been aimed primarily at extending microwave techniques to as short wavelengths as possible," Townes was able to report in the summer of 1951 only a few hundred μW at 1.1 mm from the 3rd harmonic of one of CRL's unique 3.3 mm magnetrons. This, bragged Townes soon after, is "the shortest essentially monochromatic radiation so far produced and detected electronically." Gordy conceded himself outdone by "a millimeter wave program... under development at Columbia University, which employs harmonic power from magnetrons."[90]

Gordy's concession was the more generous both inasmuch as he had no intention of withdrawing from the competition, and also as pulsed magnetrons, however satisfactory to the radar engineer as a microwave oscillator, had quite serious drawbacks for the microwave spectroscopist, particularly in comparison with klystrons. The latter are continuous wave (C.W.) oscillators, and, more important still, the oscillations of a klystron are more nearly of a single frequency, and that frequency may easily be varied – "tuned" – by varying the applied voltages. To be sure, CRL was working to remedy these deficiencies of the magnetron, and with some success. In the summer of 1951, Townes declared that "preliminary results in the development of tunable magnetrons at this laboratory appear encouraging," and stated his hope that "if these tubes become available, high resolution microwave spectroscopy may be possible at wavelengths as short as 1 mm." But at that time, and for some time afterward, Townes and his collaborators had very little to show for their efforts beyond their record breaking 1.1 mm measurement – really just "some broad gas absorption lines."[91] Clearly the magnetron was just not good enough for anything but government work.

But these millimeter magnetrons were really not good enough even for government work. As Townes was allowing publicly in 1951, "the microwave oscillators, which are now being made in small quantities, are unfortunately expensive, finicky, and short-lived." Worse, the decline in performance with decrease of wavelength – even of the relatively effective magnetron oscillators – was staggering. In halving their wavelength from 6 to 3 mm, the efficiency of CRL's magnetrons fell by a factor of 10, to just a few percent, and their average power dropped from 40 W to 2 W.[92] Worst of all, the lifetimes of these 3 mm magnetrons were wretchedly short: "2 minutes to 2 hours."[93]

The difficulties encountered in pushing the existing microwave oscillators into the millimeter region were clear to all three military services. But at this early date only the Navy had a central bureau – The Office of Naval Research – for the promotion of research and development, which, as a matter of practice as well as policy, made considerable efforts to anticipate its needs and address its problems by drawing on, and drawing in, the most capable scientists in academia and industry as expert advisors. More, the Office of Naval Research made considerable efforts to penetrate, and orientate toward its interests, the professional-scientific life in several disciplines, notably in physics. ONR pushed particular fields by sponsoring symposia at American Physical Society meetings, and also by convoking topical conferences which recipients of its funds were obliged to attend.[94]

In pursuit of this program, ONR decked itself with a panoply of advisory committees, ranging from a top-level Naval Research Advisory Committee – on which Rabi sat – reporting directly to the admiral in charge, down through divisional committees, and standing committees for particular fields – of which that for solid state physics, familiarly known as "The Chowder and Marching Society," proved particularly important.[95] Further, ONR impanelled ad hoc committees to investigate, to advise, or otherwise to promote the solution of particular technical or operational problems. ONR's "summer studies," an intensive variant of these ad hoc committees, were to be among the most famous and influential resorts to scientific advice on military matters in the post-war period.[96] Far less famous in any military context, but renowned as stimulus to Townes' conception of the maser, was ONR's Advisory Committee on Millimeter-Wave Generation. This committee was set up under Townes' chairmanship in the spring of 1950. Its charge was "to advise on the soundness and promise of research proposals which the ONR is asked to sponsor"; "to study and make available information and reports in the field of millimeter wave generation"; and "general stimulation of interest and activity in this field."[97]

Retrospectively, Townes has maintained that the ONR selected him to organize and lead this committee *not* because of his institutional connection with Columbia, its Radiation Laboratory, or the millimeter-wave magnetron development ONR was jointly sponsoring there: "In fact, I was approached as an individual." Yet in defending his individuality, Townes overlooked the inconsistency between this contention and his immediately juxtaposed statement that "there was a rather unusual arrangement that the Navy provided money to me through the Columbia Radiation Laboratory so that from the very beginning I could take care of even the administrative arrangements for the committee." This arrangement, which Townes took as evidence of the ONR's extraordinary confidence in him as an individual, seems rather to be evidence of that especially intimate institutional connection between ONR and CRL, in which Townes, as member of the Columbia Radiation Laboratory, shared.[98]

Townes called the initial meeting of his Advisory Committee on Millimeter-Wave Generation to coincide with the American Physical Society's annual spring meeting in Washington, D.C., late in April 1950.[99] It next met in November 1950, at the Navy Department, together with ONR contractors engaged in millimeter wave research.[100] A meeting in conjunction with the Physical Society's spring meeting in Washington was held once again in 1951.[101] Townes cancelled the anticipated autumn 1951 meeting, after consultation with program officers at ONR, but admonished his committee to stand by, ready to meet and to act whenever needed.[102] In the spring of 1952 the committee met for two days at Bell Telephone Laboratories.[103] I have encountered no references to further meetings of Townes' committee between spring 1952 and autumn 1953, when it met in Washington following a two-day ONR-sponsored conference on millimeter waves.[104] That the committee continued active for more than three years is testimony to Townes' ability and willingness, in this context, to identify closely with his military patrons and view the scientific-technical world through their field glasses.

7. "NEW SCHEMES FOR MILLIMETER-WAVE GENERATION"

The Directorship of the Columbia Radiation Laboratory, first filled by Rabi, devolved in the late 1940's upon a succession of physics faculty members, albeit the military sponsors never ceased to regard Rabi as really in charge.[105] Finally, at the end of 1950, the mantle fell upon Townes. Enormous energy, a capacity to concentrate, and to shift the focus of his attention instantly, equipped Townes particularly well for administrative functions. Likewise on the scientific side of the Radiation Laboratory, molecular microwave spectroscopy of gases – which until Townes' arrival at Columbia in January 1948 had been a neglected stepchild of Lamb's 'fundamental' program, falling further and further behind the rapidly advancing state of the microwave-spectroscopic art – now quickly expanded to become the largest wheel in the research mill at CRL, as Townes himself advanced into a position second to none in the world of microwave spectroscopy.[106]

It was then not due to any lack of talent and drive that Townes nonetheless remained Director of the Columbia Radiation Laboratory for only a year and a half, through June of 1952. Indeed, higher ambition was doubtless an important factor in Townes' decision in January 1952 to come forward as a candidate for the chairmanship of Columbia's Physics Department. No less important, however, as we will see in the next section, was the Signal Corps' dissatisfaction with Townes' stewardship of CRL, particularly Townes' lack of interest in the Laboratory's obligatory work on those magnetrons that he had working so hard for him.

No, Townes was not inclined to apply his energies and abilities to marginal improvements in a type of gadget even then pushing against its technological limits. There no Nobel prizes lay. Instead, Townes envisioned, and used his directorial authority to build up, a program of his own: "New schemes for millimeter wave generation."[107] Into such a program the maser, when once conceived, could and would fit very well. Without such a program Townes would probably never have conceived that "Apparatus for obtaining short microwaves from excited atomic or molecular systems" – as he titled the May 1951 statement of the maser concept in his research notebook (reproduced below, Figs. 4 and 5).[108] But even conceding the conception, without such a program Townes would almost certainly not have undertaken to *make* such a gadget, to reduce that concept to practice (as U.S. patent parlance has it).

Just how early this program took shape in Townes' mind remains unclear. Considering that Townes' responsibilities as Chairman of the ONR's Advisory Committee on Millimeter-Wave Generation, which began in the spring of 1950, provided an essential stimulus, and that his appointment as Director of the Columbia Radiation Laboratory, at the end of that year, provided the essential means, it seems unlikely that the initiation of this program would have waited beyond the spring of 1951. When the Signal Corps' representative, Israel Senitzky, visited CRL early in January of 1952, he found the Laboratory to be pursuing five programs: the two familiar technical programs of magnetron research and microwave circuit component research; the two familiar scientific programs of molecular microwave spectroscopy (Townes) and atomic microwave spectroscopy (Lamb's hydrogen fine structure studies continued by others); and, between the technical and the scientific, a fifth program: "New schemes for millimeter wave generation."

One of those "new schemes" was the maser – or, rather, as Townes called it at this time, the "molecular beam oscillator," or, more briefly, the "molecular oscillator," an appellation strongly reminiscent, as we have seen, of Lyons' "atomic oscillator."[109] But the molecular oscillator was not the first of those schemes, neither in point of conception, nor in its claim upon Townes' attention, nor upon the resources devoted to the "New schemes" program. In all three respects priority was held by a quite different scheme, namely the generation of microwaves by Cerenkov radiation.

The electromagnetic analog of the bow wave of watercraft, or the Mach wave of supersonic aircraft, Cerenkov radiation arises when a charged particle moves at velocities greater than the phase velocity of electromagnetic waves in the medium through which, or near which, the charged particle is travelling. Now familiar as the ghostly glow in "swimming-pool" reactors, this luminescence was stumbled upon in 1934/35 by Sergei Vavilov and Pavel Čerenkov at the Physical Institute of the Academy of Sciences in Moscow (FIAN), and interpreted by their colleagues there, the theorists Igor Tamm and Ilya Frank.[110] The notion that microwaves of wavelengths from one centimeter downwards to a small fraction of a millimeter could in principle be obtained by this same 'bow wave' mechanism was advanced late in 1946 by one of FIAN's younger theorists, Vitaly Ginzburg, who went beyond a general theoretical treatment of the phenomenon to estimate the microwave power from a concrete experimental arrangement, and also to suggest the more practicable experimental arrangement that Townes would eventually adopt.[111]

That proposed arrangement for production of microwave Cerenkov radiation involved an electron beam, accelerated through some kilovolts, travelling in a vacuum close along side a block of material with a high dielectric constant – i.e., a material in which electromagnetic waves propagate far more slowly than they do in a vacuum. In this way the motion of the kilovolt electrons is unimpeded, but because the electromagnetic fields originating with the moving charges penetrate into the adjacent material medium, the effect follows much as though the high-velocity electrons were moving through the medium itself. The efficiency with which the kinetic energy of the electrons is converted into Cerenkov radiation is low, with a significant reduction in the strength of this already weak effect if the beam is more distant from the surface of the dielectric than a small fraction of a free-space wavelength of the Cerenkov radiation to be generated. If, however, the electrons in the beam are gathered into "bunches," of N electrons each, regularly spaced to follow one another at frequency f, then, much as in a klystron, not only is the radiated energy, which otherwise would be spread over all frequencies, concentrated at the frequency f, but, equally important, the rate of conversion of the kinetic energy of the electrons into radiation is increased N-fold.[112]

Townes, whose drive for control of his subject included control of the literature of his field, was probably already aware of Ginzburg's ideas through an entry in *Physics Abstracts* late in 1949. This abstract accurately stated the leading results of a general discussion by Ginzburg – a paper of some 18 pages, in Russian, published early in 1947 in the *Bulletin* (*Izvestia*) of the Soviet Academy of Sciences – covering the possibility of generating shorter microwaves by beams of electrons or oscillating dipoles (atoms or molecules).[113] Presumably Townes' knowledge of such relatively inaccessible literature was further increased through the scientific intelligence services operated by the Office of Naval Research.[114]

In this particular case, however, Townes was materially assisted by Andrew Vasily Haeff, a Russian emigré on his committee. Haeff, in charge of vacuum tube research at the Naval Research Laboratory, was about to join the Hughes Aircraft Company, whose research laboratories he would soon after come to head. His particular interest was novel tube types, specifically for the generation of millimeter waves.[115] Haeff prepared a very readable translation of Ginzburg's *Izvestia* paper, which Townes then distributed to each member of his committee, in April 1950, with his initial, constitutive, letter.[116] And the committee, when it came to deliberate, soon singled out Cerenkov radiation as a – if not *the* – leading contender for "the 'right' way of generating millimeter waves." Such was the report and judgement of Bell Labs' John R. Pierce, the broadest member of that committee. And Pierce's appraisal was evidently wholly sympathetic to Townes himself, whose own considerable grasp of theoretical physics was firmest in classical electromagnetism, of which Cerenkov radiation was the most recently discovered, and still wholly unexploited, manifestation.[117]

It is thus not surprising that Cerenkov radiation was the first of the "New schemes for millimeter wave generation" to which Townes committed 'his'

Radiation Laboratory. And it was a substantial commitment: with the exception of the group Townes had assembled to surpass Gordy at the high-frequency frontier, none of Townes' other projects enjoyed so highly qualified a team, and none, without exception, so much of his personal attention.

Having in the Cerenkov radiation project a relatively well approved base for his "New schemes" program, Townes could afford the risk of extending that program with one or another speculative conception of his own. Just such was the "molecular oscillator." Heuristic in this conception were, I surmise, Lyons' concept of an "atomic oscillator," on the one hand, and on the other hand Townes' own quite original – but, his patents notwithstanding, still quite vague – conception prefacing his May 1945 memorandum: "Resonant molecules may furnish a number of circuit elements of future systems using electromagnetic waves shorter than one centimeter...." Townes apparently sought, as had Lyons, for some close analog of familiar electronic microwave oscillators, in which the macroscopic microwave-electronic resonant structure was replaced by some sort of microscopic molecular-atomic-electronic 'resonator.' Townes' scheme for generating microwave Cerenkov radiation was a 'classical', non-quantal version of such a 'resonator'; thus it was natural for Townes next to consider quantal versions.[118]

Only after such direct molecular analogs of macroscopic microwave oscillators failed to stand up to scrutiny did Townes advance to a freer, more speculative, more original idea for giving operational content to his long-standing notion of molecules as "resonant elements." And truly, the required switch in viewpoint was not so very great. Instead of the molecules being fixed, 'passive' elements, responding to the stimulus of the electron beam – which electron beam, as in the Cerenkov radiation scheme, was not only the source of the microwaves' energy, but also, by its bunching, of their periodicity as well – Townes now had the "resonant molecules" both forming the beam and bearing in themselves the energy as well as the frequency of the microwaves generated.[119]

The idea came early on the morning of Thursday, April 26, 1951, on a Washington park bench, as Townes was putting his thoughts together in anticipation of chairing a meeting of his ONR Advisory Committee later that day, the first day of the American Physical Society's annual spring meeting in Washington.[120] Townes did not speak of this idea with his committee, but upon returning to Columbia discussed it with colleagues in the Radiation Laboratory, and then, on May 11, crafted the description of his scheme reproduced here (Figs. 4 and 5). That idea, stated as Townes then conceived it, and thus with some care to avoid anachronisms, was the following: if one could contrive to remove from a stream of atoms or molecules – say, ammonia molecules – those molecules in the lower of two states that differed by a quantum of microwave energy, and then contrive that the stream (now consisting only of higher energy ammonia molecules) enter a microwave

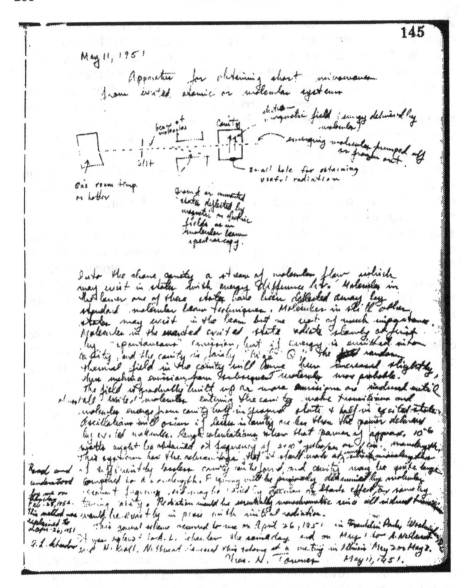

Fig. 4. Townes' original proposal for a 'maser'.[108]

cavity (i.e., a box which, on account of the low electrical resistance of its interior wall, tends to reflect and thus to accumulate, rather than to dissipate as heat, the microwave radiation that the energetic ammonia molecules emit in the course of relaxing to their lower energy state), one could expect, if the flux of excited molecules into the microwave cavity were more than great

May 11, 1985

**Apparatus for obtaining short microwaves
from excited atomic or molecular systems**

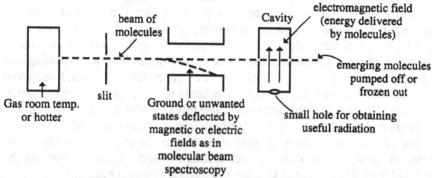

Into the above cavity a stream of molecules flow which may exist in states with energy difference hν. Molecules in the lower one of these states have been deflected away by standard molecular beam techniques. Molecules in still other states may exist in the beam but are not of much importance. Molecules in the wanted excited state radiate slowly at first by "spontaneous" emission, but if energy is emitted into cavity, and the cavity is fairly "high Q," the random thermal field in the cavity will have been increased slightly, thus making emission from subsequent molecules more probable. The field is gradually built up as more emissions are induced until almost all excited molecules entering the cavity make transitions and molecules emerge from cavity half in ground state & half in excited state. Oscillations will occur if losses in cavity are less than the power delivered by excited molecules. Rough calculations show that power of approx. 10^{-6} watts might be obtained at frequency of 3×10^{10} cycles/sec or 1 cm wavelength. This system has the advantage that it should work at much shorter wavelengths if sufficiently lossless cavity can be found, and cavity may be quite large compared to a wavelength. Frequency will be primarily determined by molecular resonant frequency, and may be varied by Zeeman or Stark effect, or some by tuning cavity. Radiation would be essentially monochromatic since all induced transitions would be exactly in phase with initial radiation.

This general scheme occurred to me on April 26, 1951 in Franklin Park, Washington, D.C. It was explained to A.L. Schawlow the same day and on May 1 to A. Nethercot and N. Kroll. Nethercot discussed this scheme at a meeting in Illinois May 2 or May 3.

Chas. H. Townes   May 11, 1951.

[Marginal note:] Read and understood by me on this day Feb. 25, 1952. This method was explained to on Apr. 26, 1951 A.L. Schawlow.

Fig. 5. Transcription of Fig. 4. (Reproduced from Forman, "Inventing the maser..."[2], p. 115. ©1992 by The University of Chicago. All rights reserved.)

enough to replenish the energy unavoidably dissipated, that such an apparatus would be a steady source, a generator, of microwave radiation whose frequency would closely approximate that which, according to the quantum theory, is connected with the difference in energy between those two states of the ammonia molecule.

Although Townes was well aware of the originality of *this* idea – the patent-preparatory form and language of his description (Figs. 4 & 5) makes that evident – six months would elapse before he gathered a few collaborators and began seriously to explore the feasibility of his proposal. And for good reason. Although unmentioned in the May 11 description – consistent with its ulterior purpose as patent disclosure – several serious obstacles to putting this scheme into practice were evident to Townes and certainly to Rabi, Kusch, and all others with knowledge of molecular beam techniques. The consideration and reconsideration of this scheme in the fall and winter of 1951 – with the assistance of postdoctoral fellow Herbert Zeiger, third-year graduate student James Gordon, and first-year graduate student George Dousmanis – led, however, to new ideas mitigating the difficulties and warranting handing the molecular oscillator over to Gordon as doctoral research project.[121]

8. PATRONS' PRESSURES & CRL'S RESPONSES

America's leap into the Korean conflict in the summer of 1950, taken with the apprehension that war with the Soviet Union itself could be imminent, led to steep increases in military R&D, and big jumps in the budgets of the several academic laboratories funded by the armed services. The annual budget of the Columbia Radiation Laboratory rose about $100 000, to roughly $350 000 – these additional resources appearing just as Townes was assuming the directorship. Fort Monmouth and the Office of Naval Research shared the "hope to see Professor Rabi's group" – i.e., the CRL – "effectively carry an increased applied load with the extra funds provided." But Rabi, with "his unswerving devotion to America," and with his recognition that mobilization put the Laboratory just then in a strong position vis-a-vis its patrons, resisted.[122] And Townes, I assume, fell readily in behind *his* patron, Rabi, for his own ideas about appropriate uses of the extra funds, though they did not coincide exactly with Rabi's, were also very different from those of CRL's sponsors.

When a year, and the direct emergency, had passed, Fort Monmouth, with less urgent need of Columbia's assistance, was better able to insist on it. Those responsible for monitoring this contract decided "to keep in closer touch" with the work at Columbia "by means of periodic visit[s]." On the first of those visits, early in January 1952, Fort Monmouth's representative, Israel Senitzky, "explained the policy" to Director Townes. And while Senitzky's report on that visit avoids any direct criticism, the Signal Corps' dissatisfactions are unmistakable in the comments and comparisons withheld: "the magnetron group has a large number of projects and few people... The molecular spectroscopy group is the largest in the laboratory."[123]

Senitzky's superiors, I suppose, found his report far from reassuring – confirming, rather, their impression that the military services' interests were being seriously neglected under Townes' administration of CRL. Perhaps Rabi received a phone call. But whatever other steps may have been taken to communicate Fort Monmouth's frown back to "Rabi's group," it was not long before Senitzky was back again, with a message: "Prof. Townes and the undersigned discussed," late in March, 1952, "the general philosophy behind the large amount of molecular spectroscopy work which is done at the Radiation Laboratory."[124]

By this date it had, however, already been decided within Columbia's Physics Department – and Senitzky surely so reported informally on his return to Fort Monmouth – that Townes would be surrendering the directorship of the Columbia Radiation Laboratory three months hence to Polykarp Kusch (and assuming, in succession to Kusch, the chairmanship of the Physics Department).[125] Senitzky's next visit to Columbia was delayed until the middle of August – that is, until Kusch had been some weeks in office as Executive Director of CRL. On this occasion, Kusch provided the Signal Corps' representative the welcome (and, presumably, fully expected) assurance that:

A new program will be initiated for a more thorough testing and better understanding of the... magnetrons developed in the past few years at the Laboratory... Prof. Kusch feels that, in order for the tubes to be useful to the Services, much more should be known about their operating characteristics. Connected with this testing program there will be a development program to improve deficiencies which might be revealed in the testing.[126]

So assured, the monitors of CRL's contract waited eight months before sending Senitzky again to Columbia, early in April, 1953, preliminary to an important, high-level joint visitation by representatives of the Signal Corps and the Navy to take place late in May. And this time Senitzky's report gave CRL good marks: "the magnetron group has been increased in personnel. They seem to have a well defined program and their morale has improved considerably compared to what it has been in the past year."[127]

If morale in the tube shop was considerably better, in CRL's research groups it was painfully depressed. To a man, those who were graduate student research assistants and post-doctoral researchers in the Columbia Radiation Laboratory in the early 1950s laud Townes' directorship. Never before had scientific life there been so active and convivial; never would it be so good again.[128] Kusch, though he possessed many fine qualities as a lecturer, writer, researcher, and mentor, was as director a martinet – exigent and righteous in petty matters (perhaps because lacking a clear conscience in substantial matters), plaguing the research assistants with ukases about unlocked doors, office assignments, and hours of attendance in the laboratory.[129]

In his role as Executive Director of the Columbia Radiation Laboratory, Kusch was prepared to stress testing and engineering of magnetron hardware to whatever extent seemed required to keep his military sponsors content, elevating the morale of the technical personnel while depressing that of the junior scientific corps. Strikingly different was the stance of *Professor* Kusch. Like Rabi and his Columbia Physics Department colleagues generally, Kusch made a great show of his commitment to purity and fundamentality. Contributing in October, 1950, on behalf of his department – of which he was then Executive Officer (chairman) – to a report on the state of the University, Kusch stressed that "The research of the Department of Physics has traditionally been characterized by an emphasis on questions in fundamental physics." Moreover, "this traditional position is still being maintained," despite "a general trend in the contrary direction" brought about "through the exigencies of the times, such as the recent war and current military preparations."[130]

The contradiction between this avowal and Kusch's carriage as Director of the Columbia Radiation Laboratory, however flagrant, should not be regarded as an indication of thoughtless hypocrisy, but rather of a deliberate strategy of 'compartmentalization.' Kusch both believed, and believed that in his own researches he acted upon, this avowal of fundamentality. The technical program of CRL was, and was meant to be, an entirely separate matter: it was not physical research, and was not to be confused with, or conducted as, research appropriate to a physics department like Columbia's. CRL was a milk cow for physics at Columbia, supporting faculty and graduate students, facilitating their research, and, in return, requiring from them various non-academic services. However intermixed the physical premises of CRL's technical work and Departmental research, their 'moral' and epistemic premises were held worlds apart. Thus Gordon Gould was surely right in his recollection that when, in November, 1957, he conceived the idea of a laser – a maser operating with light rather than with microwaves –

I realized I would have to leave Columbia to work on it, because Professor Kusch, my thesis advisor, would never let me substitute a thing like that for a research project of the very pure and basic type that was characteristic of Columbia, although Townes dabbled in such things... Columbia did not deal with applied physics.[131]

But here we have gotten ahead of our story. During the fifteen months running from spring 1952 through spring 1953, Rabi and his proxy Kusch overcame more than this one threat from CRL's sponsors to seriously curtail its technical program – and consequently reduce the resources divertible to the research that really mattered. Further threats arose, first, out of the sponsors' demand for security classification of the magnetron work; second, from the election of a Republican administration early in November, 1952, and the consequent pressure upon the sponsoring agencies to reduce their expenditures.

Pressure from the military services for classification of the research they were sponsoring on university campuses, recurrent in the late forties, became stronger and steadier in the early fifties. The Columbia Radiation Laboratory was especially vulnerable to such demands as the services were, apparently, sponsoring similar magnetron work in industry under a cloak of secrecy. These pressures reached a head in the spring of 1952, just at the time CRL was also under pressure to magnify its magnetron work.[132]

Yet as accommodating to the exigencies of military sponsorship as Rabi then was on the issue of CRL's attention to perfecting its magnetrons, just so "adamant" did the Signal Corps find him to be in refusing security classification of any part of the Laboratory's work. Even though Harold Zahl warned Rabi that "if some classification was not acceptable as an objective to work toward," then they must expect an alteration of CRL's program "so as to remove the classification question" – which "would undoubtedly lead to some reduction in contractual funds allowable for subsequent extensions" –

Dr. Rabi reaffirmed his position that the Physics Department had so thoroughly integrated the work on the subject contract into their academic program, so as to make operation under classification, even of Restricted, impossible.[133]

We are not altogether surprised by this affirmation of principle, consistent as it is with the position Rabi took at the first of Princeton's Bicentennial Conferences, in September, 1946.[134] (Indeed, it is thinkable that Rabi's compliance with Fort Monmouth's wishes for more emphasis on those magnetrons was all the fuller as he foresaw his eventual refusal to comply on the question of classification.) But we should also recognize that this issue is – as Rabi affirmed – as much one of practicability as of principle. The sticking point for Rabi was certainly not secrecy, per se: several classified contracts were in fact held by Columbia's Department of Physics in the late 1940s and early 1950s. Rather, although the work of the Radiation Laboratory was – as we have seen – very far from being "thoroughly integrated" with the "academic program" of the Physics Department, the premises and personnel of the two entities were so thoroughly intertwined that classification of any degree would unquestionably have made great practical difficulties.

Zahl himself was not a proponent of the push for classification, and so the matter was allowed to remain, ad interim, unresolved. It was not long, however, before this matter merged with, and was submerged by, yet another class of considerations that also threatened to deprive Columbia of its magnetron milk cow: Eisenhower's electoral success on a budget-cutting platform. By the middle of November, 1952, just days after the Republicans' victory, Fort Monmouth and the Office of Naval Research were seeking ways to reduce their outlays. Although barely three months had elapsed since Kusch, as newly installed Director, had promised that CRL would do much better by the services in their task of magnetron development and testing, both agencies – ONR leading, Fort Monmouth hesitantly following – now fixed upon the magnetrons as that which they preferred to eliminate from the Laboratory program. As half the CRL budget went, the funders reckoned, into "tube" work, they anticipated halving their "rate of expenditure" on that contract.[135]

The matter being so important, Zahl ignored Kusch and turned to Rabi with the proposal that CRL's technical program be cut back. Rabi – apprehending that ONR and Fort Monmouth meant this jettisoning of applied work not as an espousal of fundamental physics, but on the contrary as a way of letting Columbia down gently – temporized, urging that so serious a change in policy required consideration by the so-called Joint Services Technical Advisory Committee (JSTAC), made up of representatives of these self-same sponsoring agencies. The ploy was shrewd, for it both reminded the sponsors of the matter of 'due process' and gained CRL an indefinite delay, inasmuch as the JSTAC met only at irregular and infrequent intervals – though that irregularly would soon be rectified.[136] And so, by way of a lengthy but necessary excursion, we come again to that important, high-level meeting at Columbia late in May, 1953, in preparation for which, as we recall, Senitzky had visited CRL early in April, and had reported, reassuringly, improved morale in the tube shop.

By the time this JSTAC meeting took place the immediate post-election panic had passed. The Signal Corps was represented by Zahl, Senitzky, and G. Ross Kilgore, whose contributions to microwave generation went back into the 1920s and who was the individual to whom Senitzky reported, formally, on his visits to CRL – including this one. The Office of Naval Research was represented by Commander Gould Hunter and by Elliott Montroll, a mathematical physicist of considerable ability then serving as the Director of its Physical Sciences Division. (Both Hunter and Montroll would sit again for ONR – together with Piore – at the October 1953 meeting of Townes' Advisory Committee on Millimeter Wave Generation.) "The question of a possible change in emphasis of the Radiation Laboratory program in event of a cut in funds was raised," Senitzky's report impersonally records. "Prof. Rabi felt that it would be unwise to discard the magnetron work, and Dr. Zahl expressed a desire for the continuance of the same degree of emphasis on both the applied and the basic research."[137]

Thus CRL came through this review with its magnetron program – and its budget – largely intact: some reduction was exacted, probably about 10%, bringing the annual budget down from $344 000 to a little more than $300 000, I surmise.[138] Townes, by far the heaviest drawer upon CRL's resources for research purposes, was the chief beneficiary of this happy outcome – happy above all in the commitment it implied to the continuance of the Laboratory that housed and supported his researches and himself. Indeed, Rabi's success should probably be seen as a vindication of his tactic of shielding his relatively weakly supported Laboratory behind its technical program: while CRL dropped about 10% in this first round of Eisenhower budget cuts, its more favored big-sister laboratory at MIT, the Research Laboratory of Electronics, suffered a 20% reduction in its funds.[139] Though it seems a high price to pay for patronage, this tactic certainly had a rational basis in the late forties and early fifties when it was still uncertain that Federal funding of academic research would be a permanent feature of scientific life[140] – and when the Columbia Radiation Laboratory was still leading industrial research laboratories in millimeter magnetron work.[141]

The question of course arises, how far this victory for Rabi's magnetron-as-hostage tactic had been fairly won, i.e., to what extent the administrative staffs of the funding agencies would have supported "Rabi's group" under any circumstances – because Rabi was so well respected, and, in the Eisenhower administration especially, so well connected. Thus the continuance of the magnetron program had been set up by Senitzky's favorable report on his visit to CRL in advance of the May 1953 review. Although Senitzky, along with his superiors, certainly had reservations about that program, evidently he believed such to be better left unstated on that occasion. When he next visited CRL, in February, 1954, he gave on returning to Fort Monmouth a franker appraisal of the strengths and weaknesses of this research organization for the prosecution of the services' task: "There seems to be small possibility of the infusion of new spirit and enthusiasm into the magnetron work, but CRL is rolling along fairly well on accumulated experience."[142] Against this evidence of the military services playing softball with Rabi we must set, however, their evident intent, late in 1952, to liquidate CRL, and, further, the severe cut, which we will shortly see ONR inflict, in the summer of 1953 in the funds of Rabi's own Molecular Beams Laboratory.

9. TOWNES' SCHEMES VERSUS THE OTHERS' REALITIES

What, however, was the progress of, and the standing of, Townes' "New schemes for millimeter wave generation" – of the maser, in particular – as those principally concerned with the administration of the Columbia Radiation Laboratory performed this political-economic tango? It was of no small importance to the men at Fort Monmouth – and consequently to Rabi and Kusch, as well – that, as Senitzky noted in his minutes of the May 1953 high-level review of CRL policy, "the Cerenkov radiation and molecular oscillator experiments are still in preparation." Through all these many months the "preparation" of these "experiments" had been regularly and separately reported in CRL's Quarterly Progress Report, in the lead section, along with the magnetron work, under the rubric "The Generation of High Frequencies." In virtue of that prominence and that objective, Townes' two "experiments" had necessarily to remain in the eye of the sponsoring agencies. Indeed, from the time, late in 1951, that Fort Monmouth initiated "the policy... to keep in closer touch with all the aspects of the work" performed under its contract with Columbia, every report by its representative on his visits to the Laboratory included some words about the status of each of these two "New schemes for millimeter wave generation."[143]

The regular and specific attention given these two "experiments" was altogether exceptional, and, from Rabi's and Kusch's perspective, very risky. Far safer was the handling typical of all the numerous experiments in atomic and molecular microwave spectroscopy: unless CRL's leaders drew attention to them as significant achievements, no specific notice was taken in the reports Senitzky rendered on his visits. Townes in initiating these "New schemes" doubtless viewed its high profile as a means to lead the sponsoring agencies toward a more imaginative notion of CRL's technical role. Rabi and Kusch, however, must have seen that same high profile as puncturing CRL's technical armor. With both experiments – Cerenkov and molecular oscillator – "still in preparation" after more than two years of regularly reported effort, they could not have been very comfortable defending this program at the May 1953 review.

With another couple of months' effort, the Cerenkov radiation experiment produced detectable microwaves late in the summer of 1953. (Townes himself had been taking a large hand in this work for more than a year.)[144] For the molecular oscillator, however, this was the beginning of a bleak period. Upbeat Jim Gordon had thought that his experiment too was then on the verge of working, but soon he found himself overwhelmed by unanticipated problems. No one in the Laboratory – not even Townes – had ever been confident that this less-favored scheme could be made to succeed fully, i.e., be made to *generate* microwaves. Nor had Townes ever shown for the molecular oscillator that strong personal interest that the Cerenkov radiation experiment so clearly aroused in him.[145]

Now, with the Cerenkov radiation experiment proving a success in principle – it was, and would continue, far from a success in practice – Rabi and Kusch reckoned the juncture opportune to pressure Townes to abandon the molecular oscillator experiment as wasting CRL's reduced resources. But opportune or inopportune, apportioning CRL's resources had now become an especially urgent issue for Rabi and Kusch as ONR had recently informed them that their own contract, separate from CRL's, for the support of their molecular beams laboratory, would be reduced 40%, from $25000 to $15000 annually.[146]

Townes turned Rabi and Kusch down. He refused to pull the plug on the molecular oscillator, not merely out of unwillingness to surrender his own original scheme, and his personal autonomy as researcher, but also out of unwillingness to pull the rug from under Gordon, whom he had promised a doctoral dissertation from this "experiment." Rabi, in alliance with the military, had enlisted Townes in a complex structure of intellectual goals, material resources, social institutions, and personal obligations. In doing so, Rabi had lost control over his operative's actions. Whatever Rabi might think or want, that structure would now determine Townes' actions, and would, miraculously, conduce to an invention/discovery that more than met the objectives of every party concerned in the Columbia Radiation Laboratory.

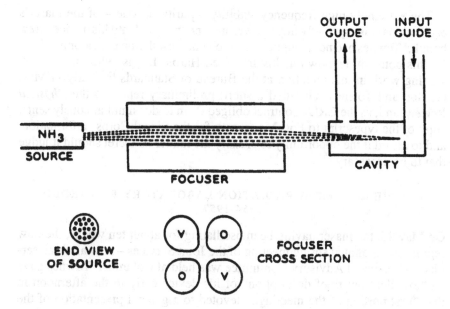

Fig. 6. Schematic illustration of the maser, spring 1954, from the first publication[151]. Note the indication of microwaves introduced into the apparatus via the "input guide," reflecting the initial 'artificial respiration' approach to maser operation.

In December, 1953, about three months after Rabi and Kusch had sought to convince Townes that he'd never succeed in breathing life into the damn thing, the maser showed its first vital signs. The respiratory metaphor is the more apt, as the approach that Townes had fixed upon, nearly two years before, to achieve a preliminary proof of principle, bears analogy with artificial respiration: reckoning it unlikely that the molecular oscillator could be brought to the point of generating microwaves by self-sustaining oscillations, Townes and his collaborators were 'breathing' microwave radiation into the device from an external source, and measuring the radiation emerging from it – hoping for an augmentation of that radiation at the resonant frequency of the ammonia molecules.[147]

This augmentation is what Gordon observed shortly before Christmas, 1953. Even six or eight weeks later, when Senitzky happened once again to be visiting the Laboratory, he found that "oscillation is not really expected," at least not by Kusch and the leaders of the magnetron work to whom he spoke.[148] But Townes had assigned a rather older, able, Chinese graduate student, Tien-Chuan Wang, to help Gordon amplify those vital signs. In March, just a few weeks after Senitzky's visit, they got the molecular oscillator to oscillate.[149]

The extremely high frequency stability – purity of tone – of the maser's oscillations was especially impressive, and unexpected. Until his device 'drew breath,' Townes had not thought of the molecular oscillator as an atomic *clock*. Of this potential he now quickly informed Harold Lyons, who lost no time starting work in this direction at the Bureau of Standards.[150] Early in May, Gordon and Townes submitted a short, preliminary report to the *Physical Review*, and, as the CRL's contract obliged them to do, simultaneously sent a copy of the typescript to Fort Monmouth for the Signal Corps' information, and to afford it the opportunity to classify the matter as secret should it judge that to be necessary.[151]

10. THE COLUMBIA RADIATION LABORATORY REWARDED, 1954-1957

On May 25, the maser having been oscillating for about ten weeks, the now regular, semi-annual spring meeting of the Joint Services – now all three services – Technical Advisory Committee was held at Columbia. Townes gave an "excellent but brief description" of his device early in the afternoon in that "first portion of the meeting... devoted to a general presentation of the research program underway." Presumably the maser was to be seen oscillating in "the later [sic] part of the afternoon... spent with visits to the various research projects where demonstrations were made and opportunity given for private discussions with Columbia personnel." Naval and Signal Corps representatives were clearly impressed by Townes' "molecular beam amplifier and/or oscillator," as indeed they had every reason to be. Here was a new gadget that could do superlatively well several different 'things' from which a military advantage was potentially to be gained: generating a microwave of unequaled frequency stability; measuring time with unequaled precision; detecting microwaves with unequaled selectivity and sensitivity.[152]

And now, in the spring of 1954, the policy and budgetary conjunctures were again propitious. The first wave of Republican budget-cutting frenzy having passed, the military services found themselves in the era of the "New Look." High-tech was the order of the day, and Federal – i.e., military – funding of the physical sciences would again be on the upswing – until the spring of 1957. Fort Monmouth, which launched just at this time a program to encourage the development of atomic clocks, let a bevy of maser R&D contracts with both commercial and academic organizations as well as initiating such work in its own laboratories. The Air Force soon followed suit, pursuing in particular the original object of Townes' invention, viz., the maser as a means to generate *much* shorter microwaves.[153]

Early in 1955 the maser was announced to the public, creating a considerable sensation.[154] In the following two years efforts in several laboratories to make a maser out of a small crystal, instead of Townes' cumbersome molecular beam, led to considerable success. Such "solid state" masers were looked to not as atomic clocks but as highly sensitive detectors of microwaves that could increase enormously the capabilities of radar and communication systems.[155]

The consequence was that over the course of these three years the maser transformed the Columbia Radiation Laboratory in the parole of its Executive Director and in the eyes of its sponsors. In effect, the maser redeemed CRL from the purgatory of applications, allowing it to enter the paradise of "pure research." Documentation of the earlier stages of this transformation has not come to hand, but a letter from Kusch to Fort Monmouth in September, 1956, shows the vast difference in posture from 1952/53.[156] Referring to the informal agreement reached at the previous meeting of the Joint Services Technical Advisory Committee,

that a considerably increased emphasis on pure research in the general field of high frequency and microwave physics at the possible expense of research and development work in microwave generators is appropriate,

Kusch spelled out the alterations in CRL's contract that he desired, and that he clearly thought had been authorized. Noting that the present and prior contracts "require a detailed and highly specific program of work in microwave generator development," Kusch now went so far as to "suggest that all of the specific requirements... be removed from the contract." In place of the hardware items and "specific requirements" on which CRL's health and wealth had previously depended, Kusch now listed the areas of investigation "with which we are presently concerned." No longer do microwave generators head the list, and when they do appear, toward the bottom, the services' interest in such devices is clearly to be of no concern to CRL:

(D) Research in microwave generators both as devices of intrinsic interest and as devices to be used in other portions of the research program of the laboratory.

What does appear at the head, and the shoulders, of Kusch's list is a research program for the Laboratory in which the only areas of application are maser related (and very vaguely stated):

(A) A study of the properties of molecules, atoms and nuclei by the use of microwave and other high frequency methods.

(B) A study of the properties of solids by an investigation of their interaction with microwave fields.

(C) Research in the general field of the utilization of atomic and molecular processes for the generation and amplification of power.

By the following spring this reordering of priorities – in effect, a redefinition of CRL – had apparently been fully accepted: of the six talks by CRL staff scheduled for the May, 1957, meeting of the Joint Services Technical Advisory Committee (JSTAC), not one dealt with a technical subject. All six described physical research, with Townes' talk on "Maser Research" allotted far more time than any of the other five presentations.[157]

Riding the coattails of this maser-induced redefinition of the Columbia Radiation Laboratory we find Kusch himself and the molecular beams laboratories. Where previously their participation in the Signal Corps' largess was silent, a tribute to Rabi, now Kusch led off the series of talks at the May, 1957, JSTAC meeting with twenty-five minutes on "Atomic and Molecular Beam Research" – second in length only to Townes' forty minutes. Now the scope of the committee's policy discussion, as well as the site inspection, was explicitly stated to be any "matters related to the work of the Columbia Radiation Laboratory and the molecular beams laboratories of Columbia University." As Kusch reported to Rabi following this review:

the response of our sponsors to the technical exposition and to the description of our business affairs is most satisfactory and, at present, our relationships continue to be most friendly. The present contract was renewed for a period of two years at $27000 per month. The money is better than in some previous years and it will be useful not to have to negotiate annually.

Where in 1952/53 Rabi's pull, and the weight of the magnetron program, had saved CRL from the rampant budget-cutting, in 1956/57 it was the maser that enabled CRL to drop its burdensome magnetron shield, and nonetheless maintain its funding level while steep reductions threatened laboratories all around the country.[158]

11. CONCLUSION: COMPARTMENTALIZATION NOTWITHSTANDING

Compartmentalization of knowers – of the bearers of different pieces or classes of information – has a long history as a means of ensuring the security of secrets. Physicists were obliged to participate in this practice during World War II, particularly in their work on the Manhattan Project, and did not like it.[159] This was not because they were, as they then supposed, categorically and inalterably opposed to compartmentalization. Indeed, ironically, the American physicists' emulation in their postwar research of the team approach, to which they had become accustomed through their war work, involved them, not by intention but in effect, in a 'compartmentalization of knowers' – as was soon recognized and often deplored among the physicists themselves.[160] What, however, they could not so readily recognize is that, as scientists, they were committed to another, different, but still more pervasive principle of compartmentalization, namely compartmentalization of knowledges.

Compartmentalization of knowledges (hereinafter compartmentalization, tout court) is the separation of different pieces or classes of information within the mind of one and the same knower, the prevention of the mental juxtaposition and cognitive integration of disparately labeled 'data.' It is, thus, a psychic mechanism rather than a social mechanism – but a psychic mechanism with important social consequences. Such compartmentalization is character-istic of Western science in all epochs prior to our present postmodern era, being inseparable from the determination to see scientific knowledge as sep-arately existent, transcending all mundane institutions, instrumentalities, and personal circumstances through which it arises and becomes established.[161] Compartmentalization allowed the scientist to employ his mental faculties and social skills for negotiating the practical complexities of his scientific life, while yet maintaining the self-image of untrammeled seeker after transcen-dent knowledge, or, alternatively, the self-image of untrammeled seeker after "fun"[162] – which two self-images, how ever much they differ in the moral di-mension, are fungible in respect of autonomy. By 'compartmentalizing' both the multifarious contingencies of science as mundane social practice, and the behaviors adopted to cope with those contingencies, the inconsistencies with the psychically required self-image were repressed.[163]

This insistence on a split between knowledge, per se, and the conditions under which knowledge is produced, is the 'original compartmentalization,' that from which most of the others derive.[164] It has been hegemonic in the West 'from Plato to NATO,' roughly speaking. However, the geo-political and techno-scientific configurations that emerged in the United States out of the Second World War gave much greater occasion for brain workers to resort to compartmentalization for 'reconciliation' of the unblinkable circumstances of their lives as scientists and the ideal images of themselves and of science that they wished to uphold. Thus compartmentalization as a psychic mecha-nism became especially characteristic of the physicist in American Cold-War culture.

The walls of these mental compartments were conceived by physicists not as altogether impermeable, but rather as "semi-permeable" – to extend the metaphor Rabi deployed at the Princeton Bicentennial Conference cited in Section 2.[165] Mere information may indeed flow in every direction – although most freely and copiously outward from fundamental science. But *influence* flows in one direction only, namely, again, *outward* from fundamental re-search to applied science, epistemology, and high culture generally. Upon such transcendent science no 'backward' influence – from the many mun-dane, often clandestine, spheres in which the physicists found themselves so deeply involved[166] – was permissible, hence none was conceivable.

The widespread reliance on the psychic mechanism of compartmentaliza-tion – indeed, the obligation to employ it – was not missed by the physicists' empathetic ironist Arthur Roberts. His refrain for "How nice to be a physicist" (1947), quoted in Section 1, is quite specifically an exhortation to compart-mentalization: "Research is long/ And time is short/...Never give/ A second thought...." I.e., in the face of the ever endless demands of research, the ide-ological stance of the postwar physicist – the stance he feels to be incumbent upon him – is the exclusion of every thought of the 'external' circumstances under which he is producing knowledge, whether that be, "You can have whatever can be bought," or, "The F.B.I.'s approval must be sought."[167]

To explore fully the Cold-War culture of compartmentalization is an undertaking far beyond the limits of the present essay. It is important, however, in concluding, to draw attention once again to the term 'gadget,' here within the framework of the concept of compartmentalization: as Oppenheimer surely anticipated, that initial compartmentalization of knowers which provided the excuse for introducing the term 'gadget' in lieu of 'bomb,' was quickly transformed into a compartmentalization in the mind of the knowing physicist, a compartmentalization that separated the scientific-technical challenge from disturbing thoughts about the consequences of its achievement. Generalized and diffused in the postwar decades to refer to any sort of hardware for "defense," the term 'gadget' was never intended to diminish the military's valuation of the physicists' contribution, but only the burden of conscience that the physicists carried for the uses 'Caesar' made of their playful inventions. In this respect, it fitted excellently with a military jargon intended to achieve much the same separation in the imaginative faculty between applications of physical force and the human consequences of such.[168]

Thus does the superego, like an ascetic stylite, make itself content with scraps drawn up from the ego's table. But if we descend to ego's realm of getting and spending we must recognize that questions of ideological stance – including both content (e.g., 'purity') and mechanisms (here, compartmentalization) – are emphatically also questions of status and control, both social and personal. This Rabi understood, perhaps only intuitively, but nonetheless extremely well, for he was putting into operation a form of compartmentalization much more deliberate and self-consciously egoistic than that of the stylitic superego. In order to retain control over the character of the knowledge produced with the military's money, Rabi relied on no mere mental compartmentalization – whose effectiveness is at best as myth, and at worst as mere illusion – but, rather, on an institutionalized compartmentalization.

The problem – the danger to be avoided – was of course clear to Rabi from the outset. As he said in his presentation at the Princeton Bicentennial Conference in September, 1946,

In the universities we are in a particularly critical time in the matter of support. It is a fact in the subject of physics now, as in many other things, that enormous amounts of money are available for the support of research. *But if it were decided to control universities and university research, there could be no better way to do this than the way it is being done now.* Distributing large funds brings the distributors to the position in a short time of being able to apply certain pressures to the universities to change their method of working, to justify their activities, and conform their policies with other broad national and governmental activities.[169]

Rabi's strategy for forestalling this danger in accepting the Signal Corps' money was to keep the technical program of the Columbia Radiation Laboratory intellectually compartmentalized, academically ostracized. The increased emphasis on the Laboratory's technical mission in the spring of 1952 in order to mollify the military sponsors was fully compatible with this strategy of compartmentalization, indeed, was an expression of it. Rabi was ready to shift the fiscal boundary between the technical program and the fundamental research supported by CRL's budget, as also to shift the amount of attention that members of CRL's staff gave to magnetrons. But the directions of effort on each side of that boundary were to remain essentially unaltered. Any blurring of the intellectual boundary between the fundamental research proper to a department of Columbia University and the technical R&D of the Laboratory was subversive of this endeavor to keep the researches of the Physics Department unswerved by national emergencies and untainted by applications. Townes' two applied physics projects under the rubric "New schemes for millimeter wave generation" violated this taboo.

In contrast with the illusory achievements of mental compartmentalization, Rabi's institutionalized compartmentalization had real successes to its credit, scientific as well as budgetary. Both Willis Lamb's and Polykarp Kusch's Nobel prize-winning researches were heavily dependent on resources siphoned from CRL in the years immediately following the war.[170] And the maintenance of CRL as a resource reservoir for fundamental physics certainly owed something to Rabi's shrewd tactic of using CRL's technical work as a hostage to shield the Laboratory from budget cuts when its sponsors entered one of their recurrent phases of financial stringency.

Two points must here, however, be emphasized: first, that the successes – both scientific and budgetary – attributable to Rabi's (and Kusch's) strategy of compartmentalization were relatively modest compared with those attributable to that gross violation of their strategy that was the maser; second, that compartmentalization notwithstanding, physical life in the postwar years was emphatically a two-way street, with ideas as well as resources flowing out of the secret committee rooms. The maser provides a clear example – it is by no means the only example – of the significance of secret discussions under national security auspices for the advancement of science in the early Cold-War decades.[171]

Smithsonian Institution's National Museum of American History, Washington

NOTES

[1] Forman, P., 1987. "Behind quantum electronics: national security as basis for physical research in the United States, 1940–1960," *Historical Studies in the Physical and Biological Sciences, 18*, pp. 149–229, cited hereinafter as "Behind Q.E." (In other references to this journal its title is abbreviated as *HSPS*.) Some of same purposes are pursued by Molina, A.H., 1989. *The social basis of the microelectronics revolution* (Edinburgh Univ. Pr., Edinburgh).

[2] Eventually a book will emerge on the emergence of the maser. Meanwhile the closest studies are: Bromberg, J.L., 1991. *The laser in America, 1950–1970* (MIT Press: Cambridge, MA), pp. 14–24, 220–223; Forman, P., 1992. "Inventing the maser in post-war America," *Osiris, 7,*

pp. 238–267.The most important archive for the present paper is the National Military Personnel Records Center, St. Louis, Missouri (here abbreviated as NMPR), where much, perhaps most, of the surviving records of the activities of the U.S. Army Signal Corps' research and development laboratories at Fort Monmouth, New Jersey, in the period 1945–65 are held. Other repositories frequently cited are: Columbia University, Butler Library, Dept. of Rare Books and Special Collections (CUSC); Library of Congress, Manuscript Division (LCMD); Niels Bohr Library of the American Institute of Physics (AIP), now located at the American Center for Physics, College Park, Maryland. I am indebted to David H. DeVorkin, Ronald E. Doel, and Robert W. Seidel for their criticism of a draft of this paper.

[3] Biographical/autobiographical sources on Townes are, most recently, Nebeker, F., 1993. "From radar bombing systems to the maser: Charles Townes as electrical engineer," ch. 3, pp. 61–92, of Nebeker, *Sparks of genius: portraits of electrical engineering excellence* (IEEE Press: Piscataway NJ), based in considerable part on an interview of Townes by Nebeker, September 14–15, 1992 (deposited at the IEEE Center for History of Electrical Engineering, Rutgers University, New Brunswick NJ). Further: Berland, T., 1962. *The scientific life* (Coward-McCann: New York), ch. 3, pp. 70–101, "Charles Hard Townes"; interviews with Townes by William V. Smith, June 18–20, 1979, by Joan L. Bromberg, January 28 and 31, 1984, and by Finn Aaserud, May 20, 1987 (all deposited at the AIP); and by P. Forman, February 12, 1988. Further, the transcript of a 35-hour oral history interview, 1991/92, covering all aspects of Townes' career, is available at/from the Regional Oral History Office, Bancroft Library, University of California, Berkeley, but not in time to be consulted for this paper.

[4] Such dimensions will be considered in an eventual book; they are touched upon in Forman, "Inventing the maser ..."[2]. Forman, "'Swords into ploughshares': breaking new ground with radar hardware and technique in physical research after World War II," *Reviews of Modern Physics*, 67, pp. 397–455 (April 1995), provides part of the physical-technical context for Townes' work.

[5] Weart, S.R. "The physics business in America, 1919–1940: a statistical reconnaissance," in: Reingold, N. (ed), 1979. *The sciences in the American context* (Smithsonian Institution Press: Washington, D.C.), pp. 295–358. Further studies focusing on rising resource levels in the 1930s: Seidel, R.W., 1978. "Physics research in California: the rise of a leading sector in American physics," Ph.D. Diss., Univ. of Calif., Berkeley (University Microfilms, order nr 79–04599 :Ann Arbor, MI, 1979), v + 596 pp.; Heilbron, J.L. and Seidel, R.W., 1989. *Lawrence and his laboratory: a history of the Lawrence Berkeley Laboratory*, vol. 1 [to 1941] (Univ. of California Pr.), pp. 207–225; Dennis, M.A., 1990. "A change of state: the political cultures of technical practice at the MIT Instrumentation Laboratory and the Johns Hopkins University Applied Physics Laboratory, 1930–1945," Ph.D. Diss., Johns Hopkins University (University Microfilms, order nr 91–13643: Ann Arbor, MI, 1991), 451 pp.; Owens, L., 1990. "MIT and the Federal 'angel': academic research and development and Federal-private cooperation before World War II," *Isis*, *81*, pp. 189–213; Pang, A.S.-K., 1990. "Edward Bowles and radio engineering at MIT, 1920–1940," *HSPS*, *20*, pp. 313–337. A contrary trend at Cornell is indicated by Schweber[7], below, p. 164.

[6] Heilbron, J.L., Seidel, R.W., Wheaton, B.R., 1981. *Lawrence and his laboratory: nuclear science at Berkeley, 1931–1961* (Office for History of Science and Technology, Univ. of California, Berkeley), pp. 46–47. More detailed background to Lawrence's shift of stance between January 1944 and May 1945 is given by Seidel, R.W., 1983. "Accelerating science: the postwar transformation of the Lawrence Radiation Laboratory," *HSPS*, *13*, pp. 375–400, on pp. 377–383. Galison, P. and Bernstein, B., 1989. "In any light: scientists and the decision to build the superbomb, 1942–1954," *HSPS*, *19*, pp. 267–344, on p. 274, paraphrase a June 16, 1945, report of the Scientific Advisory Panel (A.H. Compton, Fermi, Lawrence, and Oppenheimer) of the Secretary of War's Interim Committee calling for post-war Federal funding of the nation's nuclear activities at a billion dollars per year, a continuing rate of expenditure equal to the peak

briefly attained by the Manhattan Project.

[7] [Arthur Roberts], untitled brochure, 11 pp., containing program notes and lyrics for six songs privately recorded in Cedar Rapids, Iowa, December 23, 1947. (Brochure and phonograph records in the collections of the Division of Electricity and Modern Physics, NMAH, Smithsonian Institution, Washington D.C.) The question of how their Second World War experience altered American physicists's desires and expectations regarding the directions, organization, and resources for their work is a principal theme of the work of Dennis, M.A.[5], of Lowen, R.S., 1991. "Transforming the university: administrators, physicists, and industrial and federal patronage at Stanford, 1935–1949," *History of Education Quarterly, 31*, pp. 365–388, and of Galison, P. and Hevly, B. (eds), 1992. *Big Science: the growth of large-scale research* (Stanford Univ. Pr.), in particular the essays of Seidel, R. "The origins of the Lawrence Berkeley Laboratory," pp. 21–45; Galison, Hevly and Lowen, "Controlling the monster: Stanford and the growth of physics research, 1935–1962," pp. 46–77; Schweber, S.S. "Big science in context: Cornell and MIT," pp. 149–183; Hoddeson, L. "Mission change in the large laboratory: the Los Alamos implosion program, 1943–1945," pp. 265–289; and Needell, A.A. "From military research to big science: Lloyd Berkner and science-statesmanship in the postwar era," pp. 290–311. Pertinent also are most of the contributions to De Maria, M. and Grilli, M. (eds), 1989. *The restructuring of physical sciences in Europe and the United States, 1945–1960* (World Scientific: Singapore, etc.). Further: Pestre, D., 1992. "Les physiciens dans les sociétés occidentales de l'après-guerre: un mutation des pratiques techniques et des comportements sociaux et culturels," *Revue d'Histoire Moderne et Contemporaine, 39*, pp. 56–72. These same issues underlie, and often surface in the 700 closely printed pages of, Schweber, S.S., 1994. *Q.E.D and the men who made it* (Princeton Univ. Pr.).

[8] Klopsteg, P.E., 1956. "University responsibilities and government money," *Science, 124*, pp. 919–922, preceded by Klopsteg, P.E., 1955. "Role of government in basic research," *ibid., 121*, pp. 781–84, and followed by Klopsteg, P.E., 1956. "How shall we pay for research and education?" *ibid., 124*, pp. 965–68. See, England, J.M., 1982. *A patron for pure science: the National Science Foundation's formative years, 1945–57* (National Science Foundation: Washington, DC), pp. 134–35, 313,414. Even more clearly unhappy that "Research has become big business" was the head of the Biological Sciences Division of the Office of Naval Research, Reid, R.D., 1953. "Freedom and finance in research," *American Scientist, 41*, pp. 286–292. Agents of private foundations, indignant, turned their backs on the natural sciences ("Behind Q.E.," pp. 184–187), thus arousing concern in elder statesmen: Reingold, N., 1987. "Vannevar Bush's new deal for research: or the triumph of the old order," *HSPS, 17*, pp. 299–344, on pp. 334–35. For approbation of this new order by a major beneficiary, J.A. Stratton (1953), quoted in "Behind Q.E.," p. 218. A general and comparative view of the development of this situation at Stanford and MIT is given by Leslie, S.W., 1992. *The Cold War and American science* (MIT Press: Cambridge Mass.). Stanford is examined more closely still by Lowen, R.S., 1995. *Creating the Cold War university: patrons, administrators, and scientists at Stanford* (University of California Press). I have profited from the perceptive work of Olwell, R.B., 1991. "Science in a 'time of crisis': Project Matterhorn, sponsored research, and nuclear physics at Princeton University, 1945–53," B.A. thesis, Dept. of History, Princeton University.

[9] I.I. Rabi to George B. Pegram, 1944 Sep. 15 (Columbia University, Office of the Secretary, file "Rabi"). Rigden, J.S., 1987. *Rabi:scientist and citizen* (Basic Books: New York); pp. 71–72 treat Rabi as teacher. Cf. John A. Simpson, quoted in note 45 below. Many scientific minds were turning at just this time to the question of postwar positions and research facilities, prompted in part by Vannevar Bush's announcement that the OSRD would be terminated in the very near future: Stewart, I., 1948. *Organizing scientific research for war: the administrative history of the Office of Scientific Research and Development* (Little, Brown: Boston), ch. 21: "Demobilization of OSRD"; Schweber, "...Cornell and MIT"[7], pp. 169, 176.

[10] Antisemitism in prewar academic physics is discussed by Kevles, D.J., 1978. *The physi-

cists: the history of a scientific community in modern America (A.A. Knopf: New York), pp. 210–16, 278–79, 372; Heilbron and Seidel, *History of the Lawrence Berkeley Laboratory*, vol. 1[5], pp. 248–258; Rigden, *Rabi*[9], pp. 104, 169, 180–184. For Columbia: Trilling, D. "Lionel Trilling: a jew at Columbia," reprinted from *Commentary* (March 1979) in: Trilling, D. (ed), 1980. Lionel Trilling, *Speaking of literature and society* (Harcourt Brace Jovanovich: New York), pp. 411–429. Wechsler, H.S., 1977. *The qualified student: a history of selective college admission in America* (John Wiley: New York), pp. 139–140, 177–78, 194–211, provides valuable background on the admission of jewish students and appointment of jewish trustees at Columbia, and shows the rapid shift in public attitudes against religious and racial discrimination circa 1945. Thus, for example, Nathan Glazer recalled of the zionist-socialist student organization, Avukah, that he joined as a freshman at Columbia in 1940: "It would be impossible to list all the members of Avukah who became professors. No one expected that they would become professors before the war... (who dreamed of any job except a clerkship with the government?)" Glazer, N. "From socialism to sociology," in: Berger, B.M. (ed), 1990. *Authors of their own lives: intellectual autobiographies by twenty American sociologists* (University of California Press: Los Angeles), pp. 190–209, on p. 196. Steinberg, S., 1974. *The academic melting pot: catholics and jews in American higher education. A report prepared for the Carnegie Commission on Higher Education* (McGraw-Hill: New York, etc.), lacks the needed discussion of the appointment of catholics and jews to faculty positions *prior* to World War II, but nonetheless provides statistics showing *implicitly* how quickly jews came forward in academia after 1945. The significance of this historical phenomenon is not diminished through being ignored in works to which it is most pertinent, as, for instance, Freeland, R.M., 1992. *Academia's golden age: universities in Massachusetts, 1945–70* (Oxford Univ. Pr.: New York).

[11] That Rabi was not yet acceptable as Physics Dept. chairman in the fall of 1944 is evident from LCMD, Rabi Papers, box 20, folder "Columbia Univ. internal., Pegram, 1935–1952." For Nobel award: *New York Times*, November 10, 1944, p. 16; Dec. 11, p. 25. For chairmanship, *ibid*, July 29, 1945, p. 37.

[12] Eisenhower to Harvey S. Mudd, January 5, 1950, in: Galambos, L. (ed), 1984. *The papers of Dwight David Eisenhower*, vols 10 and 11: *Columbia University* (John Hopkins Univ. Pr.: Baltimore), p. 911. Rabi's continuing access to Eisenhower: E.R. Piore to L.A. DuBridge, June 5, 1953 (LCMD, Rabi Papers), and Kevles, *Physicists*[10], p. 377: "In 1953 Rabi spent an estimated 120 days advising the government in one capacity or another. An enthusiast of Dwight Eisenhower as President... Rabi exercised considerable power as a quiet insider." On Rabi's exceptional influence as a member of the AEC's GAC and of the Office of Defense Mobilization's Science Advisory Committee: Hewlett, R.G. and Holl, J.M., 1989. *A history of the United States Atomic Energy Commission*, Vol. 3: *Atoms for peace and war, 1953–1961* (Univ. of California Pr.: Berkeley), pp. 232, 573; Anders, R.M. (ed), 1987. *Forging the atomic shield: excerpts from the office diary of Gordon E. Dean* (Univ. of North Carolina Press: Chapel Hill), p. 56; McDougall, W.A., 1985. *The heavens and the earth: a political history of the space age* (Basic Books: New York), pp. 159–160; Rigden, *Rabi*[9], pp. 247–250; and, especially, Kevles, D.J. "$K_1 S_2$: Korea, science, and the state," in: Galison and Hevly (eds), *Big Science*[7], pp. 312–333, on pp. 324–26, 330–32. Among Rabi's less well known, but historically important, advisory posts was as chair of the so-called "Scientific Advisers to the Policy Council" of the U.S. military's Joint Research and Development Board (JRDB), 1946/47: Needell, A.A. "Rabi, Berkner, and the rehabilitation of science in Europe: the Cold War context of American support for international science, 1945–1958," in: Heller, F.H. and Gillingham, J. (eds), forthcoming. *The United States and the integration of Europe: legacies of the post war era* (St Martin's Press).

[13] In October, 1944, Polykarp Kusch, who had been (and would again be) closely associated with Rabi at Columbia, could still take for granted Rabi's antipathy to "intellectual vegeta-

tion and republicanism and professional gentlemanliness." Kusch to Rabi, October 29, 1944 (LCMD, Rabi Papers). Kusch was characterizing the industrial research environment in which he was then working. That at this date, in the week before the 1944 presidential election, physicists, as a group, stood conspicuously to the left of engineers is suggested by Edward M. Purcell's recollection of the two radar labs in wartime Cambridge, Mass., the MIT Rad Lab, staffed chiefly by physicists, and the Harvard Radio Research Lab, staffed chiefly by engineers: "You could tell the difference in the 1944 election." Purcell, E.M., 1986. "Radar and physics. Morris Loeb Lecture in Physics, Harvard University," (Videotape, Harvard University Science Center, available at the AIP).

[14] Leslie, S.W., 1990. "Profit and loss: the military and MIT in the postwar era," *HSPS, 21*, pp. 59–85, on pp. 60–66, and Leslie, *Cold war and American science*[8], pp. 25–32, 266–67. Also, Ehrmann, S.C., 1974. "Past, present, futures: a study of the MIT Research Laboratory of Electronics" (Center for Policy Alternatives, Sloan School of Management, MIT), pp. II–61, II–70, II–77. For "Armavy": DeVorkin, D.H. "Back to the future: American astronomers' response to the prospect of Federal patronage, 1947–1955: the origins of the ONR and NSF programs in astronomy," in: Reingold, N. and Van Keuren, D. (eds), forthcoming. *Science and the Federal patron: post-World-War-II government support of American science.*

[15] Hewlett, R.G. and Anderson Jr., O.E., 1962. *A history of the United States Atomic Energy Commission*, vol. 1: *the new world, 1939/1946* (Pennsylvania State Univ. Pr.; reprinted by Univ. of California Pr., 1990), p. 55; Rigden, *Rabi*[9], p. 87.

[16] Guerlac, H.E., 1987. *Radar in World War II* (Tomash Publishers: Los Angeles; American Institute of Physics: New York), p. 290; Schweber, *Q.E.D*[7], pp. 141–44; Collins, G.B. (ed), 1948. *Microwave magnetrons*, "Radiation Laboratory Series, vol. 6" (McGraw-Hill: New York, etc.), from which Fig. 19–65 on p. 793 is here reproduced as Fig. 1.

[17] "Behind Q.E.," pp. 179–180, 205, 215, 217. CRL papers in CUSC, Physics Department Papers, Box 23. The one other military-funded academic laboratory comparable in these immediate post-war years to MIT's RLE and the CRL, was Stanford University's Microwave Laboratory. Funded at a level somewhat below that of CRL, its program was more similar to MIT's RLE: Leslie, S.W., 1987. "Playing the education game to win: the miliary and interdisciplinary research at Stanford," *HSPS, 18*, pp. 55–88, esp. pp. 68–69; Leslie, S.W. and Hevly, B., 1985. "Steeple building at Stanford: electrical engineering, physics, and microwave research," Institution of Electrical and Electronic Engineers (IEEE), *Proceedings, 73*, pp. 1169–1180. And more recently, Lowen, "Transforming... Stanford"[7], and Leslie, *Cold war and American science*[8], ch. 2.

[18] Contract No. W–36–039 sc–32003, Article I, signed January 9, 1946 (LCMD, I.I. Rabi Papers, box 43, folder "OSRD, 1942–46"). Guerlac, *Radar*[16], pp. 519–522.

[19] This characterization is based on descriptions of research projects in the Columbia Radiation Laboratory, Quarterly Progress Report. On Lamb and his fundamental experiment: Schweber, S.S. *Q.E.D and the men who made it*[7], section 5.3, pp. 212–19. Further: Laurence, W.L., 1947. "Radar waves find new force in atom," *New York Times*, September 21, p. 1, col. 2; p. 62, col. 2–5. Lamb Jr., W.E., 1955. "Fine structure of the hydrogen atom. Nobel lecture, December 12," in: Nobelstiftelsen, Stockholm (ed), 1964. *Nobel lectures, including presentation speeches and laureates' biographies: physics, 1942–1962* (Elsevier: Amsterdam, etc.), pp. 281–297; "Some history of the hydrogen fine structure experiment," New York Academy of Sciences, *Transactions*, ser. 2, *38* (1977), pp. 82–86; "The fine structure of hydrogen," in: Brown, L.M. and Hoddeson, L. (eds), 1983. *The birth of particle physics* (Cambridge Univ. Pr.), pp. 311–328.

[20] For the expectation by the Army *and* the Navy that CRL would provide quite concrete technical services relative to millimeter magnetrons: G.R. Hill, Naval Research Laboratory, to CRL, March 3, 1950 (NRL Records, S67/38, Folder #47, "Sept 1949, Radio-Vacuum Tubes"), and Army Signal Corps, Fort Monmouth, to CRL, June 6, 1956 (NMPR, accession 64–434, box 47, file "Columbia U."). Although in subsequent contracts the mission statements of the

Columbia Radiation Laboratory and of MIT's Research Laboratory of Electronics were essentially identical, viz. "extending the useful range of the electromagnetic spectrum" ("Behind Q.E.," p. 217, n. 124), the latter laboratory's task was conceived on far more liberal lines, without the obligations to hardware development of the CRL.

[21] Quoted from the introduction supplied by Columbia's future President, Nicholas Murray Butler, to Friedrich Paulsen's *German universities* (1895) by Finch, J.K., 1954. *A history of the School of Engineering, Columbia University* (Columbia Univ. Pr.: New York), p. 87, who implies that Butler's attitudes were decisive for Columbia's subsequent development.

[22] Columbia University, *Handbook of information, 1950–1952*, p. 39; *Announcement of the Faculty of Pure Science, 1947–48*, p. 9.

[23] Butler, N.M., 1912. As quoted by Randall, Jr., J.H., 1954. "Introduction" to *A history of the Faculty of Philosophy, Columbia University* (Columbia Univ. Pr.: New York), pp. 3–57, on p. 5.

[24] Randall, *ibid.*, p. 6, who continued, p. 7, "The ideal of pure scholarship and pure research remained the goal of the Graduate Faculties... The final report of the Woodbridge Committee in 1938 reaffirms it once more." Nicholas Murray Butler supported these mandarin ideals as President for fortyfive years, retiring only in 1944. He was succeeded by an Acting President, F.D. Fackenthal, 1944–48, and a President, D.D. Eisenhower, 1948–51, who were throwbacks to 19th century American collegiate values – viz., character above knowledge – but were not thereby more favorably disposed to any descent from ideality. James S. Coleman, who was a graduate student in sociology at Columbia in the early 1950s, recalled, in reference to Paul Lazarsfeld's Bureau of Applied Social Research, that it "had only tenuous acceptance by the university. Such applied research... was an upstart activity with which traditional Columbia administrators were uneasy." Coleman, "Columbia in the 1950s," in: Berger (ed), *Authors*[10], pp. 75–103, on pp. 77–78.

[25] Columbia University, President's Committee on the State of the University, Pure Science Subcommittee, "Science at Columbia University: a report by the ...," December 1950 [mimeographed], on p. 2. (Columbiana Collection, Low Library.)

[26] Dunn, L.C., Merton, R.K. and Nagel, E. with Hofstadter, R., 1957. "Memorandum on the Graduate Faculties [Columbia University] ... prepared ... at the invitation of the President's Committee on the Educational Future of the University," [dittoed], on p. 13. (Columbiana Collection, Low Library).

[27] *Physical science and human values: a symposium ...* (Princeton Univ. Pr., 1947), pp. 28, 30, 50. Similarly, Rabi, 1945. "The physicist returns from the war," *Atlantic Monthly, 176*, nr. 4, 107–114, on p. 107: "Before the war these [i.e., the academic] physicists almost never had occupied themselves with problems and questions which in any direct way could be called immediately practical. They directed their whole attention to discovering and understanding the laws of the physical universe... They made it their code to communicate these results to others in the most frank, direct, and expeditious manner." Rabi's general theme in his essays for public consumption in the postwar decades was that science is culture with a big 'C,' essentially like, if not better than, painting, music, literature, etc.: *Science: the center of culture* (World Publishing Co.: New York, etc., 1970). (See note 20 for evidence that CRL was by no means so very different from commercial and government laboratories.)

[28] Indications of this devious streak in Rabi's carriage are given by Needell, A.A. "Rabi, Berkner, and ... Europe"[12]. That there was something phoney in Rabi's ideological posture was not missed by his colleagues; it is gently parodied in the lyrics that Arthur Roberts composed late in 1944 to celebrate Rabi's Nobel Prize: "It ain't the money, it's the principle of the thing."[7] (original typescript copy in LCMD, Rabi papers, box 18, folder "poem").

[29] On the plight of Columbia's Physics Department postwar: Ramsey, N.F., 1968. "Early history of Associated Universities and Brookhaven National Laboratory," *Brookhaven National Laboratory Lectures in Science, 2*, pp. 181–198 (originally issued as BNL report 992/T–421).

Details are to be found in diverse Columbia University records – e.g., I.I. Rabi to D.D. Eisenhower, November 15, 1948, 12 pp., in the files of the Office of the Secretary, Columbia University[9], and Pegram papers, box 31, folder "Growth 1904–1954" (CUSC).

[30] A clear statement of the University's plight is given in a draft of a circular letter to department heads from Provost Jacobs, A.C., 1949. 3 pp., in Pegram papers, folder "Growth"[29], and it is openly referred to in the 1950 Dec. report of the Pure Science Subcommittee[25].

[31] Columbia University, "Annual reports of the President and Treasurer to the Trustees" (privately printed, 1932–1950). Max Frankel, an undergraduate in the years of Eisenhower's presidency, laid much of the blame for Columbia's financial plight on Ike himself: Frankel, M. "Low Memorial indeed," in: First, W. (ed), 1969. *University on the Heights* (Doubleday: Garden City, New York), pp. 127–130. "Behind Q.E.," pp. 188–191, estimates the total annual U.S. Federal Government expenditure on basic research in physics at this time as roughly $15 million, based on 'supply side' figures. The increasing availability of 'demand side' figures – through researches in the archives of receiving institutions – will probably lead to a substantial upward revision of this figure. Thus R.B. Olwell's work[8] credits also Princeton's Physics Department with about a million a year in Federal funds. On Rabi's personal role: "Behind Q.E.," pp. 179–180, and Needell, A.A., 1983. "Nuclear reactors and the founding of Brookhaven National Laboratory," *HSPS, 14*, pp. 93–122. Another strategy for postwar expansion, one not relying on Federal funds with military overtones: Feffer, S.M., 1992. "Atoms, cancer, and politics: supporting atomic science at the University of Chicago, 1944–1950," *HSPS, 22*, pp. 233–261.

[32] Balogh, B., 1991. *Chain reaction: expert debate and public participation in American commercial nuclear power, 1945–1975* (Cambridge Univ. Pr.: New York), makes an attempt in his first two chapters, "Professionalization and politics in twentieth-century America" and "The promise of the proministrative state," to identify the "barriers to the marriage of professionals and the state" in the interwar period, but finds nothing to support himself upon except the uneven history of science literature and the metaphor of "maturity" (pp. 21–22).

[33] Kevles, *The physicists*[10], devoted two chapters to the work of American physicists *during* the First World War, but found the military connection *after* that war required only four pages, pp. 145–148. McBride, W.M., 1992. "The 'greatest patron of science'?: the Navy-academia alliance and U.S. naval research, 1896–1923," *Journal of Military History, 56*, pp. 7–33, points out, pp. 27–33, that although collaboration with academic scientists during World War I – especially at New London, Conn., on submarine detection – had persuaded individual naval officers that the long-hairs could be useful to the Navy, none of the proposals for a continuing relationship envisaged the Navy supporting research in academic laboratories, but contained only schemes to draw academic scientists into Navy labs for longer or shorter periods. Bruce Hevly, in his contribution to this volume, "The tools of science: radio, rockets, and the science of naval warfare," points out, however, that in the years immediately following World War I some research of direct military application was carried on at academic institutions, but apparently not research without such direct military application. Through all this exception must be made of relations falling within the long-standing tradition of assistance by the military services for scientific activities possible only by piggy-backing on military operations or employing unique military facilities. Such, for instance, was the early cosmic-ray research of Robert A. Millikan, which depended heavily upon the military facilities placed at his disposal in the immediate post-war period; his cash, however, came from private sources: De Maria, M., Ianniello, M.G. and Russo, A., 1991. "The discovery of cosmic rays: rivalries and controversies between Europe and the United States," *HSPS, 22*, pp. 165–192, on pp. 171–72.

[34] The labor force increased by half; output per 'man'-hour doubled: U.S. Bureau of the Census, *Historical statistics of the United States: colonial times to 1970* (U.S. Gov't Printing Office: Washington, D.C., 1975), pp. 126, 948, 958.

[35] U.S. Bureau of the Census, *Hist. stat.*, pp. 175–76, 198, 1115–16, 1141. This outlay includes

all "national security" expenditures (thus also the AEC's nuclear weapons facilities), but no veterans' benefits. See also "Behind Q.E.," Fig. 2, pp. 161–62.

[36] Sherry, M.S., 1977. *Preparing for the next war: American plans for postwar defense, 1941–45* (Yale Univ. Pr.: New Haven). Caraley, D., 1966. *The politics of military unification* (Columbia Univ. Pr.: New York).

[37] Quoted by Leslie, S.W., "The military and MIT"[14], p. 64. Stratton's MIT, more than any other university, would get into the midst of those plans for a Third World War: Leslie, *The Cold War and American Science*[8], and Needell, A.A., 1993. "Truth is our weapon: Project TROY, political warfare, and government-academic relations in the national security state," *Diplomatic History, 17*, pp. 399–420.

[38] Trefousse, H.L., 1955. "The failure of German intelligence in the United States, 1935–1945," *Mississippi Valley Historical Review, 42*, pp. 84–100, shows German espionage to have been astonishingly incompetent and easily thwarted. At the Oppenheimer loyalty/security hearings in 1954, Leslie R. Groves, the Army general in charge of the Manhattan Engineer District, testified that "there was never from about 2 weeks from the time I took charge of this project [Sept. 1942] any illusion on my part but that Russia was our enemy and that the project was conducted on that basis": U.S. Atomic Energy Commission, *In the matter of J. Robert Oppenheimer: transcript of hearing...*(U.S. Government Printing Office, 1954; reprinted by MIT Press: Cambridge, Mass., 1970), p. 173. In his own personal account, *Now it can be told: the story of the Manhattan Project* (1962; reprinted by Da Capo Press: New York, 1983), pp. 140–145, Groves 'justified' his preoccupation with Soviet espionage by stating that no *a priori* judgment was made about which foreign country was most to be guarded against. It seems, moreover, that "atomic secrets" were much like other technoscience in this regard, for one finds in other World War II military-scientific records, lists of active communists to be fired: National Archives, Record Group 298, "Coordinator of Naval Research," Box 3, file "A1–9." But however uniform this anti-soviet policy within the U.S. military, it was not uniformly U.S. government policy: Beardsley, E.H., 1977. "Secrets between friends: applied science exchange between the western allies and the Soviet Union during World War II," *Social Studies of Science, 7*, pp. 447–473.

[39] Quoting: Baldwin, H.W., 1945. *Life*, pp. 27–18; Furnas, C.C., 1956. quoted in *Naval Research Reviews*, pp. 24–25. Also, Army General Arthur Trudeau, quoted in "Behind Q.E.," p. 151, and Kevles, "...Korea, science, and the state"[12], p. 312, quoting Beckler, D.Z., 1955.

[40] Sapolsky, H.M., 1990. *Science and the Navy: the history of the Office of Naval Research* (Princeton Univ. Pr.: Princeton, New Jersey), p. 9 *et passim.*, is a diatribe against this scientocentric conceit, which, in its most extreme form, leads to the illusion that Federal largesse was bestowed in gratitude for the physicists' war work and in recognition of the transcendent value of their scientific discoveries: "A grateful American nation was persuaded after World War II to invest heavily in particle physics facilities for educational, cultural and technological reasons." Marshak, R.E. "Scientific and sociological contributions of the first decade of the 'Rochester' conferences to the restructuring of particle physics (1950–1960)," in: *Restructuring physical sciences ... 1945–1960*[7], pp. 745–786, on p. 745. A clear-sighted analysis is given by Seidel, R.W. "The postwar political economy of high-energy physics," in: Brown, L.M. et al. (eds), 1989. *Pions to quarks: particle physics in the 1950s* (Cambridge Univ. Pr.), pp. 497–507. The reverse, but closely related, process, in which a technological enterprise whose principal aims are national power and prestige, is 'sold' as science is most effectively portrayed by DeVorkin, D., 1989. *Race to the stratosphere: manned scientific ballooning in America* (Springer-Verlag: New York). See also: "Behind Q.E.," pp. 218–19; Needell, A.A., 1987. "Lloyd Berkner, Merle Tuve, and the Federal role in radio astronomy," *Osiris, 3*, pp. 261–288, and Needell, A.A., 1987. "Preparing for the space age: university-based research, 1946–1957," *HSPS, 18*, pp. 89–109.

[41] Such is the implication of the passage quoted from the first annual report of the Institute

for Defense Analyses by Schweber, S.S. "Some reflections on the history of particle physics in the 1950s," in *Pions to quarks*[40], pp. 668–693, on p. 681: "Such are the reasons why it is of paramount importance not only to give all necessary supports to weapons research, but also the maintenance of the most effective possible bridge between military strategy and the total of technology..." I.e., although academic scientists, especially physicists – Townes among them – played a central role in IDA, they did so qua technologists. See, also, note 69, below. Further testimony on this general point is offered in "Behind Q.E.," p. 221, n. 130 (Waterman, A. and Estermann, I.), and, especially, in D.H. DeVorkin's contribution to the present volume, "The military origins of the space sciences in the American V–2 era." Needless to say, this syllogism does not hold for non-military technologies, where the imposition of an economic criterion for viability of a technology excludes most technical capabilities. Here is the structural problem of U.S. high-technology industries in the post-Cold-War era.

[42] Zahl, H.A., 1951. "Defense Department plans for basic research: the Army," *Electronics*, pp. 82–84.

[43] For the Signal Corps' 1954 restriction, see "Behind Q.E.," p. 210. Early on there was some discussion of the likelihood and consequences of military influence on the directions and character of postwar research: "Behind Q.E.," pp. 200–229; Ridenour, L.N., 1947. "Should the scientists resist military intrusion?" *American Scholar, 16*, pp. 213–218, with comments by Philip Morrison, Vannevar Bush, Norbert Wiener, Donald Young, Seymour E. Harris, Albert Einstein, Alan T. Waterman, Robert K. Merton, and Aldous Huxley, *ibid*, pp. 218–225, 353–363. The closest studies of this matter, focusing on particular fields or programs, are Redmond K.C. and Smith, T.M., 1980. *Project Whirlwind* (Digital Press: Bedford, MA); Noble, D.F., 1984. *Forces of production* (Knopf: New York) [automatic control of machine tools]; and DeVorkin, D.H., 1992. *Science with a vengeance: the military origins of the space sciences in the American V–2 era* (Springer-Verlag: New York).

[44] E.R. Piore interviewed by R.D. Hagen, as quoted in Hagen, R.D., 1986. *Windows to the origins: the Office of Naval Research* [ONR: Alexandria, Virginia], p. 38.

[45] So I argue in "Behind Q.E.," pp. 201, 213, 216, 221. A nice example of such sociophysical conditions providing the motive for instrumental innovation is Simpson, J.A., 1947. "A precision alpha-proportional counter," *Review of Scientific Instruments, 18*, 884–893, on p. 884: "Within the past few years it has become increasingly important to provide methods for the accurate measurement of large numbers of prepared alpha-radioactive samples... It is desirable that these measurements be made by technicians, and that the measuring instrument maintain constant characteristics over a long interval of time. The instrument described in this paper fulfills these requirements, and during the past three years it has found many applications in physical, chemical, and biological research." For the tilt toward technique in the biological sciences: Wright, S., 1986. "Recombinant DNA technology and its social transformation, 1972–1982," *Osiris, 2*, pp. 303–360, esp. pp. 307, 359. Gornick, V., 1983/1990. *Women in science* (Simon and Schuster: New York), pp. 27–28, quotes one informant declaring "Thinking is out, technique is in," and a second who "brightens helplessly, 'Yes, but technique is exciting.'"

[46] Piore interviewed by B.R. Wheaton, March 7, 1977. Transcript, pp. 35–37 (AIP).

[47] D. Lazarus interviewed by L. Hoddeson and K. Szymborski, December 4, 1981. Transcript, p. 16 (AIP). Similarly, Gerald Tape interviewed by F. Nebeker, June 10, 1991: "... and Piore after the war headed up the Office of Naval Research, which turned out to be a godsend for research at universities." (Transcript published in IEEE Center for the History of Electrical Engineering, *Rad Lab: oral histories documenting World War II activities at the MIT Radiation Laboratory* (IEEE: Piscataway, N.J., 1993), on p. 404.) The point to be emphasized is the enormous exaggeration of Piore's authority and influence – exaggeration even by Tape, who himself having served in high-level positions between the Federal government and the universities, ought certainly to have known better. See also "Behind Q.E.," p. 209. The role

PAUL FORMAN

of such program managers in astronomy and biology at ONR and NSF is examined by D.H. DeVorkin and Toby A. Appel, respectively, in papers to appear in: Reingold, N. and Van Keuren, D. (eds). *Post World War II government support of American science*[14].

[48] I.I. Rabi, recommendation of Norman Ramsey for Guggenheim fellowship, enclosed with letter to H.A. Moe, November 30, 1953 (LCMD, Rabi Papers): "I am very pleased, particularly at the present time, at the thought that a man like Ramsey will be travelling around Europe. It will help to dispel the strange notions which exist in Europe of the enslavement of American science to the military forces." See, "Behind Q.E.," p. 180–182. Wolfgang Pauli, who had spent the war years at the Institute for Advanced Study in Princeton, returning soon after to Zurich, wrote to Einstein, September 19, 1946, that "For the more distant future (let's say in about 5 years) I see... the great danger of military interference [*Einmischung*] in physics (directly or roundabout, through a commission of non-physicists in civilian dress)... Under "interference [*Einmischung*]" I understand not only censorship, but also an influence on the investigatory orientation in experimental work." Translated from Pauli, *Wissenschaftlicher Briefwechsel*..., vol. III: *1940–1949*, Meyenn, K. von (ed), 1993. (Springer Verlag: Berlin, etc.), pp. 383–85. Krige, J., 1992. "The prehistory of ESRO, 1959/60," European Space Agency report ESA HSR-1, p. 6, highlights Eduardo "Amaldi's insistence that a collaborative European space effort be civilian in character... : he made a point throughout the postwar period of publicly distancing himself from the direct military applications of science, even setting down his day-to-day movements in a diary intended to 'prove' that he had not been personally involved in such activities." Just what some of those activities were with which Amaldi was at pains to prove himself uninvolved is indicated by Krige. "What is 'military' technology? Two cases of US-European scientific and technological collaboration in the 1950s," in: Heller and Gillingham (eds), *U.S. and... Europe*[12], where Needell, "Rabi, Berkner, and ... Europe"[12], provides a most revealing picture of the efforts of American physicists – preeminently Rabi – to integrate the Europeans into their military/national-security framework.

[49] *Oxford English Dictionary*, 2nd edition (1989), vol. 6, p. 305, which finds that "One of the most plausible etymological suggestions is F. [i.e., from the French] *gâchette*, which is or has been applied to various pieces of mechanism, e.g. in a lock and in a gun." Mencken, H.L., 1945. *The American language, Supplement I* (Knopf: New York), p. 518, observed that although the word first appeared in the argot of British sailors, "It seems thoroughly and even typically American today..." In his *Supplement II* (Knopf: New York, 1948), pp. 693–94, Mencken adds a 1938 use of "Gadget" in strip tease slang for "a G-string." The sexualization of the term is unquestionably real and important, but in the opposite sense: it is a commonplace that all mechanical instruments and contrivances signify preeminently the male, rather than the female, genitalia. Thus Freud, 1961. *Die Traumdeutung* (Fischer Taschenbuch Verlag: Frankfurt a.M. [which corresponds rather closely to the 1929 edition]), pp. 291–95, emphasizes that "Alle komplizierten Machinerien und Apparate der Träume sind mit grosser Wahrscheinlichkeit Genitalien – in der Regel männliche –, in deren Beschreibung sich die Traumsymbolik so unermüdlich wie die Witzarbeit erweist. Ganz unverkennbar ist es auch, dass alle Waffen und Werkzeuge zu Symbolen des männlichen Gliedes verwendet werden..." In illustration of this point: a friend recalls from her childhood in the 1940s, in which she, as eldest, was followed into her family by two brothers, that in the parlance of her household "gadget" was code for a little boy's penis.

[50] Compton, A.H., 1941. "Not knowledge alone but wisdom," *Vital speeches, 8*, pp. 51–52 (emphasis added, pf). For depression-era deprecations of technical innovation (and of science as a source of such): Kuznick, P.J., 1987. *Beyond the laboratory: scientists as political activists in 1930s America* (Univ. of Chicago Pr.), and Kevles, *Physicists*[10], ch. 16: "Revolt against science."

[51] For the reorganization of the MIT Rad Lab in the spring of 1942: Guerlac *Radar*[16], pp. 291–98. For "Luie's gadgets" my only source is Alvarez, L.W., 1987. *Alvarez: adventures*

of a physicist (Basic Books: New York), p. 101. George Greenstein, G., 1992. "Luie's gadgets: a profile of Luis Alvarez," *American Scholar, 61*, pp. 90–98, contributes nothing beyond what is to be found in Alvarez' memoirs.

[52] Whimsical names of radiation monitoring instruments appear in Kathren, R.L. and Ziemer, P.L. (eds), 1980. *Health physics: a backward glance* (Pergamon Press: New York, etc.), pp. 80, 90–91, 160–61. Names of *Winnie the Pooh* characters were given to the several beta-ray counting assemblies used in conjunction with CP-1 for measuring neutron flux by activation of metal foils. E.g., Metallurgical Laboratory Technical Note Book Nr 86, "Vital Statistics," issued to George Weil, pp. 2–3 (NARA, Federal Records Center, Chicago), and Albert Wattenberg, interview by Sarah Heald et al., April 24, 1992, at the National Museum of American History. Also photograph of counting assembly (Argonne National Laboratory, Neg. Nr. 1821).

[53] The striking difference in American physicists' attitude before and after World War II toward the christening of constructs is pointed out by Heilbron, J.L. "An historian's interest in particle physics," in: *Pions to quarks*[40], pp. 47–54, on. p. 53: "In the 1930s, the tradition of growing neologisms on Greek roots was still vigorous, and in the naming of the deuteron, for example, the several competing suggestions came forward with the certification of philologists and elaborate justification from scientists... What do we have in the 1950s? 'Strangeness,'... and latterly we have had quarks, with their colors, flavors, tops, and bottoms. Does the new terminology express cynicism or disdain by particle theorists toward their own creations?" The circumstance to which Heilbron draws attention is consistent with the striking difference in American physicists' self-image before and after World War II: Forman, P. "Social niche and self-image of the American physicist," in: *Restructuring of physical sciences... 1945–1960*[7], pp. 96–104.

[54] Rosten, L. "Harold Lasswell: a memoir," in: Rogow, A.A. (ed), 1969. *Politics, personality, and social science in the twentieth century: essays in honor of Harold D. Lasswell* (Univ. of Chicago Pr.), pp. 1–14. Smith, M.C., 1994. *Social science in the crucible: the American debate over objectivity and purpose, 1918–1941* (Duke Univ. Pr.: Durham, North Carolina, and London), pp. 243–247.

[55] MacLeish, A., 1940. *The irresponsibles: a declaration* (Duell, Sloan and Pearce: New York), 34 pp., on p. 27: "What matters now is the defense of culture – the defense truly, and in the most literal terms, of civilization as men have known it for the last two thousand years... neither the modern scholar nor the modern writer admits responsibility for the defense... The irresponsibility of the scholar is the irresponsibility of the scientist upon whose laboratory insulation he has patterned all his work." The circumstances of publication of this manifesto, and the reaction to it – especially that of MacLeish's old friend Ernest Hemingway – are described by Donaldson, S., 1992. *Archibald MacLeish: an American life* (Houghton Mifflin: Boston, etc.), pp. 333–37, who, however, makes no mention of Lasswell.

[56] Lasswell, H.D., 1941. "The garrison state," *Am. Journal of Sociology, 46*, pp. 455–468, on pp. 465–66. For background, exposition, and critique: McDougall, D., 1984. *Harold D. Lasswell and the study of international relations* (Univ. Pr. of America: Lanham, MD), ch. 4: "Civil-military relations: the garrison state construct," pp. 151–176. Lasswell's "Garrison state" was frequently reprinted through the cold-war decades: Muth, R. et al., 1990. *Harold D. Lasswell: an annotated bibliography* (New Haven Press: New Haven, Conn., and Kluwer Academic Publ.: Dordrecht), pp. 91–92. Its influence circa 1950 is indicated by Needell, A.A. "Project TROY"[37], pp. 418–19. I am indebted to Needell for bringing Lasswell to my attention by directing me to Friedberg, A., 1992. "Why didn't the United States become a garrison state?" *International Security, 16*, pp. 109–142. In addition to "garrison state," Lasswell is alleged to have originated several other frequently used phrases/concepts, including "rising expectations," "participant observer," and "world revolution of our time": Rosten, "Lasswell"[54], p. 10.

[57] Lasswell, "Garrison state," pp. 465–66.

[58] Lasswell's "Garrison state" essay was preceded by a lecture "Science and democracy,"

reported in *New York Times*, August 31, 1940, p. 12, and September 12, p. 27, and published in *Vital Speeches*, 7 (Nov. 15, 1940), pp. 85–87, a journal that Conant, along with most others then concerned with current affairs, read regularly. MacLeish was an admired personal friend of Conant, and had a large influence on the development of Conant's thinking about fascism and the defense of freedom, 1935–39: Hershberg, J.G., 1993. *James B. Conant: Harvard to Hiroshima and the making of the nuclear age* (Knopf: New York), pp. 112–116, 123, drawing upon the very thoroughly researched dissertation of Tuttle Jr., W. McC., 1967. *James B. Conant, pressure groups, and the national defense, 1933–1945*, Ph.D. diss., University of Wisconsin (University Microfilms: Ann Arbor, Mich., 1970). For Conant's mentoring of Oppenheimer: Hershberg, pp. 166–67. (Neither Tuttle nor Hershberg mention Lasswell.)

[59] Serber, R. *The Los Alamos primer: the first lectures on how to build an atomic bomb*, annotated by Serber, R., introduced by Rhodes, R. (ed), 1992 (Univ. of California Pr.: Los Angeles), p. 4. Oppenheimer himself used the term, instead of bomb, in opening the first of the series of conferences, begun immediately following Serber's course of lectures, on the task of the laboratory (presumably without the 'excuse' of workmen around): Hoddeson, L. et al., 1993. *Critical assembly: a technical history of Los Alamos during the Oppenheimer years, 1943–1945* (Cambridge Univ. Pr.), p. 75. Galison and Bernstein, "... the superbomb, 1942–1954"[6], pp. 271–72, give further examples of 'gadget' in lieu of 'bomb' from 1944. Ten years later, "the thermonuclear gadget" surfaced in Teller's testimony at the Oppenheimer hearings[38], p. 711.

[60] Hawkins, D. et al., 1961. *Manhattan District History, Project Y, the Los Alamos Project* (Los Alamos Scientific Library; reprinted by Tomash Publishers: Los Angeles, 1983), p. 197. Similarly, Hewlett and Anderson, *AEC... 1939/1946*[15], p. 311.

[61] This obliviousness is perhaps most remarkable in the case of Rhodes, R., 1986. *The making of the atomic bomb* (Simon and Schuster: New York), for this author has elsewhere flaunted his preoccupation with the employment of *his* 'gadget,' and has, moreover, surely some acquaintance with the novel of Freeling, N., 1977. *Gadget* (Heinemann: London), in which assaultive and sadistic sexuality, as act and as metaphor, is interwoven with the 'private' construction of a uranium fission bomb. Revealing in this connection are the memoirs of Bernstein, J., 1987. *The life it brings: one physicist's beginnings* (Ticknor and Fields: New York), pp. 92–93, who thirty years before, as a young postdoc, was spending a summer at Los Alamos: "Although all of us had our Q clearances, we were, effectively, divided into two classes – what I used to think of as the adult world and the child's world. The adults, of which I was not one, knew the "secret" – the secret of the hydrogen bomb. The children didn't know it and were – or so I felt – not going to know it until we became adults; i.e., until we actually began working on weapons." The pioneering work drawing attention to nuclear weapons development – and indeed atomic physics research generally – as sexual metaphor/fantasy is Easlea, B., 1981. *Science and sexual oppression* (Weidenfeld and Nicolson: London) and Easlea, 1983. *Fathering the unthinkable: masculinity, scientists and the nuclear arms race* (Pluto Press: London). More recently and 'ethnographically': Cohn, C., 1987. "Sex and death in the rational world of defense intellectuals," *Signs, 12*, pp. 687–718. It is worthy of note that for all that protesters against nuclear weapons in Britain and the United States in the 1970s and 1980s found the sexual significance of such devices obvious, and lampooned it in chant and leaflet, that dimension was largely avoided in all literature that was more than ephemeral (i.e., that got entered in library catalogs). Thus Hilgartner, S. et al., 1983. *Nukespeak* (Penguin Books: New York, etc.; orig. publ. by Sierra Club Books, 1982), does use "gadget" once, p. 29, but sees no sexual significance in it, nor in any of the other terms used or devices referred to. Among the several, scholarly, contributions to Chilton, P. (ed), 1985. *Language and the nuclear arms debate: nukespeak today* (F. Pinter: London, Eng., and Dover, NH), only one, that of Bob Hodge and Alan Mansfield, "'Nothing left to laugh at...': humour as a tactic of resistance," pp. 197–211, even grazes the subject of sexuality. The authors allow, pp. 208–210, that it is

prominent in the rhetoric of the resistors – and offer examples – but decline to grapple with that circumstance: "The intricacy of the discourse of gender and sexuality with the nuclear debate is far too complex to be discussed in this chapter."

[62] Morse, P.M., 1950. "Must we always be gadgeteers?" *Physics Today*, Dec., pp. 4–5. The difference in orientation between gadgeteers and analysts is one of the themes of Aaserud, F., 1995. "Sputnik and the 'Princeton three': the national security laboratory that was not to be." *HSPS, 25*, pp. 185–239, on pp. 186, 192, 215, 217, 230. On Morse, see Schweber, S.S. "The mutual embrace of science and the military," in: Mendelsohn, E. et al. (eds), 1988. *Science, technology and the military*, 1 vol. in 2 (Kluwer: Dordrecht), pp. 3–45, Fortun, M. and Schweber, S.S., 1993. "Scientists and the legacy of World War II: the case of operations research (OR)," *Social Studies of Science, 23*, pp. 595–642, on p. 631, Pickering, A., 1995. "Cyborg history and the World War II regime," *Perspectives on Science, 3*: 1–48, and Edgerton's essay *supra* in this volume, at his note 98. An instance of the unconscious but consistent use of the term 'gadget' for and only for contrivances conceived to have some military context or connection: Hacking, I. "The self-vindication of the laboratory sciences," in: Pickering, A. (ed), 1992. *Science as practice and culture* (University of Chicago Press), pp. 29–64, on pp. 42, 59; wherever Hacking has no military application in mind, he speaks of "apparatus," "equipment," "instruments," "devices," or "machines" – but not "gadgets."

[63] DuBridge, L.A., 1954. "Goals of research," *J. of Engineering Education, 45*, pp. 34–39. Also DuBridge (1949) quoted in "Behind Q.E.," p. 185. The "gadget-oriented" character of the post-war work of American high-energy theorists is pointed out by Schweber, S.S., in *Pions to quarks*[41], p. 673.

[64] Snow, C.P., 1961. *Science and government* (Harvard Univ. Pr.: Cambridge, Mass.), pp. 68–72. On p. 88, Snow footnotes "gadget": "I am using 'gadget' to mean any practical device, from an egg beater to a hydrogen bomb. The kind of mind which is fascinated by the one is likely to be fascinated by the other." And as if in response to Snow – to prove him wrong – in the spring of 1967 the Military Aircraft Panel of the U.S. President's Science Advisory Committee affirmed that, "A nation cannot be built with gadgets." (The nation, of course, was South Vietnam. The quotation is from Herken, G., 1992. *Cardinal choices: Presidential science advising from the atomic bomb to SDI* (Oxford Univ. Pr.: New York), pp. 146, 156–57, 273.)

[65] Snow's charge that scientists are prone to "the euphoria of gadgets" was cited by Margaret Gowing in the official history of the British nuclear program, *Britain and atomic energy, 1939–45* (Macmillan: London, 1964), p. 81, in order then, p. 82, to take issue with Snow, contending that the MAUD committee did not act like typical "gadgeteers" (when in the summer of 1941 it made a strong case for the feasibility of a uranium fission bomb). And from here the reference to scientists as gadgeteers entered the *Oxford English Dictionary*[49].

[66] Berry Jr., F.C., 1988. *Gadget warfare*, "The illustrated history of the Vietnam war" (Bantam Books: New York), pp. 19–20.

[67] Chronology prepared by George B. Pegram from Columbia University and Physics Department records (CUSC, Pegram Papers, Box 31, file "Growth 1904–1954"); Rabi to F.D. Fackenthal, Acting President, Columbia Univ., August 28, 1947, and Fackenthal to Rabi, August 29, 1947 (Columbia Univ., Physics Dept. Office, file "Townes"). Uncertainty of the continuance of the CRL contract is evident in D.P. Mitchell, Exec. Dir. CRL, to Pegram, June 4, 1947 (LCMD, Rabi Papers). Townes' own recollections of the circumstances of his 'call' to Columbia are given in tape-recorded interviews[3]. Also, Berland, "Townes"[3], p. 80. Not being one of "the Charles River boys" – i.e., not being an alumnus of the MIT Rad Lab – put Townes at a distinct disadvantage in competition with his contemporaries.

[68] Townes, "Concentration of the heavy isotope of carbon and measurement of its nuclear spin," Ph.D. thesis, California Institute of Technology, 1939 (CIT Library, accession nr. T639); "Concentration of C^{13} and measurement of its nuclear spin," *Phys. Rev. 56* (1939), p. 850,

abstract of paper presented at the meeting of the American Physical Society, Stanford, California, June 28–July 1, 1939.

[69] Townes as quoted by Berland, "Townes"[3], p. 77; Berland continues, p. 78, "This enforced gadgeteering during the war perhaps was fortunate, for it was this work he hated that actually drew him toward interests that culminated in his developing the maser. ('One has,' he says, 'a mixture of direction in his life and opportunism.')" Thus although Townes may not have been "hot for gadgeteering" in 1941, by 1959, when he took over as chief scientist of the Institute for Defense Analyses, he was seen in the office of the U.S. President's Science Advisor as one who "wants, to put it crudely, inventions." (Quoted from Aaserud, "... the national security laboratory,"[62] p. 230.)

[70] Townes' wartime work is covered well by Nebeker, F. "...Townes as electrical engineer"[3]. This general area of World War II R&D is described, without naming the individuals involved, in *A history of engineering and science in the Bell System: national service in war and peace (1925–1975)*, Fagen, M.D. (ed). (Bell Telephone Laboratories, 1978), pp. 91–93, 100–108. On Wooldridge's route: Berland, 1962. *The scientific life*, ch. 9, "Dean Everett Wooldridge," pp. 255–284; Ramo, S., 1988. *The business of science* (Hill and Wang: New York), pp. 39–44 et passim. Townes' recollections of the resistance he encountered to his requested reassignment to research are given in the interviews cited above[3]. They are supported by the recollections of his BTL research associate, A.N. Holden, interview by L. Hoddeson, June 21, 1976 (AT&T Archives, Warren, NJ).

[71] Guerlac, *Radar*[16], pp. 507–524. Becker, G.E. and Autler, S.H., 1946. "Water vapor absorption of electromagnetic radiation in the centimeter wave-length range," *Physical Review, 70*, pp. 300–307; Lamb Jr., W.E., 1946. "Theory of a microwave spectroscope," *ibid.*, pp. 308–317. Townes, C.H., 1952. "Microwave spectroscopy," *American Scientist, 40*, pp. 270–290.

[72] Townes, research notebook nr. 20464, issued by Bell Telephone Laboratories March 5, 1945 (AT&T Archives, Warren, NJ). Townes, "Applications of microwave spectroscopy," undated, mimeographed memorandum, 7 pp., stamped "Received April 8, 1946 Patent Department." A photostatic copy of this memorandum obtained from William V. Smith contains on its last page the annotation in Townes' hand "This memorandum was prepared approximately May 15, 1945/ C.H. Townes." The passage was quoted by Townes himself in the earliest of his autobiographical accounts: "Masers," in: , Overhage, C.F.J. (ed), 1962. *The age of electronics* (McGraw-Hill: New York), pp. 164–195, on p. 171.

[73] Forman, P., 1985. "Atomichron® : the atomic clock from concept to commercial product," IEEE, *Proceedings, 73*, pp. 1181–1204.

[74] Townes, "Stabilization of an electronic oscillator on a molecular absorption frequency – case 38139-4," Bell Telephone Laboratories memorandum MM–47–110–68, dated September 19, 1947, 5 pp. + 4 Figs. Townes, "Frequency stabilization of oscillators," U.S. Patent 2, 707, 231 (filed April 26, 1947, revised and extended April 27, 1948), and "Frequency selective systems," U.S. Patent 2, 707, 235 (filed April 26, 1947); both patents were assigned to BTL. This achievement of 'atomic' frequency stabilization was announced publicly in Townes, Holden, A.N. and Merritt, F.R., 1948. "Microwave spectra of some linear XYZ molecules," *Phys. Rev., 74*, pp. 1113–1133 (MS received by journal June 29, 1948).

[75] Forman, P., 1986. "The first atomic clock program: NBS, 1947–1954," *Proceedings of the Seventeenth Annual Precise Time and Time Interval (PTTI) Applications and Planning Meeting, Washington, D.C., December 3–5, 1985* (Time Service Dept., U.S. Naval Observatory, Washington, D.C. 20390–5100), pp. 1–17. Snyder, W.F. and Bragaw, C.L., 1986. *Achievement in radio: seventy years of radio science, technology, standards and measurement at the National Bureau of Standards*, "NBS Special Publication 555" (U.S. Government Printing Office: Washington, DC), pp. 292–313. On the Bureau of Standards at this time: Wang, J., 1992. "Science, security, and the Cold War: the case of E.U. Condon," *Isis, 83*, pp. 238–269, and Kellogg, N.R., 1987. "Standardizing the Bureau: the battery additive controversy and the

reorganization of the National Bureau of Standards," ms.

[76] Snyder, *Achievement in radio... At NBS*[75], pp. 403–416 on the establishment of the IRPL and CRPL, and pp. 417–618 on their work. Bromberg, *Laser*[2], p. 24, points out the importance of accurate and stable frequency standards for navigation and for missile tracking and guidance. Such was the concern of a mission from the U.S. Army Signal Corps laboratories at Fort Monmouth to Redstone Arsenal, the Army's ICBM development laboratory, January 10–13, 1955 (NMPR, accession 338-H-869, box 25, file "Proj. 142B, External Program for Frequency Control, July–December 1954"). Likewise, "Need for a *Stable Time Reference*" was placed at the top of the list of "Gaps in the long range navigation program" in the February 21, 1951 report of the Ad Hoc Working Group on Long Distance Navigation, of the Panel on Air Navigation, of the Committee on Navigation, of the Research and Development Board, U.S. Dept of Defense (copy courtesy of Allan Needell from U.S. National Archives, RG 300, entry 341, box 274). Several CRPL staff members attended the meetings of this panel as members and observers. The CRPL was also intimately involved in testing the radio transmission techniques proposed by Project Troy, among which was the locating of jamming stations by what would decades later, when adopted by astronomers, come to be known as VLBI – for the employment of which "the most serious technical problem is presented by the requirement of high-precision time standards." Wiesner, J.B., 1951. "Radio intelligence," Annex 24 to MIT, Project Troy, *Report to the Secretary of State, February 1*. I am indebted to Allan A. Needell for the opportunity to see this recently (partially) declassified report and for his illuminating discussion of it in his "Project TROY"[37].

[77] Snyder, *Achievement*[75]. Forman, "NBS, 1947–1954"[75]; "Atomichron® "[73].

[78] Rigden, *Rabi*[9], pp. 170–71, 275. Forman, "Inventing the maser"[2], at notes 7, 8, 54.

[79] Forman, *ibid*[55]; Forman, "NBS, 1947–1954"[75].

[80] Surveying this "already large and rapidly growing new field in pure physics," Gordy, W., 1948. "Microwave spectroscopy," *Reviews of Modern Physics, 20*, pp. 668–717, on p. 668, stated that "so far as I know, only one paper on microwave spectroscopy exists in prewar literature," that stemming from C.E. Cleeton's 1935 Ph.D. dissertation at the University of Michigan. Two years later Coles, D.K., 1950. "Microwave spectroscopy," *Advances in Electronics, 2*, pp. 299–362, on p. 306, could add a second Ann Arbor dissertation, that of H.S. Howe in 1940, as "so far as the writer is aware... the only attempts at microwave spectroscopy prior to the Second World War." Townes and Schawlow, A.L., 1955. *Microwave spectroscopy* (McGraw-Hill: New York) in their bibliography of well over one thousand works add only two additional prewar references directly related to microwave spectroscopy. The debt of "pure physics" to war-inspired technology is noted by Gordy, loc. cit., and again more emphatically by Gorter, C.J., 1951. "Spectroscopy at radio frequencies," *Physica, 17*, pp. 169–174, in the opening address at a conference with the same title in Amsterdam, September 1950, at which Townes was present. This point was made his own in Townes, 1952. "Microwave spectroscopy," *American Scientist, 40*, pp. 270–290, on p. 287, which paper was based on material he had presented on various campuses as Sigma Xi lecturer for 1951. The point is developed in Forman, "Swords into ploughshares"[4].

[81] Townes, *ibid*, pp. 273–74; Townes and Schawlow, *Microwave spectroscopy*[80], pp. 338–342; Townes, "Applications..."[72]; Kisliuk, P. and Townes, C.H., 1952. *Molecular microwave spectra tables*, "NBS circular 518" (US Government Printing Office: Washington, DC).

[82] S. Geschwind, telephonic interview by author, October 2, 1989.

[83] "Behind Q.E.," pp. 149–152; Seidel, "The postwar political economy"[40], p. 497; Bromberg, *Laser*[2], ch. 1; Leslie, *Cold War and American Science*[8].

[84] Harold A. Zahl to Director of Engineering, Evans Signal Laboratory, Fort Monmouth, October 21, 1947, and Zahl to J.B. Wiesner, MIT, June 2, 1953 (NMPR, accession 338-76-H-869, box 1 of 302, file "Columbia Radiation Laboratory." This file originated with G. Ross Kilgore, Chief, Microwave Tubes Section, Thermionics Branch. A docket memorandum in

the file states it to be incomplete, the primary file being that of the Chief of the Thermionics Branch, John E. Gorham. I have not encountered Gorham's file.)

[85] Harold A. Zahl, Director of Research, Fort Monmouth, to Emanuel R. Piore, ONR, February 9, 1951 (NMPR, accession 338-I-3014, box 1 of 2, file "400.112, R&D, 1951, General"). More specifically, J.J. Slattery, Chief, Radar Branch, Evans Signal Laboratory, Fort Monmouth, to Director of Engineering, January 23, 1950: "It has been determined that the development of short-range lightweight precision radar devices for use with Army Field Forces, such as tanks, tank-destroyer, etc., will require components operating in the millimeter region."[84].

[86] Bromberg, *The laser in America*[2], pp. 62–65, 263–64, provides evidence of continuing work and continuing military interest in millimeter waves after 1954, much of it focusing on the maser. Noteworthy is the appearance of the commanding officers of the Office of Naval Research, the Air Force Office of Scientific Research, and of Fort Monmouth at the opening session of the Polytechnic Institute of Brooklyn, Microwave Research Institute, "Symposium on millimeter waves," March 31, 1959. *Proceedings*, "Symposia Series, vol. 9" (Polytechnic Press: Brooklyn N.Y., 1960), pp. ix–xiv; also Skolnik, M.I., 1970. "Millimeter and submillimeter wave applications," *Proceedings of the symposium on submillimeter waves, New York, N.Y., March 31, April 1, 2, 1970*, "Symposia Series, vol. 20" (Polytechnic Press: Brooklyn, N.Y.), pp. 9–25. For earlier work at these extremely short waelengths, see: Wiltse, J.C., 1984. "History of millimeter and submillimeter waves," IEEE *Transactions on Microwave Theory and Technique*, MTT-32, pp. 1118–1127; Bryant, J.H., 1988. "The first century of microwaves," *ibid.*, MTT-36, pp. 830–858.

[87] "The basic problem with electronic amplifiers or oscillators," Townes recalled in his 1964 Nobel Prize lecture, "seemed to be that inevitably some part of the device which required careful and controlled construction had to be about as small as the wavelength generated." As the wavelength decreased from centimeters to millimeters, the microwave oscillator tubes shrank to doll-house dimensions, and the difficulty of achieving precisely controlled construction on such a diminutive scale increased greatly. Still greater is the difficulty such diminutive structures have in dissipating the heat generated by their increasingly inefficient operation. "This", Townes recalled himself reasoning twenty years before, in 1944, "set a limit to construction of operable devices." With this insight Townes stood in a consensus of informed opinion. And to this consensus Townes appealed in May 1945 when seeking to persuade his superiors to a program in microwave spectroscopy by holding out to them the (quite novel) hope that "molecules, resonant elements provided by nature... with the reproducibility inherent in molecular structure," might eventually obviate these difficulties. Townes, 1965. "Production of coherent radiation by atoms and molecules: Nobel lecture, December 11, 1964," *Science, 149*, pp. 831–841; also in: Nobelstiftelsen (ed), 1972. *Nobel lectures including presentation speeches and laureate's biographies: physics, 1963–1970* (Elzevier: New York), pp. 55–88, on p. 59. The existence of a theoretical limit to the frequency obtainable with klystrons and magnetrons, from consideration of heat dissipation, current density, transit time, etc., was demonstrated with varying degrees of formality by many investigators, e.g., Elliott, R.S., 1952. "Some limitations on the maximum frequency of coherent oscillations," *J. of Applied Physics, 23*, pp. 812–818.

[88] Pierce, J.R., 1950. "Millimeter waves," *Physics Today*, Nov., pp. 24–29; Slattery, J.J., 1950[85]. Mueller, G.E., 1955. "Millimeter tubes," in Polytechnic Institute of Brooklyn, Microwave Research Institute, *Proceedings of the symposium on modern advances in microwave techniques, ...New York, N.Y., November 8, 9, 10, 1954*, "Symposia Series, vol. 4" (Polytechnic Inst. of Brooklyn: Brooklyn, N.Y.), pp. 145–157.

[89] Smith, A.G., Gordy, W., Simmons, J.W. and Smith, W.V., 1949. "Microwave spectroscopy in the region of three to five millimeters," *Phys. Rev., 75*, pp. 260–264; Johnson, C.M., Trambarulo, R. and Gordy, W., 1951. "Microwave spectroscopy in the region from two to three millimeters, part II," *Phys. Rev., 84*, pp. 1178–1180. Loubser, J.H.N. and Townes, C.H., 1949.

"Spectroscopy between 1.5 and 2 mm wave-length using magnetron harmonics," *Phys. Rev.,* 76, p. 178 (abstract of paper presented to APS, April 28–30, 1949); Klein, J.A., Loubser, J.H.N., Nethercot, A.H. and Townes, C.H., 1952. "Magnetron harmonics at millimeter wavelengths," *Review of Scientific Instruments, 23*, pp. 78–82; Nethercot, A.H., Klein, J.A., Loubser, J.H.N. and Townes, C.H., 1952. "Spectroscopy near the boundary between the microwave and infrared regions," *Nuovo Cimento, 9*, supplement, pp. 358–363. Also important to Townes' initial success was a radiation detector recently devised by the chief scientist at one of the Signal Corps' Fort Monmouth laboratories: Golay, M.J.E., 1952. "Bridges across the infrared-radio gap," Institute of Radio Engineers (after 1960, IEEE), *Proc., 40*, pp. 1161–1165.

[90] Townes, "Magnetron harmonics"[89]; Townes, "Microwave spectroscopy,"[80], on p. 274; Gordy, W., 1952. "Microwave spectroscopy," *Physics Today*, Dec., pp. 5–9. Gordy persisted, however, and soon surpassed Townes: Burrus, C.A. and Gordy, W., 1954. "Submillimeter wave spectroscopy," *Phys. Rev. 93*, pp. 897–98 (Ms received by journal December 21, 1953); Townes and Schawlow, *Microwave spectroscopy*[80], pp. 452–462; Gordy, "Millimeter and submillimeter waves in physics," PIB, MRI, *Symposium on millimeter waves*[86], pp. 1–23.

[91] Lamb Jr., W.E., in: *Birth of particle physics*[19], p. 321; Townes, "Spectroscopy near the boundary"[89]; Townes, "Magnetron harmonics"[89].

[92] Townes, "Microwave spectroscopy"[80], p. 283; Pierce, "Millimeter waves"[88]. As we have just seen[89], even a year later, in mid–1951, Townes and collaborators couldn't get more than a fraction of a milliwatt at 1.1 mm.

[93] U.S. Navy, Office of Naval Research, Advisory Committee on Millimeter Wave Generation, "Minutes of the first meeting of the Advisory Committee... April 26, 1950, Washington, D.C.," 2 pp., signed by Chodorow, M. (Louis D. Smullin Papers, MC 216, Institute Archives and Special Collections, MIT Libraries, Cambridge, Massachusetts).

[94] Schweber, S.S. "The mutual embrace of science and the military: ONR and the growth of physics in the United States after World War II," in: Mendelsohn, E. et al. (eds), 1988. *Science, technology and the military*, "Sociology of the sciences, vol. 12," 1 vol. in 2 (Kluwer: Dordrecht), pp. 3–46; Hoch, P. "The crystallization of a strategic alliance: the American physics elite and the military in the 1940s," *ibid*, pp. 87–116; DeVorkin, "... the ONR and NSF programs in astronomy,"[14]; Forman, "Behind Q.E.," p. 210. Disappointingly uninformative on these matters but nonetheless indispensable is Sapolsky, *History of ONR*[40]. See my review in IEEE, *Annals of the History of Computing, 14*, nr 2 (1992), pp. 60–62.

[95] Weart, S.R. "The solid community," in: Hoddeson, L. et al. (eds), 1992. *Out of the crystal maze: chapters from the history of solid state physics* (Oxford Univ. Pr.: New York), pp. 617–669.

[96] Marvin, J.R. and Weyl, F.J., 1966. "The summer study," *Naval Research Reviews*, Aug., inside front cover through p.7, pp. 24–28; Schweber, "The mutual embrace"[94]; Leslie, *Cold war and American science*[8]. Needell's "Project TROY"[37] describes a consequential State-Department sponsored 'summer study.' Sapolsky's *History of ONR*[40], pp. xii, 6, 8, 112, 117, is, on the contrary, skeptical of the alleged influence of university-based scientific advisors.

[97] Townes, letter to members of ONR Advisory Committee on Millimeter Wave Generation, April 14, 1950, 3 pp. (Smullin papers[93]). A copy of this letter is one of several enclosures with a letter from Townes to Joan L. Bromberg, July 31, 1987, 28 pp. (Laser History Project records, AIP.)

[98] Townes to Bromberg, *ibid.*, p. 7; Bromberg, *Laser*[2], pp. 15–16. See, also, "Behind Q.E.," p. 215. It must be said that the contractual relations between the Navy and Columbia were by no means limited to the CRL, and that, in particular, ONR maintained with Columbia, inter alia, a blanket contract, N6-onr-271, whose various "task orders" authorized very diverse projects, including Ruth Benedict's "Research in contemporary cultures" (T.O. 3) and, most pertinent, the management of the Naval Research Advisory Committee (T.O. 12). However, Townes' committee had its own contracts: Nonr–06500 to get started, April 1 to June 30, 1950, then

Nonr–14400. (CUSC: Committee on Government Aided Research, box 1, folder "Pegram, 1949–50"; Columbia Univ., Administrative Records, Contract Corresp., 1951–52, item 15.)

[99] Townes, letter to members of millimeter wave committee, April 14, 1950[97].

[100] Townes to E.L. Chaffee, October 13, 1950, and Townes to Smullin, October 30, 1950 (Smullin papers[93]).

[101] This is the 'famous' meeting of the committee to which reference is made in every account of the origin of the maser.

[102] Townes to Smullin, November 3, 1951 (Smullin papers[93]). In the U.S. there were at this time – autumn 1951 – at least ten universities as well as several commercial laboratories working on millimeter waves: Research and Development Board, Committee on Electronics, Panel on Electron Tubes, "Report on government sponsored millimeter wave research and development in the United States," September 19, 1951. iv + 17 pp. (copy in Smullin papers[93]).

[103] April 28–29: Townes to Smullin, April 15, 1952 (Smullin papers[93]).

[104] Senitzky, I.R., memorandum for Chief, Thermionics Branch, Evans Signal Laboratory, Fort Monmouth, December 15, 1953. "Report of attendance at Office of Naval Research Millimeter Wave Advisory Committee Meeting in Washington, D.C., on October 16, 1953" (NMPR, accession 338–76–H–869, box 1, file "112B, External program for electron tubes, 1953"). Bromberg, Laser in America[2], pp. 63–64, 263, draws attention to Coleman, P.C., 1963. "State of the art: background and recent developments – millimeter and submillimeter waves," IEEE Transactions on Microwave Theory and Techniques, MTT–11, pp. 271–288, who identifies, p. 285, three special conferences on millimeter waves, 1951–1955, including the ONR conference in Washington October 15–16, 1953. About the other two – Urbana, 1951; Fort Monmouth, 1955 – I have no information.

[105] J.R. Dunning, chairman of Columbia's Committee on Government Aided Research, to Fort Monmouth, November 27, 1950 (Columbia University, Physics Department Office, file "Townes"); Townes to Bromberg[97], p. 7. Rabi regarded as boss: Zahl to Piore, February 9, 1951; Zahl, memos of conversations with Rabi, May 16, 1952, and November 17, 1952[84].

[106] Schawlow, interview, January 19, 1984, by J.L. Bromberg (AIP); Geschwind, interview[82]; Stanley Autler, interview, November 12, 1989, by S.S. Schweber and author. From 1946 through 1955 the Quarterly Report of the Columbia Radiation Laboratory listed an average of two publications per year on the "Lamb shift." From 1950 onward, i.e. two years after Townes arrival, 10–15 publications were listed annually in molecular microwave spectroscopy. As a measure of Townes' prominence: in volume 76 of the Physical Review, issued July through September, 1949, Townes had ten papers, letters, abstracts, or titles of invited talks. Early in the following year the Departmental Executive Officer, Polykarp Kusch, in recommending Townes' promotion to Professor, stated that "He is now, without any qualification, the outstanding man in this country and abroad in the important field of microwave spectroscopy." (Kusch to Pegram, February 22, 1950, in Columbia Unversity, Physics Department Office, file "Townes".)

[107] Senitzky, January 31, 1952, report on January 11 visit to CRL[84].

[108] Entry for May 11, 1951, on p. 145, in Townes, Columbia University notebook 1, "Used from January 2, 1948, to July 1951" (in possession of C.H. Townes, Department of Physics, University of California, Berkeley).

[109] The "molecular beam oscillator" project first appeared in the Quarterly Report of the Columbia Radiation Laboratory for October–December, 1951, pp. 7–8, and regularly thereafter, appearing under the moniker 'maser' only in 1955. See Forman, "Inventing the maser"[2].

[110] Čerenkov, P.A., 1937. "Sichtbares Leuchten der reinen Flüssigkeiten unter der Einwirkung von harten β-Strahlen," Comptes Rendus (Doklady) de l'Académie des Sciences de l'URSS, 14, pp. 101–108; Frank, I.M. and Tamm, I.E. "Coherent visible radiation of fast electrons passing through matter," ibid., pp. 109–114. Frank, I.M., 1984. "A conceptual history of the Vavilov-Cherenkov radiation," Soviet Physics-Uspekhi, 27, pp. 385–395.

[111] The idea was put forward by Ginzburg in a series of four papers – the first dated November 25, 1946, the last February 10, 1947 – in vol. 56 (1947) of the *Doklady* (i.e., *Comptes Rendus, Proceedings) of the Soviet Academy of Sciences*, pp. 145–48, 353–54, 583–86, 699–702. (Ginzburg, V.L. et al. "Use of Cherenkov effect to generate high radio frequencies. Four papers... Translated from the Russian by Hope, E.R., 1954. " Canada, Defense Research Board, Defense Scientific Information Service, report T103R, 24pp., available from the U.S. Dept. of Defense, Defense Technical Information Center, as document AD610899.) The latter two of these four papers carried, as second author, Ilya M. Frank, Ginzburg's more senior colleague at FIAN, who, with the still more senior Igor Tamm, had provided the interpretation of the effect discovered by Vavilov and Čerenkov. The last of these four papers, "Izluchenii elektrona i atoma... " Akademiia Nauk SSSR, *Doklady*, 56 (1947), pp. 699–702, opens with an acknowledgement that Leonid I. Mandelstam († 1944), FIAN's intellectual leader, had pointed out, sometime before 1940, that Vavilov-Cerenkov radiation should also result from the motion of a charged particle not through, but near to, a dielectric.

[112] Lashinsky, H., 1961. "Cerenkov radiation at microwave frequencies," *Advances in Electronics and Electron Physics, 14*, pp. 265–297; Jelley, J.V., 1958. *Cerenkov radiation and its applications* (Pergamon Press: London, etc.), pp. 44–47, 55–56, 76–77, 95–96, 292.

[113] *Physics Abstracts, 52* (1949), p. 766, being an abstract of Ginzburg, "Ob izluchenii mikroradiovoln...," *Izvestiya Akad. Nauk SSSR, Seriya Fizicheskaya, 11* (1947), pp. 165–182. Earlier in 1949 *Physics Abstracts* (52, p. 76) had carried a summary of the fourth of Ginzburg's papers in the *Doklady*[111] – not, however, of the three preceding – translating the title as "Radiation of an electron or atom moving along the axis of a channel in a dense medium," but gave no indication that the radiation investigated was in the microwave, rather than the optical, range.

[114] An important function of ONR offices abroad was as listening posts (see Needell[12]). The readiness of ONR staff to procure scientific documents for its contractors is emphasized in its *Contractors' manual* (NAVEXOS P–454, rev. February 1948), p. 4. Their intent to make special efforts on behalf of Townes' committee is stated in Townes, circular letter of April 4, 1950[97].

[115] Haeff, A.V., 1948/1949. "The electron wave tube: a novel method of amplification and generation of microwave energy," Naval Research Laboratory, *Report*, nr R–3306, and "The electron wave tube," Institute of Radio Engineers, *Proceedings, 37*, pp. 4–10.

[116] Townes letter to millimeter wave committee members, April 14, 1950[97]; a copy of Haeff's translation of Ginzburg's *Izvestia* paper[113], "Radiation of microwaves and their absorption in air," is also in the Smullin papers, MIT Archives. The very typescript – prepared by the "Ditto" stencil process – distributed by Townes to his committee, was subsequently marketed by Reilly Translations (RT–399) and was available from the National Translation Service, Library of Congress, while that service existed.

[117] Pierce, "Millimeter waves"[88]. Townes, his committee, and ONR were much excited and activated late in 1951 by the rumor that J. Bernier, at the French electronics firm CSF, had produced 1 mm wavelength Cerenkov radiation: Townes to Smullin and committee, November 3, 1951 and June 25, 1952 (Smullin papers[93]). Nebeker, "...Townes as electrical engineer,"[3] as well as Townes' own recollections[3] stress the grounding in classical electromagnetic theory that he had obtained at Caltech as assistant to William R. Smythe.

[118] Townes, "Applications of microwave spectroscopy."[72] An indication of such direction of Townes' thoughts is given in Townes' 1979 interview by W.V. Smith[3].

[119] It is thinkable that Townes here found suggestive Ginzburg's[111] generalized treatments of the Cerenkov effect, including 'Cerenkov radiation' by an oscillating dipole. Cf., Frank, I.M., 1964. "Optics of light sources moving in refractive media. Nobel lecture, December 11, 1958," in: *Nobel lectures, including presentation speeches and laureates' biographies: physics, 1942–1962* (Elsevier for the Nobel Foundation: New York), pp. 442–469, on pp. 448–452.

[120] The park-bench epiphany appears in all of Townes's recollections and in nearly every

secondary account of the origin of the maser. The source closest in time to the actual event is
Townes, 1962. "Masers"[72], on p. 172.

[121] Forman, "Inventing the maser"[2], section III; Bromberg, *Laser in America*[2], pp. 19, 21. By
Berland's account[3] (p. 83), "For two years, he [Townes] and his helpers built gadgets..."

[122] For budgets: Leslie, "The military and MIT"[14], and Ehrmann "MIT RLE"[14], and "Behind
Q.E.," pp. 173–177, 206–209. For CRL and Rabi: Zahl to Piore, February 9, 1951[85]; Zahl to
Director, Evans Signal Laboratory, Fort Monmouth, November 17, 1952[84]. For guns *and* basic
research: "Behind Q.E.," pp. 158–59; Needell, "Preparing for the space age"[40]; and England,
...the NSF[40], ch. 8, "Defining a program."

[123] I.R. Senitzky January 31, 1952 report on January 11 visit to CRL[84]. Senitzky, five years
Townes' junior, and with a Ph.D. in physics awarded him by Columbia less than two years
previous, could hardly have been entirely self-assured vis-a-vis these Columbia professors. He
had, however, been working at Fort Monmouth for ten years, and was both an accomplished
experimentalist and a capable theorist.

[124] Senitzky, April 3, 1952 report on March 28 visit to CRL[84].

[125] Rabi later stated (*New York Times*, October 30, 1964, p. 23) that Townes had *insisted*
on becoming chairman (Executive Officer) of the Physics Department, and it indeed seems
unlikely that Townes would have been *drafted* after only four years at Columbia, and only one
year directing CRL. Dissatisfaction with the management of CRL's magnetron development
work is clearly implied by the reports I.R. Senitzky, Microwave Tubes Section, rendered to
the Chief, Thermionics Branch, Evans Signal Laboratory, Fort Monmouth, on his visits to
Columbia University Radiation Laboratory on January 11, March 28, and August 14, 1952[84].

[126] Senitzky, September 15, 1952 report on August 14 visit to CRL[84].

[127] Senitzky, memorandum to Chief, Thermionics Branch, Evans Signal Lab., April 21, 1953
(April 15), "Report of trip to Columbia University Radiation Laboratory, on 6 April 1953..."
2 pp.; same to same, undated, "Report of trip to Columbia University... on 26 May 1953... to
review, together with Office of Naval Research representatives, Radiation Laboratory policy,"
2 pp.[84]

[128] Author's interviews of S. Autler (November 12, 1989), M. Danos (April 18, 1988), S.
Geschwind (October 2, 1989), J.P. Gordon (November 4, 1983), M. Stitch (August 3, 1989),
and H. Zeiger (August 15, 1989).

[129] P. Kusch, memoranda to staff, December 4, 1950, September 15, 1952, October 30, 1953,
March 18, 1954 (CUSC, Physics Department papers, Box 23). In contrast with the position
of "Executive Officer" of the Physics Department – which, following Rabi's resignation of it
after five years' tenure, was fixed at a term of three years – that of "Executive Director" of
CRL was without term and with far more dictatorial authority in minor matters.

[130] Kusch, P., 1950. "Report of the Physics Department as contributory to a report on the state
of the University," Oct., 22 pp, on p. 3 (LCMD, Rabi papers). The report to which Kusch's
was contributory is, probably, that cited in note 25. Columbia's stress upon fundamentality
found expression at this time also in "Recommendations for the Final PhD Examinations,"
May 30, 1951, by Kusch's colleagues, the experimentally capable theorists Henry Foley and
Norman Kroll, both – and most especially Kroll – closely involved in CRL: "It is apparent that
in numerous cases a candidate who has performed creditably a piece of significant research
lacks the theoretical background necessary for an intelligent evaluation and analysis of his
own work. We believe that a candidate should have a thorough mnastery of the elementary
aspects and fundamental principles of his own field and should be acquainted with the current
status of work in this field. It is our proposal that an additional oral examination be given..."
(CUSC, Physics Department Papers, Box 1, folder "Graduate Committee, 1951–53.)

[131] Interview by Jeff Hecht, September 19, 1984, as published in Hecht et al., 1985. *Laser
pioneer interviews* (High Tech Publications: Torrance CA), pp. 73–84, on p. 74. Michael Danos
recalled, April 18, 1988, in an interview by the author, a conversation circa 1953 with Jack

Steinberger, an able young elementary particle physicist recently elevated to faculty status, that made postdoc Danos aware that there was some doubt in Columbia's Physics Department about the appropriateness of Townes' microwave generation schemes as research projects.

[132] "Behind Q.E.," pp. 205–209, for the growing pressure toward classification of academic research in the early 1950s. G.R. Kilgore's CRL file[84] shows the question of classifying the CRL Quarterly Report was repeatedly raised, and although the decision was always negative, it was understood that the QR was not to be sent beyond the borders of the U.S. by CRL or its staff, but only by the military sponsor. The anomaly – that CRL's work remained unclassified while similar work in industrial labs was classified – is stressed in Executive Officer, Evans Signal Laboratory, for Commanding Officer, Fort Monmouth, to Office of the Chief Signal Officer, October 27, 1953.[84]

[133] Zahl, H.A., Director of Research, memorandum to Director, Evans Signal Laboratory, Fort Monmouth, May 19, 1952, reporting telephone conversation with Rabi[84].

[134] Cf.[27].

[135] Zahl, H.A., Memorandum to Director, Evans Signal Laboratory, Fort Monmouth, Nov. 17, 1952, 1 p.[84]: "It is ONR's belief that a reduction of the tube effort should be shown in the proposed one year extension [beginning Dec. 1]... The undersigned called Dr. Rabi at Columbia specifically to ask if he felt that the magnetron work should be reduced."

[136] Loc. cit. In the following years, with the addition of the Air Force, the Joint Services (Technical) Advisory Committee – in fact, more nearly an executive than an advisory committee – began to meet regularly, semiannually.

[137] Senitzky, "Report of trip to Columbia University... on May 26, 1953"[84]. For Senitzky's minutes of the October 16, 1953 meeting of Townes' committee see above[104]. Zahl's appreciation of Rabi is evident in his autobiography, cited in "Behind Q.E.," note 50. For Kilgore's early work see Bryant, "Microwaves"[86].

[138] An undated docket memorandum in the file[84] states that the CRL contract for the twelve months December 1, 1952–November 30, 1953 was in the amount of $343883, but that there were modifications to this contract, the fourth being "Price is reduced." As the third modification is "Change in delivery schedule," and this clearly refers to correspondence in this file between Kusch and Fort Monmouth in June 1953 regarding the schedule for delivery of the Laboratory's quarterly reports[144], the unspecified reduction in "price" was effected in or shortly after June 1953. My supposition that the reduction was only about 10% is based on the 'inelasticity' of the magnetron program and on the tone of Kusch to Rabi [1957 June 2][158].

[139] Zahl to J.B. Wiesner, Director, Research Laboratory of Electronics, MIT, June 2, 1953. Informing him of reduction from $750000 to $600000 annually[84]; Ehrmann, "... the MIT Research Laboratory of Electronics"[14], pp. II–61, 62.

[140] Evidence of this uncertainty 1947–1949: Mitchell to Pegram, June 4, 1947[67]; "Behind Q.E.,"[13]; England... the NSF[8], pp. 98–101; and Kevles, "... Korea, science"[12], p. 319.

[141] CRL's leadership in millimeter magnetrons circa 1951 seems pretty clear. Thus, the September 19, 1951, "Report on government sponsored millimeter wave research and development in the United States"[102] lists several commercial laboratories working on klystrons, etc., but only one, Bomac, working on magnetrons. In the spring of 1950 CRL was still authoritative, Bomac's 5-mm tubes being taken there to have their power output checked: Naval Research Laboratory records, S67/38, folder n. 47. Mueller, 1954. "Millimeter tubes"[88], testifies to CRL's continuing leadership in millimeter magnetrons, but it remains unclear how far this appeared to be true because other programs were then under wraps.

[142] I.R. Senitzky, memorandum to Chief, Thermionics Branch, Evans Signal Laboratory, Fort Monmouth, March 4, 1954, "Report of visit to... Columbia Radiation Laboratory... February 16, 1954," 1 p.[84]

[143] Senitzky, "Report... May 26, 1953", and earlier reports[84].

[144] Danos, M., Geschwind, S., Lashinsky, H. and Trier, A. van, 1953. "Cerenkov effect at mi-

crowave frequencies," *Phys. Rev. 92*, pp. 828–29, MS received by journal September 8, 1953. Success was attained after the preparation of the *Second Quarterly Report* on the then-current contract, dated June 30, 1953, pp. 18–19, but before the preparation of the *Third Quarterly Report*, dated August 31, 1953, pp. 11–12. That this brackets the achievement within a period of two, rather than three, months is a consequence of CRL, on instruction from Fort Monmouth, having switched from calendar quarters to three-month periods initialized by the starting date of the contract: Kusch, letter June 10, 1953, to Director Evans Signal Laboratory, attn. Zahl; Zahl, letter June 30, 1953, to Kusch[84].

[145] Still optimistic in his report on the molecular beam oscillator for the *Second Quarterly Report* (June 30, 1953), pp. 18–19, Gordon submitted none for the next (August 31) quarterly report. An account of the difficulties with which he was struggling is given in the *Fourth Quarterly Report* (October 30, 1953), pp. 8–9. The universally low level of expectation for the molecular oscillator experiment emerges from interviews[82].

[146] The visitation by Rabi and Kusch is recalled with distinctness and feeling by Townes in his 1979 interview by W.V. Smith[3], and is already described in Townes, 1962, "Masers"[72], p. 175. The Rabi/Kusch ONR contract crisis is documented by an undated, handwritten note by Kusch to Rabi, "What do you suggest?... it is your formal responsibility" (LCMD, Rabi pap., box 19, folder "Columbia Univ.-Internal-Government Contracts, 1946–66").

[147] Forman, "Inventing the maser"[2], section III; Gordon, J.P., 1954. "Molecular beam oscillator," CRL, [*Quarterly*] *Progress Report*, Jan. 30, p. 6; Gordon, interview, November 4, 1983, by Forman (tape recording and transcript at AIP).

[148] Senitzky, "Report of visit... February 16, 1954"[84]. Senitzky lists Kusch, [Norman] Kroll, and [Melvin J.] Bernstein as interlocutors. This is, however, the first of Senitzky's reports in which his characterization of the status of the molecular oscillator *precedes* that of the Cerenkov radiation experiment.

[149] Gordon, J.P., "Molecular beam oscillator," CRL, *Quarterly [Progress] Report* (April 30, 1954), pp. 8–12; Townes, entry dated March 22, 1954, on pp. 51–53 of Columbia Radiation Laboratory Notebook #225, numbered by Townes as "III," used by Townes from July 25, 1951 to July 1, 1957 (in possession of C.H. Townes, Dept. of Physics, Univ. of Calif., Berkeley).

[150] Bromberg, *Laser in America*[2], p. 25; Townes, letter April 9, 1954, to H. Lyons (in possession of H. Lyons, Pacific Palisades, Calif); Lyons, letter November 2, 1954, to Townes (loc. cit.).

[151] Gordon, J.P., Zeiger, H.J. and Townes, C.H., 1954. "Molecular microwave oscillator and new hyperfine structure in the microwave spectrum of NH_3," *Phys. Rev., 95*, pp. 282–84, MS received by editor May 5, 1954. H.A. Zahl, memorandum June 7, 1954, to Director, Physical Sciences Division, attn.: Chief, Thermionics Branch [Evans Signal Laboratory], Fort Monmouth[84]. Townes' request for Army Signal Corps permission to publish was rather, although not wholly, exceptional; there are one or two other instances of such a request by Townes for permission to publish in the Fort Monmouth files I have located.

[152] Zahl, memorandum June 28, 1954, to Commanding Officer, Signal Corps Laboratories, Fort Monmouth, reporting on meeting of Joint Services Technical Advisory Committee, Columbia University, May 25, 1954[84]; E.A. Tucker, Acting Head, Electronics Branch, ONR, letter July 9, 1954, to Townes (in possession of C.H. Townes). In October Townes would tell Fort Monmouth that even in its present experimental form the maser was ready for employment as a frequency standard in large, liquid oxygen fueled missiles: D.D. Babb, "Report of trip to Columbia University... on 15 October 1954," memorandum to Chief, Frequency Control Branch, Squier Signal Laboratory, Fort Monmouth, October 21, 1954 (NMPR, accession 338–76–H–869, box 24, folder "Project 142A... 1954").

[153] Bromberg, *Laser in America*[2], pp. 25–27, 30, 62–64, 254–56.

[154] Forman, "Inventing the maser"[2], section I.

[155] Bromberg, *Laser in America*[2], pp. 31–41, 46–61.

[156] Kusch, as Executive Director, CRL, to Commanding General, Fort Monmouth, to attention of E.M. Reilley, Director, Electron Devices Division, Evans Signal Laboratory, September 14, 1956 (NMPR, accession 64–434, box 47, file "Columbia Univ.").

[157] Agenda for "Joint Services Technical Advisory Committee Meeting, Columbia Radiation Laboratory, May 28, 1957," with ms notes on points discussed in the hand of Edward M. Reilley, now Asst. Dir. of Research, Fort Monmouth[156]. At this meeting tentative approval was given to Kusch's proposal – Kusch to Reilley, April 17, 1957[156] – to consolidate in a single quarterly report the obligatory progress reports on the several contracts from the several services held by CRL and molecular beams faculty. This too was a further acquiescence in the redefinition of CRL that Kusch was pushing.

[158] Kusch, letter to Rabi, undated [1957 June 2] (LCMD, Rabi pap., box 21, fldr "Columbia Univ.-Internal-Physics-CRL, 1946–49"). $27000 per month is $324000 per year; I surmise that at no time since 1953 had the money been significantly better, but have data only for CRL's contract SC–64630, which entered into effect June 30, 1955: base amount $300000, augmented by $30000 in November 1955[156]. For the general, alarming, downturn in military funding of academic research in the summer of 1957 see "Behind Q.E.," at notes 28 and 29.

[159] "Behind Q.E.," p. 182; Groves, Now it can be told[38]; Goldberg, S., 1995. "Groves and the scientists: compartmentalization and the building of the bomb," Physics Today, August, pp. 38–43.

[160] Capshew, J.H. and Rader, K.A., 1992. "Big science: Price to the present," Osiris, 7, pp. 3–25, on p. 12.

[161] Forman, P., 1991. "Independence, not transcendence, for the historian of science," Isis, 82, pp. 71–86; Forman, P. "Inventing the maser"[2], section VI: "Candor and compartmentalization"; Forman, P., 1994. "Fisica, modernidad y nuestra evasion de la responsabilidad," Arbor, 147, 51–74.

[162] Forman, P. "Social niche and self-image of the American physicist," in: Restructuring of physical sciences ... 1945–60[7], pp. 96–104.

[163] The concept of 'compartmentalization' here proposed is thus the reverse – but also the complement – of the concept of 'cognitive dissonance': Festinger, L., 1957. A theory of cognitive dissonance (Row, Peterson: Evanston, IL); Boring, E.G., 1964. "Cognitive dissonance: its use in science," Science, 145, 680–85.

[164] Philosophers, of course have not failed to appreciate the difficulty of justifying this split, which in general they have instead attempted to avoid by idealization of the conditions of knowledge production as 'the scientific method,' i.e., by regarding the conditions of knowledge production as sharing with the produced knowledge in the quality of transcendence.

[165] Rabi in Physical science and human values[27], pp. 29–30.

[166] See, for the clandestine, Needell, "Project TROY"[37], and Needell, "... Rabi, Berkner, ...,"[12]; Doel, R.E., 1992. "Evaluating Soviet lunar science in Cold War America," Osiris, 7, pp. 238–264; Ziegler, C.A. and Jacobson, D., 1995. Spying without spies: origins of America's secret nuclear surveillance system (Praeger Publishers: Westport, Conn.).

[167] Roberts, "How nice ..."[7]. Thus the wild exaggeration of the authority and influence of 'their man' Manny Piore in the Office of Naval Research[47] should perhaps be understood as a form of compartmentalization, preserving the physicists from having to recognize and come to terms with the fact that the Navy's purposes in providing support for their researches were very different than their own in pursuing those researches. Similarly, by choosing to view that support as largesse, the physicists preserved themselves from recognizing the relationship as quid pro quo, and sustained their self-image as untrammeled seekers after truth and/or fun.

[168] See the works on "nukespeak"[61] and Pyenson's paper in this volume, at his note 2.

[169] Rabi, "... research in universities ..."[27], p. 31. Similarly, George Pegram, writing May 20, 1948, as Chairman, Committee on Government Aided Research, to Philip M. Hayden, Secretary of the University, regarding an inquiry from the Brookings Institution about the effects of

government funding, offerred a rather disingenuous denial of any negative consequences, but concluded by pointing out the disparity in power, "so that no matter how generous the terms of the contract, the government never deals with contractors on even terms," and acknowledging "an unavoidable tendency on the part of those who draft government contracts to... make the contract more and more one-sided..." (Columbia University Secretary's Files, folder "Pegram"). See, further, Gruber, C., 1995. "The overhead system in government-sponsored academic science: origins and early development," *HSPS, 25*, pp. 241–268.

[170] For Lamb see the works cited in note 19. For descriptions of the work that gained Kusch his ill-deserved half of that 1955 physics prize: Kusch, "The magnetic moment of the electron. Nobel lecture, December 12, 1955," in: *Nobel lectures...: physics,*[19] pp. 298–311, and Trigg, G.L., 1975. *Landmark experiments in twentieth century physics* (Crane, Russak & Co.: New York), pp. 121–134. Abragam, A., 1989. *Time reversal: an autobiography* (Oxford Univ. Pr.: New York), p. 161, gives an anecdote expressing the consensus among physicists that Kusch's Nobel Prize was undeserved.

[171] The point is made by David DeVorkin[41] in his contribution to this volume (pp. 251–52), citing also the work of S.W. Leslie,[8] and by Allan Needell in his study of Project TROY[37] with respect to ionospheric scattering; see, also, the discussion of VLBI[76]. It is probable that the Project TROY deliberations played some role in Edward Purcell's search for the 21 cm microwave emission line from interstellar hydrogen: Forman, "... radar... in physical research after World War II"[4], in section on signal/noise. The forthcoming work of Finn Aaserud on "Jason," the elite group of physicists serving as consults to the U.S. Department of Defense under the auspices of the Institute for Defense Analyses, will help to illuminate such relationships between secret committees and scientific discoveries.

LIST OF CONTRIBUTORS

David DeVorkin is curator of the history of astronomy and the space sciences at the National Air and Space Museum, Smithsonian Institution. He is at work on a biography of the astronomer Henry Norris Russell. Address: Department of Space History MRC 311, Smithsonian Institution, Washington DC 20560, U.S.A. Email: nasdsh03@sivm.si.edu.

Michael Eckert, formerly collaborator on the International Project in the History of Solid State Physics, is now editor of the scientific correspondence of Arnold Sommerfeld at the Institut für Geschichte der Naturwissenschaften der Universität München, Postfach, D-80306 München, Germany. Email: ug301an@sunmail.lrz-muenchen.de.

David Edgerton, formerly at the University of Manchester, is Reader and Head of the Centre for History of Science, Technology and Medicine in the Imperial College of Science, Technology and Medicine, London SW7 2AZ, U.K. Email: d.edgerton@ic.ac.uk.

Paul Forman is curator for modern physics in the Smithsonian Institution's National Museum of American History. Address: MRC 631, Washington DC 20560, U.S.A. Email: mah0h84@sivm.si.edu.

Bruce Hevly teaches the history of science and technology at the University of Washington. Address: Department of History DP-20, Seattle, WA 98195, U.S.A. Email: bhevly@u.washington.edu. From 1985–1987 he was Associate Historian of the Naval Research Laboratory.

Helge Kragh has published widely in the history of physics, including a biography of P.A.M. Dirac. In 1990–94 he did research in history of technology at Roskilde University, Denmark. He now has the Chair of the History of Science at the University of Oslo. Address: Centre for Technology and Culture, Gaustadall, 0371 Oslo, Norway.

Herbert Mehrtens is Professor of Modern History at the Technical University of Braunschweig. Address: Historisches Seminar, Technische Universität, Postfach 33 29, D-38023 Braunschweig, Germany. Telefax: (0531) 391-8162. Email: h.mehrtens@tu-bs.de.

P. Forman and J. M. Sánchez-Ron (eds.), National Military Establishments and the Advancement of Science and Technology. Studies in the 20th Century History 327-328

Javier Ordoñez is Profesor of the Logic and Philosophy of Science in the Facultad de Filosofia, Universidad Autónoma de Madrid, Canto Blanco, 28049 Madrid, Spain. Email: javiero@ccuam3.sdi.uam.es. He has produced Spanish editions of works of Sadi Carnot and Ludwig Boltzmann.

Eduardo L. Ortiz, Foreign Fellow of the National Academy of Science, Argentina, and of the Royal Academy of Sciences, Spain, is Professor of Mathematics and of History of Mathematics, Imperial College, University of London, London SW7 2AZ, U.K. Email: e.ortiz@ic.ac.uk.

Lewis Pyenson, a Fellow of the Royal Society of Canada, has pursued extensive researches on the exact sciences as vehicle for cultural imperialism. Formerly Profesor in the Département d'histoire, Faculté des arts et des sciences, Université de Montréal, Canada, he is now Dean of the Graduate School at the University of Louisiana, Lafayette, LA 70504–4610, U.S.A.

José M. Sánchez-Ron has the Chair of History of Science in the Faculty of Science of the Universidad Autónoma de Madrid. Address: Departamento de Fisica Teorica, Canto Blanco, 28049 Madrid, Spain. Email: sron@ccuam3.sdi.uam.es. He is at work on a biography of the Spanish physicist Blas Cabrera.

INDEX OF NAMES

INDEX OF SUBJECTS

Boston Studies in the Philosophy of Science

Editor: Robert S. Cohen, *Boston University*

1. M.W. Wartofsky (ed.): *Proceedings of the Boston Colloquium for the Philosophy of Science, 1961/1962.* [Synthese Library 6] 1963
   ISBN 90-277-0021-4

2. R.S. Cohen and M.W. Wartofsky (eds.): *Proceedings of the Boston Colloquium for the Philosophy of Science, 1962/1964.* In Honor of P. Frank. [Synthese Library 10] 1965
   ISBN 90-277-9004-0

3. R.S. Cohen and M.W. Wartofsky (eds.): *Proceedings of the Boston Colloquium for the Philosophy of Science, 1964/1966.* In Memory of Norwood Russell Hanson. [Synthese Library 14] 1967
   ISBN 90-277-0013-3

4. R.S. Cohen and M.W. Wartofsky (eds.): *Proceedings of the Boston Colloquium for the Philosophy of Science, 1966/1968.* [Synthese Library 18] 1969
   ISBN 90-277-0014-1

5. R.S. Cohen and M.W. Wartofsky (eds.): *Proceedings of the Boston Colloquium for the Philosophy of Science, 1966/1968.* [Synthese Library 19] 1969
   ISBN 90-277-0015-X

6. R.S. Cohen and R.J. Seeger (eds.): *Ernst Mach, Physicist and Philosopher.* [Synthese Library 27] 1970
   ISBN 90-277-0016-8

7. M. Čapek: *Bergson and Modern Physics.* A Reinterpretation and Re-evaluation. [Synthese Library 37] 1971
   ISBN 90-277-0186-5

8. R.C. Buck and R.S. Cohen (eds.): *PSA 1970.* Proceedings of the 2nd Biennial Meeting of the Philosophy and Science Association (Boston, Fall 1970). In Memory of Rudolf Carnap. [Synthese Library 39] 1971
   ISBN 90-277-0187-3; Pb 90-277-0309-4

9. A.A. Zinov'ev: *Foundations of the Logical Theory of Scientific Knowledge (Complex Logic).* Translated from Russian. Revised and enlarged English Edition, with an Appendix by G.A. Smirnov, E.A. Sidorenko, A.M. Fedina and L.A. Bobrova. [Synthese Library 46] 1973
   ISBN 90-277-0193-8; Pb 90-277-0324-8

10. L. Tondl: *Scientific Procedures.* A Contribution Concerning the Methodological Problems of Scientific Concepts and Scientific Explanation.Translated from Czech. [Synthese Library 47] 1973   ISBN 90-277-0147-4; Pb 90-277-0323-X

11. R.J. Seeger and R.S. Cohen (eds.): *Philosophical Foundations of Science.* Proceedings of Section L, 1969, American Association for the Advancement of Science. [Synthese Library 58] 1974   ISBN 90-277-0390-6; Pb 90-277-0376-0

12. A. Grünbaum: *Philosophical Problems of Space and Times.* 2nd enlarged ed. [Synthese Library 55] 1973   ISBN 90-277-0357-4; Pb 90-277-0358-2

13. R.S. Cohen and M.W. Wartofsky (eds.): *Logical and Epistemological Studies in Contemporary Physics.* Proceedings of the Boston Colloquium for the Philosophy of Science, 1969/72, Part I. [Synthese Library 59] 1974
   ISBN 90-277-0391-4; Pb 90-277-0377-9

Boston Studies in the Philosophy of Science

14. R.S. Cohen and M.W. Wartofsky (eds.): *Methodological and Historical Essays in the Natural and Social Sciences*. Proceedings of the Boston Colloquium for the Philosophy of Science, 1969/72, Part II. [Synthese Library 60] 1974
ISBN 90-277-0392-2; Pb 90-277-0378-7

15. R.S. Cohen, J.J. Stachel and M.W. Wartofsky (eds.): *For Dirk Struik*. Scientific, Historical and Political Essays in Honor of Dirk J. Struik. [Synthese Library 61] 1974
ISBN 90-277-0393-0; Pb 90-277-0379-5

16. N. Geschwind: *Selected Papers on Language and the Brains*. [Synthese Library 68] 1974
ISBN 90-277-0262-4; Pb 90-277-0263-2

17. B.G. Kuznetsov: *Reason and Being*. Translated from Russian. Edited by C.R. Fawcett and R.S. Cohen. 1987
ISBN 90-277-2181-5

18. P. Mittelstaedt: *Philosophical Problems of Modern Physics*. Translated from the revised 4th German edition by W. Riemer and edited by R.S. Cohen. [Synthese Library 95] 1976
ISBN 90-277-0285-3; Pb 90-277-0506-2

19. H. Mehlberg: *Time, Causality, and the Quantum Theory*. Studies in the Philosophy of Science. Vol. I: *Essay on the Causal Theory of Time*. Vol. II: *Time in a Quantized Universe*. Translated from French. Edited by R.S. Cohen. 1980
Vol. I: ISBN 90-277-0721-9; Pb 90-277-1074-0
Vol. II: ISBN 90-277-1075-9; Pb 90-277-1076-7

20. K.F. Schaffner and R.S. Cohen (eds.): *PSA 1972*. Proceedings of the 3rd Biennial Meeting of the Philosophy of Science Association (Lansing, Michigan, Fall 1972). [Synthese Library 64] 1974
ISBN 90-277-0408-2; Pb 90-277-0409-0

21. R.S. Cohen and J.J. Stachel (eds.): *Selected Papers of Léon Rosenfeld*. [Synthese Library 100] 1979
ISBN 90-277-0651-4; Pb 90-277-0652-2

22. M. Čapek (ed.): *The Concepts of Space and Time*. Their Structure and Their Development. [Synthese Library 74] 1976
ISBN 90-277-0355-8; Pb 90-277-0375-2

23. M. Grene: *The Understanding of Nature*. Essays in the Philosophy of Biology. [Synthese Library 66] 1974
ISBN 90-277-0462-7; Pb 90-277-0463-5

24. D. Ihde: *Technics and Praxis*. A Philosophy of Technology. [Synthese Library 130] 1979
ISBN 90-277-0953-X; Pb 90-277-0954-8

25. J. Hintikka and U. Remes: *The Method of Analysis*. Its Geometrical Origin and Its General Significance. [Synthese Library 75] 1974
ISBN 90-277-0532-1; Pb 90-277-0543-7

26. J.E. Murdoch and E.D. Sylla (eds.): *The Cultural Context of Medieval Learning*. Proceedings of the First International Colloquium on Philosophy, Science, and Theology in the Middle Ages, 1973. [Synthese Library 76] 1975
ISBN 90-277-0560-7; Pb 90-277-0587-9

27. M. Grene and E. Mendelsohn (eds.): *Topics in the Philosophy of Biology*. [Synthese Library 84] 1976
ISBN 90-277-0595-X; Pb 90-277-0596-8

28. J. Agassi: *Science in Flux*. [Synthese Library 80] 1975
ISBN 90-277-0584-4; Pb 90-277-0612-3

Boston Studies in the Philosophy of Science

Boston Studies in the Philosophy of Science

Boston Studies in the Philosophy of Science

Boston Studies in the Philosophy of Science

87. R.S. Cohen and T. Schnelle (eds.): *Cognition and Fact.* Materials on Ludwik Fleck. 1986
ISBN 90-277-1902-0

88. G. Freudenthal: *Atom and Individual in the Age of Newton.* On the Genesis of the Mechanistic World View. Translated from German. 1986
ISBN 90-277-1905-5

89. A. Donagan, A.N. Perovich Jr and M.V. Wedin (eds.): *Human Nature and Natural Knowledge.* Essays presented to Marjorie Grene on the Occasion of Her 75th Birthday. 1986
ISBN 90-277-1974-8

90. C. Mitcham and A. Hunning (eds.): *Philosophy and Technology II.* Information Technology and Computers in Theory and Practice. [*Also* Philosophy and Technology Series, Vol. 2] 1986
ISBN 90-277-1975-6

91. M. Grene and D. Nails (eds.): *Spinoza and the Sciences.* 1986
ISBN 90-277-1976-4

92. S.P. Turner: *The Search for a Methodology of Social Science.* Durkheim, Weber, and the 19th-Century Problem of Cause, Probability, and Action. 1986.
ISBN 90-277-2067-3

93. I.C. Jarvie: *Thinking about Society.* Theory and Practice. 1986
ISBN 90-277-2068-1

94. E. Ullmann-Margalit (ed.): *The Kaleidoscope of Science.* The Israel Colloquium: Studies in History, Philosophy, and Sociology of Science, Vol. 1. 1986
ISBN 90-277-2158-0; Pb 90-277-2159-9

95. E. Ullmann-Margalit (ed.): *The Prism of Science.* The Israel Colloquium: Studies in History, Philosophy, and Sociology of Science, Vol. 2. 1986
ISBN 90-277-2160-2; Pb 90-277-2161-0

96. G. Márkus: *Language and Production.* A Critique of the Paradigms. Translated from French. 1986
ISBN 90-277-2169-6

97. F. Amrine, F.J. Zucker and H. Wheeler (eds.): *Goethe and the Sciences: A Reappraisal.* 1987
ISBN 90-277-2265-X; Pb 90-277-2400-8

98. J.C. Pitt and M. Pera (eds.): *Rational Changes in Science.* Essays on Scientific Reasoning. Translated from Italian. 1987
ISBN 90-277-2417-2

99. O. Costa de Beauregard: *Time, the Physical Magnitude.* 1987
ISBN 90-277-2444-X

100. A. Shimony and D. Nails (eds.): *Naturalistic Epistemology.* A Symposium of Two Decades. 1987
ISBN 90-277-2337-0

101. N. Rotenstreich: *Time and Meaning in History.* 1987   ISBN 90-277-2467-9

102. D.B. Zilberman: *The Birth of Meaning in Hindu Thought.* Edited by R.S. Cohen. 1988
ISBN 90-277-2497-0

103. T.F. Glick (ed.): *The Comparative Reception of Relativity.* 1987
ISBN 90-277-2498-9

104. Z. Harris, M. Gottfried, T. Ryckman, P. Mattick Jr, A. Daladier, T.N. Harris and S. Harris: *The Form of Information in Science.* Analysis of an Immunology Sublanguage. With a Preface by Hilary Putnam. 1989   ISBN 90-277-2516-0

Boston Studies in the Philosophy of Science

105. F. Burwick (ed.): *Approaches to Organic Form*. Permutations in Science and Culture. 1987   ISBN 90-277-2541-1
106. M. Almási: *The Philosophy of Appearances*. Translated from Hungarian. 1989   ISBN 90-277-2150-5
107. S. Hook, W.L. O'Neill and R. O'Toole (eds.): *Philosophy, History and Social Action*. Essays in Honor of Lewis Feuer. With an Autobiographical Essay by L. Feuer. 1988   ISBN 90-277-2644-2
108. I. Hronszky, M. Fehér and B. Dajka: *Scientific Knowledge Socialized*. Selected Proceedings of the 5th Joint International Conference on the History and Philosophy of Science organized by the IUHPS (Veszprém, Hungary, 1984). 1988   ISBN 90-277-2284-6
109. P. Tillers and E.D. Green (eds.): *Probability and Inference in the Law of Evidence*. The Uses and Limits of Bayesianism. 1988   ISBN 90-277-2689-2
110. E. Ullmann-Margalit (ed.): *Science in Reflection*. The Israel Colloquium: Studies in History, Philosophy, and Sociology of Science, Vol. 3. 1988
                      ISBN 90-277-2712-0; Pb 90-277-2713-9
111. K. Gavroglu, Y. Goudaroulis and P. Nicolacopoulos (eds.): *Imre Lakatos and Theories of Scientific Change*. 1989   ISBN 90-277-2766-X
112. B. Glassner and J.D. Moreno (eds.): *The Qualitative-Quantitative Distinction in the Social Sciences*. 1989   ISBN 90-277-2829-1
113. K. Arens: *Structures of Knowing*. Psychologies of the 19th Century. 1989
                                   ISBN 0-7923-0009-2
114. A. Janik: *Style, Politics and the Future of Philosophy*. 1989
                                   ISBN 0-7923-0056-4
115. F. Amrine (ed.): *Literature and Science as Modes of Expression*. With an Introduction by S. Weininger. 1989   ISBN 0-7923-0133-1
116. J.R. Brown and J. Mittelstrass (eds.): *An Intimate Relation*. Studies in the History and Philosophy of Science. Presented to Robert E. Butts on His 60th Birthday. 1989   ISBN 0-7923-0169-2
117. F. D'Agostino and I.C. Jarvie (eds.): *Freedom and Rationality*. Essays in Honor of John Watkins. 1989   ISBN 0-7923-0264-8
118. D. Zolo: *Reflexive Epistemology*. The Philosophical Legacy of Otto Neurath. 1989   ISBN 0-7923-0320-2
119. M. Kearn, B.S. Philips and R.S. Cohen (eds.): *Georg Simmel and Contemporary Sociology*. 1989   ISBN 0-7923-0407-1
120. T.H. Levere and W.R. Shea (eds.): *Nature, Experiment and the Science*. Essays on Galileo and the Nature of Science. In Honour of Stillman Drake. 1989
                                   ISBN 0-7923-0420-9
121. P. Nicolacopoulos (ed.): *Greek Studies in the Philosophy and History of Science*. 1990   ISBN 0-7923-0717-8
122. R. Cooke and D. Costantini (eds.): *Statistics in Science*. The Foundations of Statistical Methods in Biology, Physics and Economics. 1990
                                   ISBN 0-7923-0797-6

Boston Studies in the Philosophy of Science

123. P. Duhem: *The Origins of Statics*. Translated from French by G.F. Leneaux, V.N. Vagliente and G.H. Wagner. With an Introduction by S.L. Jaki. 1991
ISBN 0-7923-0898-0

124. H. Kamerlingh Onnes: *Through Measurement to Knowledge*. The Selected Papers, 1853-1926. Edited and with an Introduction by K. Gavroglu and Y. Goudaroulis. 1991
ISBN 0-7923-0825-5

125. M. Čapek: *The New Aspects of Time: Its Continuity and Novelties*. Selected Papers in the Philosophy of Science. 1991
ISBN 0-7923-0911-1

126. S. Unguru (ed.): *Physics, Cosmology and Astronomy, 1300-1700*. Tension and Accommodation. 1991
ISBN 0-7923-1022-5

127. Z. Bechler: *Newton's Physics on the Conceptual Structure of the Scientific Revolution*. 1991
ISBN 0-7923-1054-3

128. É. Meyerson: *Explanation in the Sciences*. Translated from French by M-A. Siple and D.A. Siple. 1991
ISBN 0-7923-1129-9

129. A.I. Tauber (ed.): *Organism and the Origins of Self*. 1991
ISBN 0-7923-1185-X

130. F.J. Varela and J-P. Dupuy (eds.): *Understanding Origins*. Contemporary Views on the Origin of Life, Mind and Society. 1992
ISBN 0-7923-1251-1

131. G.L. Pandit: *Methodological Variance*. Essays in Epistemological Ontology and the Methodology of Science. 1991
ISBN 0-7923-1263-5

132. G. Munévar (ed.): *Beyond Reason*. Essays on the Philosophy of Paul Feyerabend. 1991
ISBN 0-7923-1272-4

133. T.E. Uebel (ed.): *Rediscovering the Forgotten Vienna Circle*. Austrian Studies on Otto Neurath and the Vienna Circle. Partly translated from German. 1991
ISBN 0-7923-1276-7

134. W.R. Woodward and R.S. Cohen (eds.): *World Views and Scientific Discipline Formation*. Science Studies in the [former] German Democratic Republic. Partly translated from German by W.R. Woodward. 1991
ISBN 0-7923-1286-4

135. P. Zambelli: *The Speculum Astronomiae and Its Enigma*. Astrology, Theology and Science in Albertus Magnus and His Contemporaries. 1992
ISBN 0-7923-1380-1

136. P. Petitjean, C. Jami and A.M. Moulin (eds.): *Science and Empires*. Historical Studies about Scientific Development and European Expansion.
ISBN 0-7923-1518-9

137. W.A. Wallace: *Galileo's Logic of Discovery and Proof*. The Background, Content, and Use of His Appropriated Treatises on Aristotle's *Posterior Analytics*. 1992
ISBN 0-7923-1577-4

138. W.A. Wallace: *Galileo's Logical Treatises*. A Translation, with Notes and Commentary, of His Appropriated Latin Questions on Aristotle's *Posterior Analytics*. 1992
ISBN 0-7923-1578-2
Set (137 + 138) ISBN 0-7923-1579-0

Boston Studies in the Philosophy of Science

139. M.J. Nye, J.L. Richards and R.H. Stuewer (eds.): *The Invention of Physical Science*. Intersections of Mathematics, Theology and Natural Philosophy since the Seventeenth Century. Essays in Honor of Erwin N. Hiebert. 1992
ISBN 0-7923-1753-X

140. G. Corsi, M.L. dalla Chiara and G.C. Ghirardi (eds.): *Bridging the Gap: Philosophy, Mathematics and Physics*. Lectures on the Foundations of Science. 1992
ISBN 0-7923-1761-0

141. C.-H. Lin and D. Fu (eds.): *Philosophy and Conceptual History of Science in Taiwan*. 1992
ISBN 0-7923-1766-1

142. S. Sarkar (ed.): *The Founders of Evolutionary Genetics*. A Centenary Reappraisal. 1992
ISBN 0-7923-1777-7

143. J. Blackmore (ed.): *Ernst Mach – A Deeper Look*. Documents and New Perspectives. 1992
ISBN 0-7923-1853-6

144. P. Kroes and M. Bakker (eds.): *Technological Development and Science in the Industrial Age*. New Perspectives on the Science–Technology Relationship. 1992
ISBN 0-7923-1898-6

145. S. Amsterdamski: *Between History and Method*. Disputes about the Rationality of Science. 1992
ISBN 0-7923-1941-9

146. E. Ullmann-Margalit (ed.): *The Scientific Enterprise*. The Bar-Hillel Colloquium: Studies in History, Philosophy, and Sociology of Science, Volume 4. 1992
ISBN 0-7923-1992-3

147. L. Embree (ed.): *Metaarchaeology*. Reflections by Archaeologists and Philosophers. 1992
ISBN 0-7923-2023-9

148. S. French and H. Kamminga (eds.): *Correspondence, Invariance and Heuristics*. Essays in Honour of Heinz Post. 1993
ISBN 0-7923-2085-9

149. M. Bunzl: *The Context of Explanation*. 1993
ISBN 0-7923-2153-7

150. I.B. Cohen (ed.): *The Natural Sciences and the Social Sciences*. Some Critical and Historical Perspectives. 1994
ISBN 0-7923-2223-1

151. K. Gavroglu, Y. Christianidis and E. Nicolaidis (eds.): *Trends in the Historiography of Science*. 1994
ISBN 0-7923-2255-X

152. S. Poggi and M. Bossi (eds.): *Romanticism in Science*. Science in Europe, 1790–1840. 1994
ISBN 0-7923-2336-X

153. J. Faye and H.J. Folse (eds.): *Niels Bohr and Contemporary Philosophy*. 1994
ISBN 0-7923-2378-5

154. C.C. Gould and R.S. Cohen (eds.): *Artifacts, Representations, and Social Practice*. Essays for Marx W. Wartofsky. 1994
ISBN 0-7923-2481-1

155. R.E. Butts: *Historical Pragmatics*. Philosophical Essays. 1993
ISBN 0-7923-2498-6

156. R. Rashed: *The Development of Arabic Mathematics: Between Arithmetic and Algebra*. Translated from French by A.F.W. Armstrong. 1994
ISBN 0-7923-2565-6

Boston Studies in the Philosophy of Science

157. I. Szumilewicz-Lachman (ed.): *Zygmunt Zawirski: His Life and Work.* With Selected Writings on Time, Logic and the Methodology of Science. Translations by Feliks Lachman. Ed. by R.S. Cohen, with the assistance of B. Bergo. 1994   ISBN 0-7923-2566-4

158. S.N. Haq: *Names, Natures and Things.* The Alchemist Jabir ibn Ḥayyān and His *Kitāb al-Aḥjār* (Book of Stones). 1994   ISBN 0-7923-2587-7

159. P. Plaass: *Kant's Theory of Natural Science.* Translation, Analytic Introduction and Commentary by Alfred E. and Maria G. Miller. 1994
ISBN 0-7923-2750-0

160. J. Misiek (ed.): *The Problem of Rationality in Science and its Philosophy.* On Popper vs. Polanyi. The Polish Conferences 1988–89. 1995
ISBN 0-7923-2925-2

161. I.C. Jarvie and N. Laor (eds.): *Critical Rationalism, Metaphysics and Science.* Essays for Joseph Agassi, Volume I. 1995   ISBN 0-7923-2960-0

162. I.C. Jarvie and N. Laor (eds.): *Critical Rationalism, the Social Sciences and the Humanities.* Essays for Joseph Agassi, Volume II. 1995   ISBN 0-7923-2961-9
Set (161–162) ISBN 0-7923-2962-7

163. K. Gavroglu, J. Stachel and M.W. Wartofsky (eds.): *Physics, Philosophy, and the Scientific Community.* Essays in the Philosophy and History of the Natural Sciences and Mathematics. In Honor of Robert S. Cohen. 1995
ISBN 0-7923-2988-0

164. K. Gavroglu, J. Stachel and M.W. Wartofsky (eds.): *Science, Politics and Social Practice.* Essays on Marxism and Science, Philosophy of Culture and the Social Sciences. In Honor of Robert S. Cohen. 1995   ISBN 0-7923-2989-9

165. K. Gavroglu, J. Stachel and M.W. Wartofsky (eds.): *Science, Mind and Art.* Essays on Science and the Humanistic Understanding in Art, Epistemology, Religion and Ethics. Essays in Honor of Robert S. Cohen. 1995
ISBN 0-7923-2990-2
Set (163–165) ISBN 0-7923-2991-0

166. K.H. Wolff: *Transformation in the Writing.* A Case of Surrender-and-Catch. 1995   ISBN 0-7923-3178-8

167. A.J. Kox and D.M. Siegel (eds.): *No Truth Except in the Details.* Essays in Honor of Martin J. Klein. 1995   ISBN 0-7923-3195-8

168. J. Blackmore: *Ludwig Boltzmann, His Later Life and Philosophy, 1900–1906.* Book One: A Documentary History. 1995   ISBN 0-7923-3231-8

169. R.S. Cohen, R. Hilpinen and Q. Renzong (eds.): *Realism and Anti-Realism in the Philosophy of Science.* Beijing International Conference, 1992. 1995
ISBN 0-7923-3233-4

170. I. Kuçuradi and R.S. Cohen (eds.): *The Concept of Knowledge.* The Ankara Seminar. 1995   ISBN 0-7923-3241-5

Boston Studies in the Philosophy of Science

171. M.A. Grodin (ed.): *Meta Medical Ethics*: The Philosophical Foundations of Bioethics. 1995   ISBN 0-7923-3344-6
172. S. Ramirez and R.S. Cohen (eds.): *Mexican Studies in the History and Philosophy of Science*. 1995   ISBN 0-7923-3462-0
173. C. Dilworth: *The Metaphysics of Science*. An Account of Modern Science in Terms of Principles, Laws and Theories. 1995   ISBN 0-7923-3693-3
174. J. Blackmore: *Ludwig Boltzmann, His Later Life and Philosophy, 1900–1906* Book Two: The Philosopher. 1995   ISBN 0-7923-3464-7
175. P. Damerow: *Abstraction and Representation*. Essays on the Cultural Evolution of Thinking. 1996   ISBN 0-7923-3816-2
176. G. Tarozzi (ed.): *Karl Popper, Philosopher of Science.*   (in prep.)
177. M. Marion and R.S. Cohen (eds.): *Québec Studies in the Philosophy of Science.* Part I: Logic, Mathematics, Physics and History of Science. Essays in Honor of Hugues Leblanc. 1995   ISBN 0-7923-3559-7
178. M. Marion and R.S. Cohen (eds.): *Québec Studies in the Philosophy of Science.* Part II: Biology, Psychology, Cognitive Science and Economics. Essays in Honor of Hugues Leblanc. 1996   ISBN 0-7923-3560-0
                    Set (177–178) ISBN 0-7923-3561-9
179. F. Dainian and R.S. Cohen (eds.): *Chinese Studies in the History and Philosophy of Science and Technology*. 1995   ISBN 0-7923-3463-9
180. P. Forman and J.M. Sánchez-Ron (eds.): *National Military Establishments and the Advancement of Science and Technology*. Studies in 20th Century History. 1996   ISBN 0-7923-3541-4
181. E.J. Post: *Quantum Reprogramming*. Ensembles and Single Systems: A Two-Tier Approach to Quantum Mechanics. 1995   ISBN 0-7923-3565-1
182. A.I. Tauber (ed.): *The Elusive Synthesis: Aesthetics and Science*. 1996
                    ISBN 0-7923-3904-5

Also of interest:
R.S. Cohen and M.W. Wartofsky (eds.): *A Portrait of Twenty-Five Years Boston Colloquia for the Philosophy of Science, 1960-1985*. 1985   ISBN Pb 90-277-1971-3

Previous volumes are still available.

KLUWER ACADEMIC PUBLISHERS – DORDRECHT / BOSTON / LONDON